AF901159

POSTHARVEST QUALITY ATTRIBUTES OF FRESH VEGETABLES AND GRAINS

Postharvest Biology and Technology Series

POSTHARVEST QUALITY ATTRIBUTES OF FRESH VEGETABLES AND GRAINS

Riadh Ilahy, PhD
Imen Tlili, PhD
Thouraya R'him, PhD
Muhammad Azam, PhD
Editors

AAP | APPLE ACADEMIC PRESS

First edition published 2027

Apple Academic Press Inc.
1265 Goldenrod Circle, NE,
Palm Bay, FL 32905 USA

760 Laurentian Drive, Unit 19,
Burlington, ON L7N 0A4, Canada

CRC Press
2385 NW Executive Center Drive,
Suite 320, Boca Raton FL 33431

4 Park Square, Milton Park,
Abingdon, Oxon, OX14 4RN UK

© 2027 by Apple Academic Press, Inc.

Apple Academic Press exclusively co-publishes with CRC Press, an imprint of Taylor & Francis Group, LLC

Reasonable efforts have been made to publish reliable data and information, but the authors, editors, and publisher cannot assume responsibility for the validity of all materials or the consequences of their use. The authors, editors, and publishers have attempted to trace the copyright holders of all material reproduced in this publication and apologize to copyright holders if permission to publish in this form has not been obtained. If any copyright material has not been acknowledged, please write and let us know so we may rectify in any future reprint.

Except as permitted under U.S. Copyright Law, no part of this book may be reprinted, reproduced, transmitted, or utilized in any form by any electronic, mechanical, or other means, now known or hereafter invented, including photocopying, microfilming, and recording, or in any information storage or retrieval system, without written permission from the publishers.

For permission to photocopy or use material electronically from this work, access www.copyright.com or contact the Copyright Clearance Center, Inc. (CCC), 222 Rosewood Drive, Danvers, MA 01923, 978-750-8400. For works that are not available on CCC please contact mpkbookspermissions@tandf.co.uk

Trademark notice: Product or corporate names may be trademarks or registered trademarks and are used only for identification and explanation without intent to infringe.

For Product Safety Concerns and Information please contact our EU representative GPSR@taylorandfrancis.com Taylor & Francis Verlag GmbH, Kaufingerstraße 24, 80331 München, Germany

Library and Archives Canada Cataloguing in Publication

Title: Postharvest quality attributes of fresh vegetables and grains / Riadh Ilahy, PhD, Imen Tlili, PhD, Thouraya R'him, PhD, Muhammad Azam, PhD, editors.
Names: Ilahy, Riadh, editor. | Tlili, Imen, editor. | R'him, Thouraya, editor. | Azam, Muhammad (Lecturer in horticultural sciences), editor.
Series: Postharvest biology and technology book series.
Description: First edition. | Series statement: Postharvest biology and technology book series | Includes bibliographical references and index.
Identifiers: Canadiana (print) 2026018196X | Canadiana (ebook) 20260181986 | ISBN 9781779642844 (hardcover) | ISBN 9781779642851 (ebook)
Subjects: LCSH: Vegetables in human nutrition. | LCSH: Vegetables—Nutrition. | LCSH: Grain in human nutrition. | LCSH: Grain—Nutrition.
Classification: LCC QP144.V44 P67 2026 | DDC 613.2—dc23

Library of Congress Cataloging-in-Publication Data

Names: Ilahy, Riadh editor | Tlili, Imen editor | R'him, Thouraya editor | Azam, Muhammad (Lecturer in horticultural sciences) contributor

Title: Postharvest quality attributes of fresh grains and vegetables / Riadh Ilahy, Imen Tlili, Thouraya R'him, Muhammad Azam, editors.

Other titles: Postharvest biology and technology book series

Description: First edition. | Palm Bay, FL : Apple Academic Press, 2026. | Series: Postharvest biology and technology | Includes bibliographical references and index. | Summary: "The consumption of fruits and vegetables is necessary for a healthy and balanced diet as outlined on recent research studies. In fact, their consumption is positively associated with the prevention of various disorders and diseases. This new book, Postharvest Quality Attributes of Fresh Vegetables and Grains, explores the beneficial aspects of commonly consumed fresh vegetables and grains. Each chapter focuses on one particular vegetable or grain, covering the health benefits, nutritional and antinutritional composition, antioxidant phytochemicals, antioxidant potential, and factors influencing the antioxidant content, such as genotype, environmental variation and agronomic conditions. The chapters comprehensively cover the bioactive compounds, mineral composition, vitamins, carotenoids, phenolics, flavonoids, tocopherols, anthocyanins, glycosylates, etc., that contribute the functional quality and help prevent various chronic diseases. Key features: Provides nutritional and antinutritional composition data for various vegetables and grains Presents advanced information about antioxidant properties of vegetables and grains Discusses the factors influencing antioxidant content, such as genotype, environmental variation and agronomic conditions Includes detailed information about the beneficial health effects This volume will be valuable for academic researchers, scientists, and food processors interested in food chemistry, nutrition, nutraceuticals, and functional foods. Additionally, the book is suitable for both undergraduate and postgraduate students in the field of food science and technology, food chemistry, and nutrition"-- Provided by publisher.

Identifiers: CCN 2026021473 (print) | LCCN 2026021474 (ebook) | ISBN 9781779642844 hardcover | ISBN 9781779642851 ebook

Subjects: LCSH: Vegetables in human nutrition | Vegetables--Nutrition | Grain in human nutrition | Grain--Nutrition

Classification: LCC QP144.V44 P67 2026 (print) | LCC QP144.V44 (ebook)

LC record available at https://lccn.loc.gov/2026021473

LC ebook record available at https://lccn.loc.gov/2026021474

ISBN: 978-1-77964-284-4 (hbk)
ISBN: 978-1-77964-285-1 (ebk)

DOI: 10.1201/9781779642851

About the Book Series: Postharvest Biology and Technology

Editor-in-Chief

Mohammed Wasim Siddiqui, PhD
Scientist-cum-Assistant Professor, Department of Food Science and Technology, Bihar Agricultural University, Sabour, Bhagalpur, Bihar, India
AAP Acquisitions Editor | Horticultural Science
Founding/Managing Editor | Journal of Postharvest Technology

As we know, preserving the quality of fresh produce has long been a challenging task. In the past, several approaches were used for postharvest management of fresh produce, but due to continuous technological advancements, increased consumer health consciousness, and environmental concerns, these approaches have been modified and enhanced to address these issues.

The Postharvest Biology and Technology series features edited books that address several aspects of postharvest technology for fresh produce. The series presents existing and novel management systems in use today or with great potential to maintain the postharvest quality of fresh produce, including microbiological safety, nutrition, and sensory quality.

The books are aimed at professionals, postharvest scientists, academicians researching postharvest problems, and graduate-level students. This series is intended to be a comprehensive venture that provides up-to-date scientific and technical information focusing on postharvest management for fresh produce.

Books in the series address the following themes:

- Nutritional composition and antioxidant properties of fresh produce
- Postharvest physiology and biochemistry
- Biotic and abiotic factors affecting maturity and quality
- Preharvest treatments affecting postharvest quality
- Maturity and harvesting issues
- Nondestructive quality assessment
- Physiological and biochemical changes during ripening
- Postharvest treatments and their effects on shelf life and quality

- Postharvest operations such as sorting, grading, ripening, de-greening, curing, etc.
- Storage and shelf-life studies
- Packaging, transportation, and marketing
- Vase life improvement of flowers and foliage
- Postharvest management of spice, medicinal, and plantation crops
- Fruit and vegetable processing waste/byproducts: Management and utilization
- Postharvest diseases and physiological disorders
- Minimal processing of fruits and vegetables
- Quarantine and phytosanitary treatments for fresh produce
- Conventional and modern breeding approaches to improve the postharvest quality
- Biotechnological approaches to improve postharvest quality of horticultural crops

BOOKS IN THE SERIES

Postharvest Biology and Technology of Horticultural Crops: Principles and Practices for Quality Maintenance
Editor: Mohammed Wasim Siddiqui, PhD

Insect Pests of Stored Grain: Biology, Behavior, and Management Strategies
Ranjeet Kumar, PhD

Postharvest Management of Horticultural Crops: Practices for Quality Preservation
Editors: Mohammed Wasim Siddiqui, PhD, and Asgar Ali, PhD

Innovative Packaging of Fruits and Vegetables: Strategies for Safety and Quality Maintenance
Editors: Mohammed Wasim Siddiqui, PhD,
Mohammad Shafiur Rahman, PhD, and Ali Abas Wani, PhD

Advances in Postharvest Technologies of Vegetable Crops
Editors: Bijendra Singh, PhD, Sudhir Singh, PhD, and
Tanmay K. Koley, PhD

Plant Food By-Products: Industrial Relevance for Food Additives and Nutraceuticals
Editors: J. Fernando Ayala-Zavala, PhD, Gustavo González-Aguilar, PhD, and Mohammed Wasim Siddiqui, PhD

Emerging Technologies for Shelf-Life Enhancement of Fruits
Editors: Basharat Nabi Dar, PhD, and Shabir Ahmad Mir, PhD

Sensor-Based Quality Assessment Systems for Fruits and Vegetables
Editors: Bambang Kuswandi, PhD, and Mohammed Wasim Siddiqui, PhD

Emerging Postharvest Treatment of Fruits and Vegetables
Editors: Kalyan Barman, PhD, Swati Sharma, PhD, and Mohammed Wasim Siddiqui, PhD

Packaging and Storage of Fruits and Vegetables: Emerging Trends
Editor: Tanweer Alam, PhD

Postharvest Physiological Disorders of Fruits
Editors: Ghulam Khaliq, PhD, and Sajid Ali, PhD

Bioengineered Fruit and Vegetables: Trends in Postharvest Quality Improvement
Editors: Mohammad Irfan, PhD, and Pankaj Kumar, PhD

Novel Packaging Systems for Fruits and Vegetables
Editors: Kirtiraj K. Gaikwad, PhD, and Suman Singh, PhD

Ozone Applications in the Food Industry: Trends and Future Prospects
Editors: Ravi Pandiselvam, PhD, Anjineyulu Kothakota, PhD, and Kaavya Rathnakumar, PhD

Nonthermal Food Processing Technologies: Impact on Color Profile
Editors: Kaavya Rathnakumar, PhD, and R. Pandiselvam, PhD

Ready-to-Eat Snacks: Emerging Technologies for Production and Safety
Editors: R. Pandiselvam, PhD and Gülşah Çalışkan Koç, PhD

Handbook of Research on Mango Fruit: Postharvest Science, Production, Nutrition, and Processing Technology
Editors: Sajjad Hussain, PhD, Kamrun Nahar, PhD, Ishtiaq A. Rajwana, PhD, Sezai Ercisli, PhD, and Shakeel Ahmad, PhD

Sustainable Approaches in Postharvest Handling and Management
Editor: Pankaj B. Pathare, PhD

Physiological Disorders of Fruits and Vegetables: Causes and Remedies
Editors: Ghulam Khaliq, PhD, Sajid Ali, PhD, and Mohammed Wasim Siddiqui, PhD

About the Series Editor

Mohammed Wasim Siddiqui, PhD
Scientist-cum-Assistant Professor, Department of Food Science and Technology, Bihar Agricultural University, Sabour, Bhagalpur, Bihar, India

Mohammed Wasim Siddiqui, PhD, is an Assistant Professor and Scientist in the Department of Food Science and Post-Harvest Technology, Bihar Agricultural University, Sabour, India, and author or co-author of peer-reviewed research articles, book chapters, two manuals, and conference papers. He has over 20 books to his credit, edited and authored, published by Elsevier, CRC Press, Springer, and Apple Academic Press. Dr. Siddiqui has established the international peer-reviewed *Journal of Postharvest Technology.* He is Editor-in-Chief of two book series *(Postharvest Biology and Technology and Innovations in Horticultural Science)*, published by Apple Academic Press. Dr. Siddiqui is also a Senior Acquisitions Editor for Apple Academic Press, USA, for Horticultural Science. He has been serving as an editorial board member and an active reviewer for several international journals.

Dr. Siddiqui has received several grants and awards from a number of organizations for his research. He is an active member of the organizing committees of several national and international seminars, conferences, and summits. He is one of the key members in establishing the World Food Preservation Center (WFPC), LLC, USA, and is currently an active associate and supporter.

Dr. Siddiqui acquired a BSc (Agriculture) degree from Jawaharlal Nehru Krishi Vishwa Vidyalaya, Jabalpur, India. He received his MSc (Horticulture) and PhD (Horticulture) degrees from Bidhan Chandra Krishi Viswavidyalaya, Mohanpur, Nadia, India, with specialization in Postharvest Technology. He is a member of the Core Research Group at the Bihar Agricultural University (BAU), which provides appropriate direction and assistance in prioritizing research.

About the Editors

Riadh Ilahy, PhD
Associate Professor, Department of Horticulture, National Agricultural Research Institute of Tunisia (INRAT), Tunisia

Riadh Ilahy, PhD, is currently working as an Associate Professor at the National Agricultural Research Institute of Tunisia (INRAT), Department of Horticulture. He is specialized in breeding for higher-nutritional-quality crops, mainly tomatoes and potatoes. He has published three book chapters, 17 communications/posters, and more than 33 scientific articles. He has acted as an external examiner as well as a paper reviewer for many internationally indexed journals. He was awarded the Amitom Young Researcher Prize from HM Clause for his research on chloroplasts/chromoplasts of high-pigment tomato cultivars. He earned his PhD after collaborating with Professor Marcello Salvatore Lenucci and Professor Giuseppe Dalessandro at the University of Salento, Lecce, Italy. His research interests are in horticultural crop breeding, phytochemistry, natural antioxidants, food chemistry, and postharvest management.

Imen Tlili, PhD
Researcher, Department of Horticulture, National Agricultural Research Institute of Tunisia (INRAT), Tunisia

Imen Tlili, PhD, is currently working as a Researcher in the Department of Horticulture at the National Agricultural Research Institute of Tunisia (INRAT). She specializes in breeding for higher-nutritional-quality crops, mainly cucurbits. She has published two book chapters, 17 communications/posters, and more than 30 scientific articles. She has acted as an external examiner and paper reviewer for many international, indexed, and impact journals. Her research interests are in horticultural crop breeding, phytochemistry, natural antioxidants, and agriculture. She was awarded her PhD in phytochemistry in 2011.

Thouraya R'him, PhD
Researcher, Department of Horticulture, National Agricultural Research Institute of Tunisia (INRAT), Tunisia

Thouraya R'him, PhD, is currently working as a permanent Researcher in the Department of Horticulture at the National Agricultural Research Institute of Tunisia (INRAT), Tunisia. She specializes in selection and breeding, plant physiology, and environmental stress, mainly of pepper. She has published communications/posters and more than 30 scientific articles. She has acted as an external examiner and paper reviewer for many international indexed and impact factor journals. She earned her baccalaureate degree (Experimental Sciences) from the secondary school in Monastir. She also earned a Diploma of Engineering in Horticultural Sciences from the Higher School of Horticulture and Landscape of Chott Meriem, Sousse, Tunisia. Afterwards, she was awarded a diploma in in-depth studies in plant eco-physiology from the Faculty of Science, Tunis, Tunisia. She received her PhD (Plant Production) from the National Agricultural Institute of Tunisia (INAT). Her research interests include horticultural crop breeding, environmental stress, and plant physiology.

Muhammad Azam, PhD
Associate Professor, Institute of Horticultural Sciences, Faculty of Agriculture, University of Agriculture, Faisalabad, Punjab, Pakistan

Muhammad Azam, PhD, is an Associate Professor at the Institute of Horticultural Sciences, University of Agriculture, Faisalabad, Pakistan. He has more than 14 years of experience in research and teaching. He completed his research on citrus volatiles during his PhD at Zhejiang University in China. He worked as a consultant on a Citrus GlobalGAP certification project in Bhalwal, Sargodha, Pakistan. He is actively working and participating in different national and international projects. He is serving as the principal investigator and co-principal investigator for six projects and has been collaborating with local and international institutions. He has supervised 32 MSc students, and four PhD students are working under his supervision on different projects. He has published research papers in peer-reviewed journals as well as several book chapters, popular brochures, abstract books, and popular articles in Urdu and English, some of which are aimed at the farming community.

Contents

Contributors

Yassmine Abidi
Laboratory of Field Crops, National Agricultural Research Institute of Tunisia (INRAT), University of Carthage, Tunis, Ariana Governorate, Tunisia

Tanveer Ahmad
Department of Horticulture, MNS University of Agriculture, Multan, Punjab, Pakistan

Muhammad Amjad
Institute of Horticultural Sciences, University of Agriculture, Faisalabad, Punjab, Pakistan

Jihen Ben Amor
Research Laboratory "Management of Horticultural Species in Organic and Conventional System," High Agronomic Institute of Chott Mariem, University of Sousse, Sousse, Sousse Governorate, Tunisia

Muhammad Anwar-Ul-Haq
Department of Horticulture, MNS University of Agriculture, Multan, Punjab, Pakistan

Chadha Ayed
Agrobiodiversity and Ecotoxicology, High Agronomic Institute of Chott-Mariem, University of Sousse, Chott Mariem, Sousse, Tunisia

Muhammad Azam
Laboratory of Pomology, Institute of Horticultural Sciences, Faculty of Agriculture, University of Agriculture, Faisalabad, Punjab, Pakistan

Mokhtar Baraket
National Research Institute of Rural Engineering, Water and Forests (INRGREF), Ariana Governorate, Tunisia

Beenish Chaudhary
Laboratory of Pomology, Institute of Horticultural Sciences, Faculty of Agriculture, University of Agriculture, Faisalabad, Punjab, Pakistan

Hela Chikh-Rouhou
Regional Research Center on Horticulture and Organic Agriculture (CRRHAB), Production and Protection for a Sustainable Horticulture, University of Sousse, Chott Mariem, Sousse, Tunisia

Marwa Chouikhi
Laboratory of Horticulture, National Agricultural Research Institute of Tunisia (INRAT), University of Carthage, Ariana Governorate, Tunisia

Bouthaina Dridi
Agrobiodiversity and Ecotoxicology, High Agronomic Institute of Chott-Mariem, University of Sousse, Chott Mariem, Sousse, Tunisia

Rym Fekih
Food Resources Education and Research Center, Graduate School of Agricultural Science, Kobe University, Kasai, Hyōgo Prefecture, Japan

Muhammad Awais Ghani
Institute of Horticultural Sciences, University of Agriculture, Faisalabad, Punjab, Pakistan

Marwa Hassine
Laboratory of Genetics and Cereal Breeding, National Agronomic Institute of Tunisia (INAT), Ariana Governorate, Tunisia

Chafik Hdider
Laboratory of Horticulture, National Agricultural Research Institute of Tunisia (INRAT), University of Carthage, Ariana Governorate, Tunisia

Imen Henane
Laboratory of Horticulture, National Agricultural Research Institute of Tunisia (INRAT), University of Carthage, Ariana Governorate, Tunisia

Hammad Hussain
Department of Horticulture, MNS University of Agriculture, Multan, Punjab, Pakistan

Riadh Ilahy
Laboratory of Horticulture, National Agricultural Research Institute of Tunisia (INRAT), University of Carthage, Ariana Governorate, Tunisia

Qumer Iqbal
Fiblast, LLC, Tuskegee, Alabama, USA

Faysal Ben Jeddi
Laboratory of Horticultural and Environmental Sciences, National Institute of Agronomy of Tunisia (INAT), University of Carthage, Ariana Governorate, Tunisia

Raoudha Khanfir Ben Jenana
Department of Horticultural Science, Higher Institute of Agronomy of Chott Mariem, University of Sousse, Chott Mariem, Sousse, Tunisia

Karima Kacem
Regional Research Center on Horticulture and Organic Agriculture (CRRHAB), Production and Protection for a Sustainable Horticulture, University of Sousse, Chott Mariem, Sousse, Tunisia

Nouri Khamassi
Laboratory of Horticulture, National Agricultural Research Institute of Tunisia (INRAT), University of Carthage, Ariana Governorate, Tunisia

Hayet Ben Haj Koubaier
Laboratory of Innovation and Valorization for a Sustainable Food Industry, Higher School of Food Industries of Tunis, University of Carthage, Tunis, Tunisia

Yosr Laayouni
Laboratory of Horticulture, National Agricultural Research Institute of Tunisia (INRAT), University of Carthage, Ariana Governorate, Tunisia

Hela Mahmoudi
Laboratory of Natural Substances, National Institute for Research and Physico-Chemical Analysis (INRAP), Biotechpole of Sidi Thabet, Ariana Governorate, Tunisia

Sonia Mansouri
Laboratory of Field Crops, National Agricultural Research Institute of Tunisia (INRAT), University of Carthage, Tunis, Ariana Governorate, Tunisia

Mouna Mechri
National Institute of Field Crops (INGC), Bousalem, Jendouba Governorate, Tunisia

Sana Medimagh
Laboratory of Field Crops, National Agricultural Research Institute of Tunisia (INRAT), University of Carthage, Tunis, Ariana Governorate, Tunisia

Najla Mezghani
National Gene Bank of Tunisia, Tunis, Tunis Governorate, Tunisia; Research Laboratory of Horticultural Crop, High Institute of Agronomy of Chott Mariem, University of Sousse, Chott Mariem, Sousse, Tunisia; Research Laboratory "Management of Horticultural Species in Organic and Conventional System," High Agronomic Institute of Chott Mariem, University of Sousse, Sousse, Sousse Governorate, Tunisia

Shiza Muqadas
Laboratory of Pomology, Institute of Horticultural Sciences, Faculty of Agriculture, University of Agriculture, Faisalabad, Punjab, Pakistan

Khaoula Nefzi
National Research Institute of Rural Engineering, Water and Forests (INRGREF), Ariana Governorate, Tunisia

Rashad Qadri
Laboratory of Pomology, Institute of Horticultural Sciences, Faculty of Agriculture, University of Agriculture, Faisalabad, Punjab, Pakistan

Anissa Riahi
Laboratory of Horticulture, National Agricultural Research Institute of Tunisia (INRAT), University of Carthage, Ariana Governorate, Tunisia; Higher School of Agriculture of Kef, University of Jendouba, Kef Governorate, Tunisia

Ons Riahi
Laboratory of Field Crops, National Agricultural Research Institute of Tunisia (INRAT), University of Carthage, Tunis, Ariana Governorate, Tunisia

Amani Romdhane
Laboratory of Horticulture, National Agricultural Research Institute of Tunisia (INRAT), University of Carthage, Ariana Governorate, Tunisia

Neji Tarchoun
Research Laboratory "Management of Horticultural Species in Organic and Conventional System," High Agronomic Institute of Chott Mariem, University of Sousse, Sousse, Sousse Governorate, Tunisia

Muhammad Tariq
Department of Horticulture, MNS University of Agriculture, Multan, Punjab, Pakistan

Imen Tlili
Laboratory of Horticulture, National Agricultural Research Institute of Tunisia (INRAT), University of Carthage, Ariana Governorate, Tunisia

Muhammad Ali Umer
Laboratory of Pomology, Institute of Horticultural Sciences, Faculty of Agriculture, University of Agriculture, Faisalabad, Punjab, Pakistan

Imen Zaouali
National Gene Bank of Tunisia, Tunis, Tunis Governorate, Tunisia

Jiaoke Zeng
Jiangxi Key Laboratory for Postharvest Preservation and Non-Destruction Testing of Fruits and Vegetables, College of Agriculture, Jiangxi Agricultural University, Nanchang, Jiangxi, China

Hatem Zgallai
National Agricultural Research Institute of Tunisia (INRAT), University of Carthage, Ariana Governorate, Tunisia

Jiukai Zhang
Agro-Product Safety Research Center, Chinese Academy of Inspection and Quarantine (CAIQ), Ronghua Nanlu, Yi Zhuang, Beijing, China

Abbreviations

2-VD	2-vinyl-4H-1,3-dithiin
AA	ascorbic acid
ABTS	2,2′-azino-bis (3-ethylbenzothiazoline-6-sulfonate)
ACE	angiotensin-converting enzyme
ACSO	S-alk(en)yl-l-cysteine sulfoxides
ADP	adenosine-5-diphosphate
AGE	aged garlic extract
ALP	alkaline phosphatase
ALT	alanine transaminase
AMPK	adenosine monophosphate-activated protein kinase
AMThs	allyl methyl thiosulfinates
AST	aspartate transaminase
ATP	adenosine triphosphate
ATPThs	allyl trans-1-propenyl thiosulfinates
BG	black garlic
BHA	butylated hydroxyanisole
CAS	controlled atmosphere storage
CNO	canola oil
CNS	central nervous system
CoQ	coenzyme Q
COX-2	cyclooxygenase-2
CRC	colorectal cancer
CRP	C-reactive protein
CUPRAC	cupric ion reducing antioxidant capacity
CVD	cardiovascular disease
DADS	diallyl disulfide
DAS	diallyl sulfide
DATS	diallyl trisulfide
DHAA	dehydroascorbic acid
DMC	dry matter content

DNA	deoxyribonucleic acid
DPPH	2,2-diphenyl-1-picryl-hydrazyl-hydrate
EO	essential oil
EPM	elevated plus-maze
ERK1/2	extracellular signal-regulated kinase 1/2
FAO	Food and Agriculture Organization
FDA	Food and Drug Administration
FOS	fructooligosaccharides
FRAP	ferric reducing antioxidant power
FV	fruits and vegetable
GAE	gallic acid equivalent
GI	glycemic index
GL	glucosinolate
GNLD	golden neo-life dynamite
GRAS	generally recognized as safe
H2S	hydrogen sulfide
I/R	ischemia-reperfusion
IBD	inflammatory bowel disease
IgA	immunoglobulin A
IL	interleukin
INF-γ	interferon gamma
iNOS	inducible nitric oxide synthase
ITC	isothiocyanate
KA	kainic acid
LDL	low-density lipoprotein
LDL-C	low-density lipoprotein-cholesterol
LDLP	low-density lipid protein
LPO	lipid peroxidation
LPS	lipopolysaccharide
MAPK	mitogen-activated protein kinase
MDA	malondialdehyde
MGMT	O-6-methylguanine-DNA-methyltransferase
MMP	matrix metalloproteinase
MYS	myrosinase enzyme
NF-κB	nuclear factor-kappa B
NK	natural killer

NO	nitric oxide
NSC	nonstructural carbohydrates
OP	open-pollinated
ORAC	oxygen radical absorbance capacity
OSC	organo-sulfur compounds
PC-8	plastochromanol-8
PUFA	polyunsaturated fatty acids
PVN	paraventricular nucleus
RDA	recommended daily amount
RNS	reactive nitrogen species
ROS	reactive oxygen species
SAC	S-allyl cysteine
SAMC	S-allyl-mercaptocysteine
SCFA	short-chain fatty acid
SFA	saturated fatty acids
STAT-1	signal transducer and activator of transcription-1
TAA	total ascorbic acid
TDF	total dietary fiber
TEAC	Trolox equivalent antioxidant capacity
TGF-β1	transforming growth factor-beta 1
TLR4	toll-like receptor 4
TNF-α	tumor necrosis factor-α
TSS	total soluble solids
WCRF	World Cancer Research Fund

Preface

The consumption of fresh produce is essential for a healthy and balanced diet, as outlined in recent research studies. In fact, their intake is positively associated with the prevention of various disorders and diseases.

The beneficial effects are often attributed to the specific bioactive compounds that possess potent antioxidant properties and can target and neutralize free radicals, thereby regulating the oxidant/antioxidant balance in human cells. In addition to a balanced mineral composition, a wide array of bioactive compounds from various classes are increasingly being identified in fresh produce, including vitamins, carotenoids, phenolics, flavonoids, tocopherols, anthocyanins, and glucosinolates. These compounds contribute to the functional quality of different fresh produce and help prevent various chronic diseases. Based on the above, numerous studies have been conducted—and others are ongoing—aiming to explore new bioactive compounds, understand their mechanisms of action, and promote non-traditional or underutilized sources of bioactives. This has led to a large number of scientific publications. However, there remains a lack of a comprehensive book compiling detailed information related to the nutritional composition, antioxidant properties, and health outcomes of the most commonly consumed fresh produce worldwide. The main objective of this book is to simplify and increase the availability of information on the nutritional content of fruits and vegetables, as well as the health outcomes of various agricultural products.

The present book is suitable for academic researchers, scientists, and food processors interested in food chemistry, nutrition, nutraceuticals, and functional foods. Additionally, the book is suitable for both undergraduate and postgraduate students in the fields of food science and technology, food chemistry, and/or nutrition. Moreover, with its accessible and straightforward content, the book is appealing to general readers seeking information on the link between the antioxidant composition of agricultural products and their health benefits.

This book is comprehensive in nature and includes a range of commonly consumed fresh produce. The book includes 11 chapters in total, with each chapter dedicated to a specific agricultural product. It has a simple layout and is divided into four parts:

- **Part I:** Inflorescence Flowers/Flower Buds (Cauliflower, Cabbage, Onion, Garlic)
- **Part II:** Leafy Vegetables (Lettuce, Spinach, Basil)
- **Part III:** Roots and Tubers (Carrot, Potato)
- **Part IV:** Cereals (Barley)
- **Part V:** Industrial Crops (Rapeseed/Canola)

Each chapter addresses various aspects of fresh produce, including health benefits, nutritional and antinutritional composition, antioxidant phytochemicals and their antioxidant potential, and factors influencing antioxidant content, including genotype, environmental variation, and agronomic conditions.

I would like to express my deep gratitude to all the experts and leaders from different research institutions worldwide who have provided valuable contributions to this book. This book would not have been possible without your admirable efforts and those of your teams. I am also thankful to Apple Academic Press for the successful production of this book. I acknowledge the Apple Academic Press editorial team for their prompt assistance and valuable advice.

I hope this book meets readers' expectations. Constructive comments and suggestions to further improve the book's content are welcome.

PART I

Inflorescence Flowers/ Flower Buds

CHAPTER 1

Cauliflower *(Brassica oleracea)*

SHIZA MUQADAS,[1] MUHAMMAD AZAM,[1,*] RASHAD QADRI,[1] JIUKAI ZHANG,[2] JIAOKE ZENG,[3] BEENISH CHAUDHARY,[1] and MUHAMMAD ALI UMER[1]

[1]*Laboratory of Pomology, Institute of Horticultural Sciences, Faculty of Agriculture, University of Agriculture, Faisalabad, Punjab, Pakistan*

[2]*Agro-Product Safety Research Center, Chinese Academy of Inspection and Quarantine (CAIQ), Ronghua Nanlu, Yi Zhuang, Beijing, China*

[3]*Jiangxi Key Laboratory for Postharvest Preservation and Non-Destruction Testing of Fruits and Vegetables, College of Agriculture, Jiangxi Agricultural University, Nanchang, Jiangxi, China*

*Corresponding author

ABSTRACT

Cauliflower (*Brassica oleracea* var. *botrytis*) is a major cruciferous vegetable with significant global production. This chapter reviews its botany, phytochemistry, and health-promoting properties. It is a rich source of bioactive compounds, including glucosinolates, their breakdown products (isothiocyanates), polyphenols, carotenoids, vitamins, and fiber. These compounds are responsible for cauliflower's potent antioxidant, anti-inflammatory, and anticancer activities. Specifically, isothiocyanates like sulforaphane modulate detoxification enzymes and induce apoptosis, offering chemopreventive benefits against gastrointestinal, prostate, and breast cancers. The vegetable also contributes to cardiovascular and digestive health. The chapter advocates for increased dietary consumption of

Postharvest Quality Attributes of Fresh Vegetables and Grains.
Riadh Ilahy, Imen Tlili, Thouraya R'him, and Muhammad Azam (Eds.)
© 2027 Apple Academic Press, Inc. Co-published with CRC Press (Taylor & Francis)

DOI:10.1201/9781779642851-2

cauliflower and related brassicas as a strategic, food-based approach to reducing chronic disease risk and promoting overall wellness.

1.1 CAULIFLOWER

Cauliflower (*Brassica oleracea*) is a widely grown brassica species, and it is cultivated as a vegetable for its distinctive aroma, super-white color, firm texture, and nutritional value.

1.2 INTRODUCTION

1.2.1 HISTORY

Cauliflower (*Brassica oleracea*) was introduced to the island of Cyprus in the middle of the 12th and 13th centuries by Arab botanists [1, 2]. Then it spread to Western Europe, where it was generally regarded as Cyprus colewort. The seeds of cauliflower are also very important, as in the 16th century, extensive trade in cauliflower seeds was conducted from Cyprus to Western Europe [3, 4]. In the middle of the 16th century, it was very popular due to its unique taste and crispy texture, with a fully starry white appearance [5]. It was spread from Genoa to France and has a specific place on grand tables due to its delicacy [6]. In 1822, the British adapted it in India. The name cauliflower originated from the Italian and Latin words "*cavolfiore*" and "*caulisflos*," meaning "cabbage flower" [7, 8]. Cyma is well known across *Brassica* species for its pleasant taste and flowering head [9]. In English, it is known as "broccoli" or "heading broccoli"; in Spanish, it is called "coliflor"; in French, it is called "choufleur"; and in Chinese, it is known as "*Hua ye cai*," but it has the preferred common name "cauliflower" [10]. It was used in Germany in the 18th century and includes the old cultivars Erfurt and Snowball. In Asia, tropical cauliflower was cultivated in the 19th century, including old varieties such as Early Benaras and Early Patna [7, 11].

1.2.2 PRODUCTION

According to FAOSTAT (Food and Agriculture Organization of the United Nations), world cauliflower production exceeded 20 million tons in 2020. The production of cauliflower globally increased, while the cultivated land area of cauliflower increased by 750,000 ha to 1,260,000 ha between the years 1994 and 2020. Cauliflower is mainly produced in China, the world's largest producer, followed by India, the second-largest producer. China and India together produce more than 70% of the world's total cauliflower. With 1,247,490 tons of production per year, the United States of America is the third largest producer of cauliflower and broccoli. In 2020, China ranked first among countries in cauliflower production at 37.21%, followed by India, the United States, Mexico, and Spain (Figures 1.1 and 1.2).

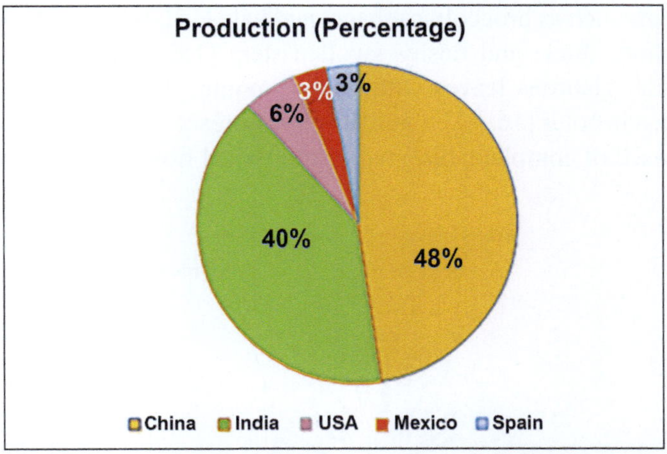

FIGURE 1.1 Production share of cauliflower by the top five countries in 2020.
Source: FAOSTAT (2021).

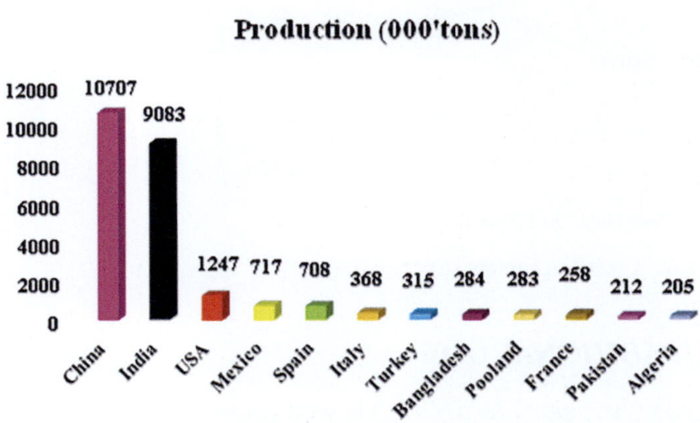

FIGURE 1.2 Top 10 countries in the production (1000 tons) of cauliflower (2021).
Source: FAOSTAT (2021).

1.2.3 BOTANICAL DESCRIPTION

Cauliflower is one of the well-known vegetables of the *Brassica* genus that belongs to the Brassicaceae, or Mustard family [12]. It is reproduces through seed and generally considered an annual plant. The cauliflower head is comprised of white inflorescence meristems [13]. The cauliflower differs from the broccoli head, as both have different flower buds that are regarded as edible [14]. Normally, the edible white flesh of cauliflower curd has the same appearance as cheese curd. It has different cultivar groups, termed Cole crops, such as broccoli, cabbage, and kale [15]. It has a taproot system with a short, thick, and fleshy, swollen stem [15]. The cauliflower plant has sessile, glabrous leaves with waxy, smooth, and curly edges, mostly blue-green in color [16, 17]. Cauliflower inflorescence, known as a raceme, is composed of complete, bisexual, symmetrical flowers (Figure 1.3).

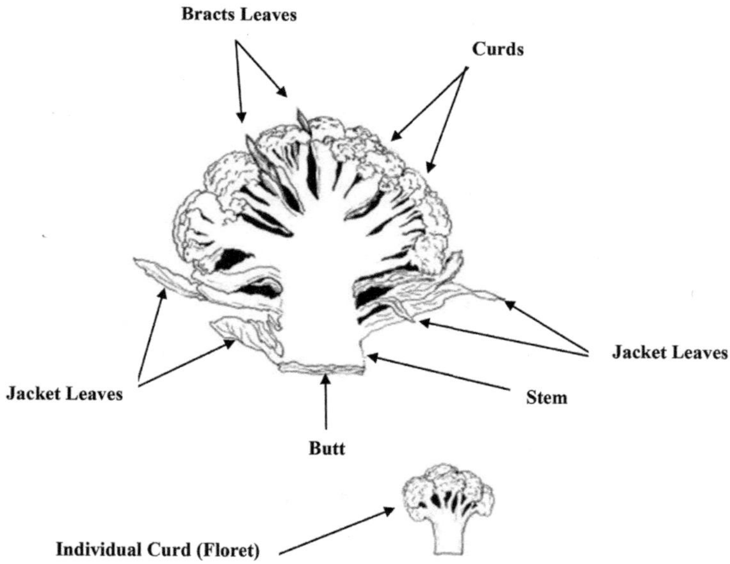

FIGURE 1.3 Cauliflower plant structure.

1.2.4 TRADITIONAL USES

Cauliflower (*Brassica oleracea* L.) is well-known among other vegetables for its unique taste and high nutritional value. Cauliflower has about 92% water, 5% carbohydrates, and 2% protein. A 100-gram serving of cauliflower has 25 kilocalories and a great content of vitamin C, and a modest amount of vitamins B and K, as shown in Table 1.1 [18]. It has low levels of dietary minerals and contains carotenoids and antioxidants, such as glucosinolates, which have anticancer properties [3]. It is widely used because it has high fiber content, several vitamins (i.e., B, C, E, and K), dietary fiber, and folic acid, omega-3 fatty acids, proteins, phosphorus, potassium, iron, magnesium, and manganese [19].

Besides, cauliflower leaves are a good source of beta-carotenoids, calcium, and iron. Generally, it has a high-water index. Cauliflower has about 3 grams of fiber, Vitamin C, and Vitamin K, with 77% and 20% of the Daily Value, respectively [20]. There are also 11% Vitamin B6, 7% pantothenic acid, 9% potassium, 8% manganese and magnesium, and 4% phosphorus present in cauliflower [18]. It is an excellent source of minerals, vitamins, antioxidants, and phytochemicals [21]. It improves health by reducing the burden of chronic diseases and regulating the immune system to prevent and reduce the prevalence of conditions such as cardiovascular disease, cancer, and tumors [22]. It can also be considered a value-added product for the treatment of anemia and micronutrient deficiencies. Cauliflower is a highly nutritious vegetable, and its head or curd is often used as food or for treatment. It stimulates the urination process and acts as a preventive measure against bladder cancer [23]. It is very helpful against anemia, menopause, scurvy, breast cancer, heart disease, diabetes, stroke, lung cancer, osteoporosis, and prostate cancer [16]. It is very low in calories yet high in vitamins, and it is used for weight loss. Cauliflower is mainly composed of glucosinolates and isothiocyanates (ITCs); these two groups of antioxidants help slow the growth of cancer cells [24].

TABLE 1.1 Dietary Structure of Cruciferous Vegetables

	Energy	Water Content (g)	Carbohydrates (g)	Fat (g)	Protein (g)	Fiber (g)
Broccoli	34	89.3	6.6	0.37	2.8	2.6
Brussels Sprouts	43	86	9	0.3	3.4	3.8
Cabbage	25	92.2	5.8	0.1	1.3	2.5
Cauliflower	25	92	5	0.3	1.9	2
Kale	49	84	8.8	0.9	4.3	3.6
Radish	16	95.3	3.4	0.1	0.7	1.6

Source: Wang et al. [18].

1.2.5 CAULIFLOWER AND BIOACTIVE COMPOUNDS

The flowering plant of the Brassicaceae family contains over 300 genera and 4000 species and is regarded as the largest plant family [20, 25, 26]. The Brassicaceae family is also known as the mustard family. Brassicaceae vegetables such as cauliflower are a rich source of biologically active compounds, that is, minerals, phytic acid, soluble sugars, tocopherols, carotenoids, folic acid, and vitamins C and E, present in the aerial parts of cauliflower, as demonstrated in Table 1.2 [16, 27–29]. Moreover, *Brassica oleracea* is rich in phosphorus, sulfur, chlorine, and potassium [29, 30].

TABLE 1.2 Total Phenolic, Flavonoid, and Chlorophyll Contents in the Aerial Parts of Cauliflower, Expressed as Mean ± SD ($n = 9$)

	Florets	Stems	Leaves
TPC (mg GAE/g DM)	2.76 ± 0.34^b	2.65 ± 0.52^b	4.40 ± 0.22^a
TFC (mg RUT/g DM)	0.29 ± 0.02^b	0.22 ± 0.02^b	0.64 ± 0.06^a
CHa (µg/g DM)	0.38 ± 0.10^b	0.34 ± 0.09^b	14.08 ± 0.63^a
CHb (µg/g DM)	0.66 ± 0.10^b	0.66 ± 0.10^b	8.36 ± 1.64^a
TCH (µg/g DM)	1.04 ± 0.14^b	1.00 ± 0.13^b	22.43 ± 2.10^a

Abbreviations: TPC: total phenolic content; TFC: total flavonoid content; CH_a: chlorophyll a; CH_b: chlorophyll b; TCH: total chlorophyll.

Note: Different letters in a row indicate a statistically significant (p-value < 0.05) differences between the samples.

Source: Drabińska et al. [23].

1.3 PHYTOCHEMICALS

1.3.1 ANTIOXIDANTS

Phenolic compounds are the reason for the highest antioxidant level in cauliflower, as shown in Table 1.3 [1]. Fruits and vegetables are a rich source of different antioxidant components like vitamin C, vitamin E, carotenoids, flavonoids, flavones, isoflavones, flavanones, anthocyanins, catechin, and isocatechin [31]. Free radicals (reactive oxygen or nitrogen molecules) cause damage to DNA, lipid membranes, and proteins [32]. These damages may cause cancer, inflammation, and heart disease. Antioxidants such as vitamins, carotenoids, dietary glutathione, endogenous metabolites, phenolic acids, and flavonoids can scavenge free radicals and quench singlet and triplet oxygen [33]. Oxygen, an element essential for life, has an adverse effect on the human body. The oxidation process is important to living organisms [34].

TABLE 1.3 HPLC Analysis of Phenolic Compound (mg/100 g DW) of Raw and Processed Cauliflower

Phenolic Compounds (mg/100 g DW)	Raw	Steam Blanching	Water Blanching	Steam Boiling	Water Boiling	Microwaving	Stir-Frying
Pyrogallol	18.9	13.71	6.84	11.54	4.84	14.52	12.47
Catechin	1.91	0.85	–	0.71	–	–	–
Protocatechuic	192.4	78.78	37.23	66.10	23.04	67.41	52.38
Catechol	4.90	0.70	0.60	0.60	0.34	0.45	–
Chlorogenic acid	25.83	11.36	–	8.14	–	12.36	10.45
Rosmarinic acid	1.73	–	–	–	–	–	–
Rutin	2.57	0.24	0.01	0.01	ND	0.12	–
Caffeic acid	0.77	0.44	0.21	–	–	–	1.644
Vanillic acid	11.90	0.031	–	–	–	–	–
Quercetin	202.46	85.20	45.13	54.19	22.35	61.78	58.74
Naringenin	1.81	–	–	–	–	–	–
Syringic acid	1.91	–	–	–	–	–	–
Coumaric acid	6.94	3.69	2.94	3.48	1.54	2.62	2.10
Cinnamic acid	3.10	0.35	ND	0.25	0.24	–	–
Kaempferol	25.91	15.12	10.91	9.21	7.34	11.36	11.84

Source: Ahmed and Ali [1].

The production of reactive oxygen species (ROS) and the uncontrolled structure of antioxidant protection cause many diseases. ROS are a class of highly reactive molecules established in aerobic life in living organisms and include superoxide anion radicals ($O_2{\bullet}^-$), hydroxyl radicals (OH•), and non-free-radical species [35]. The generation of ROS and their inactivation by the antioxidant system in organisms are emphasized by a balanced pathway [23]. However, differences in ROS and antioxidant defense mechanisms means that ROS contribute to oxidative alterations in cellular membranes and intracellular molecules. In immunological situations or under oxidative stress, ROS are produced during peroxidation of membrane lipids, resulting in lipid peroxides [36], which can be controlled by an antioxidant defense system. Antioxidants are known for their protective property and lower the oxidative damage from ROS, slowing down the progress of many chronic diseases, as well as lipid peroxidation [37].

Antioxidants are believed to inhibit or delay the oxidation of other molecules, as they slow down the initiation or propagation of oxidizing chain reactions, such as butylated hydroxyanisole (BHA), butylated hydroxytoluene (BHT), propyl gallate, and tert-butylhydroquinone, which are the most important antioxidants due to their toxicity and possible carcinogenicity [33, 37]. Cauliflower contains abundant amounts of carotenoids, flavonoids, and antioxidants, which have anticancer properties and reduce the risk of numerous other illnesses, including heart disease [32]. A high vitamin C content in cauliflower supports the immune system, reducing the risk of heart disease and cancer [38]. Pigmented curd cauliflower is different from white head cauliflower due to different contents of chlorophyll, anthocyanins, and carotenoids, as demonstrated in Table 1.4 [39]. So, chlorophyll is responsible for the green color, purple color is due to anthocyanins, and the yellowish-orange color is due to carotenoids [40].

TABLE 1.4 Composition of Phenolic Compounds

Compounds	White Cauliflower	Purple Cauliflower
	Phenolic Acid	
Ferulic acid	53.20 ± 0.15	2450.10 ± 16.40
Sinapic acid	436.21 ± 1.84	555.84 ± 4.69
Vanillic acid	224.56 ± 1.38	9103.71 ± 32.49
Total Phenolic Acids	1347.33 ± 8.99	21153.59 ± 97.20
	Flavonols (μg/g d.w.)	
Quercetin	8.51 ± 0.06	32.63 ± 0.08
Isorhamnetin	0.59 ± 0.01	4.59 ± 0.02
Kaempferol	8.97 ± 0.02	13.06 ± 0.05
Total flavanols	18.07 ± 0.06	50.27 ± 0.086
Anthocyanins (μg/g d.w.)	–	–
Cy-3-diGlu-5-Glu	–	99.29 ± 5.23
Cy-3-Glu-5-Glu	–	9.67 ± 0.60
Total Anthocyanins		3923.05 ± 112.97

Note: Results represent ± SD from independent determinations, where, WCau (white cauliflower) and PCau (purple cauliflower).

Source: Koss-Mikołajczyk et al. [39].

High levels of solar radiation and temperature fluctuations mainly increase anthocyanin levels in cauliflower. It has been noted that climatic conditions allowed the maximum expression of color fluctuation, as the color expression is mainly achieved by variations in temperature between day and night [1]. Both anthocyanins and carotenoids are antioxidant compounds that exhibit antiradical activity. In addition to these compounds, the green and violet curd tissues also contain significant levels of ascorbic acid (AA) and glucosinolates. Many studies have classified cauliflower as a functional food due to its nutritive and health benefits, as it contains a certain amount of amino acids, as represented in Table 1.5 [16]. Glucosinolates (GLs), atypical β-thioesters, comprise more than 100 compounds found in Brassicaceae species. A significant fluctuation has been observed in their amount and composition, depending on several factors [38, 41–43].

TABLE 1.5 The Content of Individual Amino Acids in the Aerial Parts of Cauliflower Expressed as Mean ± SD ($n = 9$) in mmol/ g DM

Amino Acids	Florets	Leaves	Stems	Category of Amino Acids
Valine	106.54 ± 23.81^{a}	60.66 ± 11.15^{b}	100.01 ± 24.52^{a}	Essential
Leucine	62.7 ± 11.11^{a}	10.30 ± 2.93^{c}	23.93 ± 6.2^{b}	Essential
Isoleucine	64.2 ± 15.64^{a}	16.43 ± 2.07^{c}	41.25 ± 7.23^{b}	Essential
Threonine	54.47 ± 6.56^{b}	75.14 ± 8.74^{a}	50.4 ± 14.39^{b}	Essential
Tryptophan	15.32 ± 5.66^{a}	9.13 ± 2.1^{b}	$12.66 \pm 5.25^{a,b}$	Essential
Alanine	432.07 ± 53.24^{a}	332.01 ± 46.53^{b}	341.32 ± 68.92^{b}	Non-essential
Glycine	29.54 ± 4.08^{a}	22.25 ± 5.01^{b}	33.02 ± 5.51^{a}	Non-essential
Serine	224.1 ± 32.93^{b}	276.51 ± 28.78^{a}	268.31 ± 48.04^{a}	Non-essential
Proline	21.27 ± 3.07^{a}	22.62 ± 4.07^{a}	21.18 ± 3.99^{a}	Non-essential

Note: Different letters in a row indicate a statistically significant (p-value < 0.05) difference.

Source: Drabińska et al. [16].

1.3.2 GLUCOSINOLATES

Glucosinolates are well-known secondary metabolites present in cauliflower, broccoli, cabbage, and Brussels sprouts, and are found in the aerial parts of cauliflower, as shown in Table 1.6 [44]. Glucosinolates are water-soluble anions, but glucosinolates in plants are biologically inactive until they are hydrolyzed through myrosinase, either by processing or chewing, as it releases the β-thioglucosidase, or by gut microbiota action. Breakdown processing occurs based on pH level, generally involving nitriles, epithionitriles, thiocyanates, and epithioalkanes [45].

TABLE 1.6 The Content of Individual Glucosinolates in the Aerial Parts of Cauliflower Expressed as Mean ± SD ($n = 9$) in mmol/ g DM

Glucosinolate	Florets	Leaves	Stems
Progoitrin	6.8 ± 1.58^{c}	9.69 ± 2.02^{b}	14.83 ± 2.61^{a}
Sinigrin	10.54 ± 2.57^{b}	7.48 ± 3.22^{b}	17.46 ± 6.71^{a}
4-Hydroxyglucobrassicin	13.78 ± 3.32^{b}	23.48 ± 4.65^{a}	16.84 ± 6.13^{b}
Glucobrassicin	32.21 ± 3.49^{b}	59.85 ± 11.96^{a}	16.09 ± 2.55^{c}
4-Methylglucobrassicin	10.22 ± 2.16^{c}	51.88 ± 9.27^{a}	16.61 ± 2.3^{b}
Neoglucobrassicin	8.17 ± 3.54^{c}	62.7 ± 13.6^{a}	25.49 ± 2.96^{b}
Total Glucosinolates	94.02 ± 20.71^{b}	225.2 ± 48.88^{a}	125.05 ± 25.39^{b}

Note: Different letters in a row indicate statistically significant (p-value < 0.05) differences. The total glucosinolate content is presented in bold.

Source: Drabińska et al. [23].

Glucosinolates are used in the food industry to maintain the organoleptic characteristics of food products and act as antimicrobials, facilitating the inactivation of food-borne pathogens when combined with conventional processing and non-conventional preservation technologies [46, 47]. These are generally present in medicine due to their anticancer properties. Cauliflower is a nutritious vegetable due to its antioxidant properties, which help protect body cells from harmful free radicals and inflammation. In cruciferous vegetables, cauliflower is particularly rich in glucosinolates and ITCs, which slow the growth and spread of cancer cells, as shown in Table 1.7 [15, 48]. In several studies, it was reported that glucosinolates and ITCs have properties that act against colon, lung, breast, and prostate cancer [49].

However, GLS is hydrolyzed by the myrosinase enzyme (MYS) into ITCs upon absorption in humans. These compounds are responsible for curing and treating human diseases [10, 50]. Glucoraphanin is present in cauliflower in the pigmented curds, but white head cultivars do not have these compounds [4]. On the other hand, it is present in low amounts, indicating a toxic effect on animal organisms, including humans. Despite this, many experiments showed that colored curds have more antioxidant and

TABLE 1.7 Individual and Total Glucosinolates (GLs) Content in Various Plant Food Materials

Food Matrix	Sinigrin	Glucoibervirin	Glucoiberin	Gluconapin	Progoitrin	Glucoerucin
Broccoli	0.0–1.6	0.0	1.4–11.2	0.0–0.2	0–3.7	1.1–3.7
Brussels sprouts	3.9–22.7	0–0.2	2.4–18.9	0.5–12.2	1.0–25.4	0.4–1.2
Cauliflower	2.9–14.6	1.6–7.4	0.0–22.8	0.0–0.1	0.0	0.2–1.3
Kale	0–28.7	0.0	0.0–67.8	0.0–4.1	0.0–158.1	0.0–0.9
Mustard green	693.9–779.4	0.0	0.0	2.7–8.5	0.0	0.0
Kohlrabi	0.0	9.7	2.3	0.0	0.3	33.6
Vegetable Seeds						
Broccoli	0.01–0.49	0.0–0.37	0.0–1.45	0.05–0.88	0.0–2.66	0.98–2.84
Brussels sprouts	0.20–2.24	0.02–0.35	0.0–1.13	0.49–1.55	1.70–6.02	0.86–1.28
Callards	2.55–6.30	0.10–0.50	0.11–1.97	0.60–1.00	2.39–5.96	0.31–0.47
Kale	0.1–5.4	0.0–0.30	0.0–0.73	1.19–4.10	3.97–5.08	0.15–0.64
Mustard green	12.5	0.05	0.20	0.15	0.0	0.04
Kohlrabi	0.05	0.40	2.74	0.05	0.19	2.41

Note: Individual GLs were measured: (a) content of glucosinolates (GLs) (μmol/100 g fresh wt.); (b) content of GLs (mmol/100 g defatted seed).

Source: Deng et al. [15].

antiradical activity than white curd [51]. It has been reported that different levels of isothiocyanates in cauliflower contribute to its health benefits. Several epidemiological studies have shown that it reduces cancer risk, gastrointestinal disorders, and heart diseases [52]. Hence, cauliflower has a suitable range of glucosinolate precursors, which are mainly converted in ITCs and have been developed [37, 47].

1.3.3 ISOTHIOCYANATES

Isothiocyanates are known as hydrolytic products of glucosinolates due to their anticarcinogenic properties [12]. Allylisothiocyanate is considered a hydrolytic product of glucoraphanin and sulforaphane, which is a hydrolytic product of the glucosinolate sinigrin [53]. They are the dominant ITCs present in cauliflower, as represented in Table 1.8 [39]. ITCs are very helpful in regulating the metabolic enzymes present in the liver and epithelial cells, thereby adjusting the balance of the metabolic pathways of carcinogenic compounds [54]. Meanwhile, phenethyl isothiocyanate is used to stimulate the pathways of genes that trigger cancer development, either through cell cycle regulation or antioxidant response [17].

TABLE 1.8 Isothiocyanates (ITCs) Found in Food

Compounds	White Cabbage	Red Cabbage	White Cauliflower	Purple Cauliflower
3-(Methyl sulfinyl)-propyl-ITC	0.29 ±0.036	–	0.094 ± 0.006	0.216 ± 0.019
4-(Methyl sulfinyl)-butyl-ITC	0.179 ± 0.008	0.071 ± 0.006	–	0.051 ± 0.002
2-Propenyl-ITC	0.259 ± 0.019	–	0.070 ± 0.003	0.272 ± 0.009
3-(Methylthio)-propyl-ITC	0.018 ± 0.001	–	–	–

Note: Results represent ± SD from independent determinations.

Source: Koss-Mikołajczyk et al. [38].

1.3.1.1 HEALTH BENEFITS AND BIOLOGICAL ACTIVITIES

Cauliflower belongs to the cruciferous family of vegetables, which is well known for its effectiveness against certain forms of cancer [14]. Cauliflower contains several phytochemicals and has a high nutritive value of riboflavin, niacin, vitamin A, vitamin C, calcium, iron, phosphorus, fat, and thiamine [55]. Cauliflower is well-suited as a vegetable or as an ingredient in salads, soups, and other dishes. The dietary quality of cauliflower has a contiguous structure of polyphenols, total carotenoids, flavonoids, and vitamin C, which confer defensive properties against several diseases, such as cardiovascular diseases (CVDs), cancer, and tumors, as shown in Table 1.9 [16, 47].

TABLE 1.9 The Proximate Composition of the Aerial Parts of Cauliflower, Expressed in %, as the Mean ± SD ($n = 9$)

	Florets	Leaves	Stems
Ash	$9.47 \pm 0.53^{b,1}$	10.13 ± 0.64^{b}	11.78 ± 0.74^{a}
Protein	23.28 ± 1.46^{a}	20.74 ± 1.68^{b}	19.12 ± 1.57^{b}
Fat	1.94 ± 0.24^{b}	4.68 ± 0.35^{a}	4.74 ± 0.40^{a}
Carbohydrates[2]	65.31 ± 2.31^{a}	64.45 ± 2.80^{a}	64.36 ± 2.91^{a}

Note: 1. Different letters in a row indicate statistically significant (p-value < 0.05) differences between the samples.

2. Calculated from the difference.

Source: Drabińska et al. [16].

1.3.4 CAULIFLOWER AND HEART DISEASES

Cauliflower is known for its antioxidant activity , which helps protect against harmful free radicals. Cauliflower is rich in glucosinolates and ITCs that retard the growth and spread of cancer cells. It is demonstrated that glucosinolates and ITCs act against colon, lung, breast, and prostate cancer [39, 56]. Cauliflower is responsible for a high amount of carotenoid and flavonoid antioxidants as well, and has anticancer properties, and it reduces the risk of numerous further disorders, like heart disease. It has an anti-inflammatory effect on the immune system, reducing the risk of heart disease and cancer [22, 30, 57].

CVD is the main cause of death in the world every year. According to the Dietary Guidelines for Americans and the American Heart Association, it is recommended to consume 5 servings of fruits and vegetables (FVs) daily to reduce the incidence of heart disease [58]. Still, FVs act as carriers of pesticide residues, as several pesticides, such as organophosphates, organochlorines, pyrethroids, and atrazine, are known to stimulate cellular oxidative stress and increase apoptosis [3, 6]. The risk factors involved in the onset and progression of cardiovascular events [59, 60]. Cauliflower has a high fiber content; one cup provides 3 grams of fiber, which fulfill about 10% of our daily fiber intake [61]. Fiber is an important part of our diet, as it feeds the healthy bacteria in our gut, which help fight inflammation and promote digestive health [62]. Using adequate amounts of fiber improved gastrointestinal situations, that is, constipation, diverticulitis, and inflammatory bowel disease (IBD). Many experiments have shown that a high fiber intake in the diet is associated with a lower risk of several illnesses, such as CVD, as fiber also plays a vital role in obesity prevention by reducing overall calorie intake [5, 13, 57, 63].

1.3.5 CAULIFLOWER AND CANCER

ROS produced by normal cellular metabolism are considered to play substantial roles in stimulating signaling pathways in plant and animal cells, counteracting changes in intra- and extracellular environmental conditions [47]. Cell structure may be deliberately damaged by ROS produced over a prolonged period under stress conditions. It may induce somatic mutations and neoplastic transformation [30]. Pro-tumorigenic effects of oxidative stress lead to cancer induction and development, such as expanding DNA mutation rate, inducing DNA damage, genome instability, and cell proliferation [52].

Glucosinolates and other dietary sources present in cauliflower are used to protect from chronic diseases such as colorectal cancer (CRC), prostate cancer, breast cancer, and myocardial infarction [64]. Glucosinolates are known as sulfur- and nitrogen-containing compounds found in many plants and vegetables and are derived from amino acids and glucose [65]. They are converted into ITCs of hydrolysis. Glucosinolates, ITCs, and indole products have been observed to induce apoptosis, slowing tumor growth by blocking the phase 1 enzyme, altering estrogen metabolism, and inhibiting the cell cycle. They also play an anti-inflammatory role by blocking the release of histamine, which leads to the formation of inflammation [55].

ITCs, the sulfurafane hydrolytic product of glucoraphanin (4-methylsulfinylbutyl), exhibit strong antioxidant activity and protective effects against several types of cancer, including prostate cancer. The anticancer properties of cauliflower in cancers and other tumors are associated with bioactive compounds, such as glucosinolates [26]. It has low levels of systemic oxidative stress by constraining angiogenesis and cell cycle progression and promoting apoptosis of cancer cells [66]. Glucosinolates have preventive and therapeutic properties against cancer and other tumor types, such as prostate cancer and breast cancer [29, 64, 67]. Erucin is a usual isothiocyanate found in cauliflower, as it is specific in the treatment of cancer [53]. It is used on a pancreatic adenocarcinoma cell line that exhibits antiproliferative and cytotoxic properties and inhibits cell migration.

A specific gene characterizes these cells: the KRAS gene mutation, the protein kinase pathway, and the Akt/protein kinase B pathway [14, 53]. Cauliflower contains a consistent number of bioactive compounds that can affect the occurrence and development of cancer cells through multiple mechanisms of action. Glycosylation of triterpenes, which are known as defense mechanisms in cauliflower, has properties against inflammation and microbial action [68]. It inhibits the spread of cancer growth and triggers genes involved in inflammation. These include chemokines, cytokines such as IL-1, IL-6, IL-12, and TNF-α, and adhesion molecules [53]. An experiment examined sulforaphane's effect on receptor oligomerization, which reduced cytokine-stimulating production [63, 65]. Substantially, sulforaphane and DIM suppress the phosphorylation of IKK/IkB and reduce the translocation of the nuclear subunit, impacting key inflammatory facilitators [17]. It is also capable of initiating activation through the *Nrf2* gene. It can restrain the NF-κB pathway between two transcriptional factors [18].

1.3.6 CAULIFLOWER AND OTHER HEALTH BENEFITS

Cauliflower is a good source of polyphenols, namely flavonoids, phenylpropanoids, hydroxycinnamic acids, and lignans [29, 30, 69]. These polyphenolic phytoproducts have anticancer and antioxidant properties, as they can prevent oxidative stress and induce apoptosis [28, 70]. Generally, quercetin, a flavonoid present in high amounts in cauliflower, can initiate mitochondrial apoptotic-supported cancer stem cell growth [71]. Carotenoids and tocopherols are present in plentiful amounts in cauliflower [15]. It is highly pigmented and has antioxidant properties due to its yellow, orange, and carotenoid pigments, including lycopene, α-carotene, β-carotene, and γ-carotene [49, 68]. Lipid-soluble tocopherols are considered to be primarily found in Brassicaceae vegetables [45, 29]. β-carotene has anticancer properties against many GI cancers [53, 62]. A recent experiment demonstrated that pre-treatment of human gastric cells with β-carotene can reduce *Helicobacter pylori*-1 by inducing ROS production, activating NADPH oxidase, and increasing *TRAF1* and *TRAF2* gene expression in gastric cells [67, 72].

1.4 CONCLUSION AND PROSPECTS

Cauliflower plays a role in for digestion, absorption, excretion, and protection. The GI lining epithelium is a continuous process with rapidly regenerating tissue. It is well known for its restorative and regenerative abilities, but it is quite susceptible to cancer-causing agents that can transform it into tumor cells. GI cancers are categorized as colorectal, liver, and stomach cancers and are considered a main cause of cancer-related deaths, with a 5% survival rate. The standard treatment for GI cancers is surgery, radiotherapy, and chemotherapy. The use of phytocompounds acts as a preventive and healing agent. Cauliflower has a high range of polyphenols, carotenoids, tocopherols, and vitamins. Glucosinolates are an important phytochemical group in cauliflower due to their cancer-antagonistic activity through their products, ITCs and indoles, but are used sparingly. For this, the intake of cauliflower should be promoted for its health benefits and as an effective practice for treating cancerous cells. Now is the time to modify to improve the use and intake of foods rich in glucosinolates, polyphenols, and ITCs to improve digestive conditions. We should pursue these efforts *in vitro* and *in vivo* that are responsible for improving the phytochemical content of cauliflower, thereby lowering the risk of chronic diseases. It has been observed that chemotherapeutic agents obtained from Cruciferae vegetables have the ability to enhance their anticancer activity when used clinically, either alone or in combination with chemotherapeutic agents.

KEYWORDS

- **cauliflower**
- **cruciferous vegetables**
- **gastrointestinal cancers**
- **glucosinolates**
- **isothiocyanates**
- **phytochemicals**

REFERENCES

1. Ahmed, F. A., & Ali, R. F. (2013). Bioactive compounds and antioxidant activity of fresh and processed white cauliflower. *BioMed Research International, 2013*, 1–9.
2. Aked, J. (2002). Maintaining the postharvest quality of fruits and vegetables. In *Fruit and Vegetable Processing: Improving Quality* (pp. 119–149). CABI.
3. Ares, A. M., Nozal, M. J., & Bernal, J. (2013). Extraction, chemical characterization and biological activity determination of broccoli health promoting compounds. *Journal of Chromatography A, 1313*, 78–95.
4. Bhattacharya, A., Li, Y., Wade, K. L., Paonessa, J. D., Fahey, J. W., & Zhang, Y. (2010). Allyl isothiocyanate-rich mustard seed powder inhibits bladder cancer growth and muscle invasion. *Carcinogenesis, 31*(12), 2105–2110.
5. Chen, Q. H., Wu, B. K., Pan, D., Sang, L. X., & Chang, B. (2021). Beta-carotene and its protective effect on gastric cancer. *World Journal of Clinical Cases, 9*, 6591–6607.
6. Deer, E. L., González-Hernández, J., Coursen, J. D., Shea, J. E., Ngatia, J., Scaife, C. L., Firpo, M. A., & Mulvihill, S. J. (2010). Phenotype and genotype of pancreatic cancer cell lines. *Pancreas, 39*, 425–435.
7. Bell, L., Oloyede, O. O., Lignou, S., Wagstaff, C., & Methven, L. (2018). Taste and flavor perceptions of glucosinolates, isothiocyanates, and related compounds. *Molecular Nutrition & Food Research, 62*, 17001990.
8. Egner, P. A., Chen, J. G., & Wang, J. B. (2011). Bioavailability of sulforaphane from two broccoli sprout beverages: Results of a short-term, cross-over clinical trial in Qidong, China. *Cancer Prevention Research, 4*, 384–395.
9. Berrang, M. E., Brackett, R. E., & Beuchat, L. R. (1990). Microbial, color and textural qualities of fresh asparagus, broccoli, and cauliflower stored under controlled atmosphere. *Journal of Food Protection, 53*, 391–395.
10. Dhall, R. K., Sharma, S. R., & Mahajan, B. V. C. (2010). Effect of packaging on storage life and quality of cauliflower stored at low temperature. *Journal of Food Science and Technology, 47*(1), 132–135.
11. Avato, P., & Argentieri, M. P. (2015). Brassicaceae: A rich source of health improving phytochemicals. *Phytochemistry Reviews, 14*, 1019–1033.
12. Borek, C. (2004). Dietary antioxidants and human cancer. *Integrative Cancer Therapies, 3*(4), 333–341.
13. Chaudhary, A., Sharma, U., Vig, A. P., Singh, B., & Arora, S. (2014). Free radical scavenging, antiproliferative activities and profiling of variations in the level of phytochemicals in different parts of broccoli (*Brassica oleracea italica*). *Food Chemistry, 148*, 373–380.
14. Branca, F., & Ferrari, M. (2002). Impact of micronutrient deficiencies on growth: The stunting syndrome. *Annals of Nutrition and Metabolism, 46*, 8–17.
15. Deng, Q., Zinoviadou, K. G., Galanakis, C. M., Orlien, V., Grimi, N., Vorobiev, E., Lebovka, N., & Barba, F. J. (2015). The effects of conventional and non-conventional processing on glucosinolates and its derived forms, isothiocyanates: Extraction, degradation, and applications. *Food Engineering Reviews, 7*(3), 357–381.
16. Drabińska, N., Jeż, M., & Nogueira, M. (2021). Variation in the accumulation of phytochemicals and their bioactive properties among the aerial parts of cauliflower. *Antioxidants, 10*(10), 1597–1619.

17. Sturm, C., & Wagner, A. E. (2017). *Brassica*-derived plant bioactive as modulators of chemopreventive and inflammatory signaling pathways. *International Journal of Molecular Sciences, 18*, 1890–1916.
18. Wang, Z., Xu, G., Ma, P., Lin, Y., Yang, X., & Cao, C. (2017). Isolation and characterization of a phosphorus-solubilizing bacterium from rhizosphere soils and its colonization of Chinese cabbage (*Brassica campestris* ssp. *chinensis*). *Frontiers in Microbiology, 8*, 1270–1288.
19. Florkiewicz, A., Filipiak-Florkiewicz, K., Topolska, E., Cieślik, R. B., & Kostogrys, R. B. (2014). The effect of technological processing on the chemical composition of cauliflower. *Italian Journal of Food Science, 26*, 275–281.
20. Cheng, A. C., Shen, C. J., Hung, C. M., & Hsu, Y. C. (2019). Sulforaphane decrease of SERTAD1 expression triggers G1/S arrest in breast cancer cells. *Journal of Medicinal Food, 22*, 444–450.
21. Ekström, A. M., Serafini, M., Nyrén, O., Wolk, A., Bosetti, C., & Bellocco, R. (2011). Dietary quercetin intake and risk of gastric cancer: Results from a population-based study in Sweden. *Annals of Oncology: Official Journal of the European Society for Medical Oncology, 22*, 438–443.
22. Nugrahedi, P. Y., Verkerk, R., Widianarko, B., & Dekker, M. (2014). A mechanistic perspective on process-induced changes in glucosinolate content in *Brassica* vegetables: A review. *Critical Reviews in Food Science and Nutrition, 55*(6), 823–838.
23. López-Hernández, A. A., Ortega-Villarreal, A. S., Rodríguez, J. A. V., Lomelí, M. L. C., & González-Martínez, B. E. (2022). Application of different cooking methods to improve nutritional quality of broccoli (*Brassica oleracea* var. *italica*) regarding its compounds content with antioxidant activity. *International Journal of Gastronomy and Food Science, 28*, 10–51.
24. Lo Scalzo, R., Picchi, V., Migliori, C. A., Campanelli, G., Leteo, F., Ferrari, V., & Di Carolo, L. F. (2013). Variations in the phytochemical contents and antioxidant capacity of organically and conventionally grown Italian cauliflower (*Brassica oleracea* L. subsp. *botrytis*): Results from a three-year field study. *Journal of Agricultural and Food Chemistry, 61*(43), 10335–10344.
25. Niranjana, R., Gayathri, R., Nimish, M. S., Sugawara, T., Hirata, T., Miyashita, K., & Ganesan, P. (2015). Carotenoids modulate the hallmarks of cancer cells. *Journal of Functional Foods, 18*, 968–985.
26. Park, Y., Lee, H., Lim, J. W., & Kim, H. (2019). Inhibitory effect of β-carotene on *Helicobacter pylori*-induced TRAF expression and hyper-proliferation in gastric epithelial cells. *Antioxidants, 8*, 637–661.
27. Pappa, G., Lichtenberg, M., Iori, R., Barillari, J., Bartsch, H., & Gerhäuser, C. (2006). Comparison of growth inhibition profiles and mechanisms of apoptosis induction in human colon cancer cell lines by isothiocyanates and indoles from *Brassicaceae*. *Mutation Research/Fundamental and Molecular Mechanisms of Mutagenesis, 599*, 76–87.
28. Park, J. E., Sun, Y., Lim, S. K., Tam, J. P., Dekker, M., Chen, H., & Sze, S. K. (2017). Dietary phytochemical PEITC restricts tumor development via modulation of epigenetic writers and erasers. *Scientific Reports, 7*, 405–469.

29. Zhou, Q., Tang, H., Jia, X., Zheng, C., Huang, F., & Zhang, M. (2018). Distribution of glucosinolate and pungent odors in rapeseed oils from raw and microwaved seeds. *International Journal of Food Properties, 21*, 2296–2308.
30. Ishida, M., Hara, M., Fukino, N., Kakizaki, T., & Morimitsu, Y. (2014). Glucosinolate metabolism, functionality and breeding for the improvement of *Brassicaceae* vegetables. *Breeding Science, 64*, 48–59.
31. Raiola, A., Errico, A., Petruk, G., Monti, D. M., Barone, A., & Rigano, M. M. (2018). Bioactive compounds in *Brassicaceae* vegetables with a role in the prevention of chronic diseases. *Molecules, 23*, 15–26.
32. Mandrich, L., & Caputo, E. (2020). *Brassicaceae*-derived anticancer agents: Towards a green approach to beat cancer. *Nutrients, 12*, 868–878.
33. Peñas, E., Zielinska, D., Gulewicz, P., Zielinski, H., & Frias, J. (2018). Vitamin C, phenolic compounds, and antioxidant capacity of broccoli florets grown under different nitrogen treatments combined with selenium. *Polish Journal of Food and Nutrition Sciences, 68*(2), 1–11.
34. Franzke, A., Lysak, M. A., Al-Shehbaz, I. A., Koch, M. A., & Mummenhoff, K. (2011). Cabbage family affairs: The evolutionary history of *Brassicaceae*. *Trends in Plant Science, 16*, 108–116.
35. Krinsky, N. I. (1998). The antioxidant and biological properties of the carotenoids. *Annals of the New York Academy of Sciences, 854*(1), 443–447.
36. Long, J., Guan, P., Hu, X., Yang, L., He, L., Lin, Q., Luo, F., Li, J., He, X., & Du, Z. (2021). Natural polyphenols as targeted modulators in colon cancer: Molecular mechanisms and applications. *Frontiers in Immunology, 12*, 435–484.
37. Li, H., Deng, Z., Zhu, H., Hu, C., Liu, R., Young, J. C., & Tsao, R. (2012). Highly pigmented vegetables: Anthocyanin compositions and their role in antioxidant activities. *Food Research International, 46*(1), 250–259.
38. Soundararajan, P., & Kim, J. S. (2018). Anticarcinogenic glucosinolates in cruciferous vegetables and their antagonistic effects on prevention of cancers. *Molecules, 23*(11), 2983–3212.
39. Koss-Mikołajczyk, I., Kusznierewicz, B., Wiczkowski, W., Płatosz, N., & Bartoszek, A. (2019). Phytochemical composition and biological activities of differently pigmented cabbage (*Brassica oleracea* var. *capitata*) and cauliflower (*Brassica oleracea* var. *botrytis*) varieties. *Journal of the Science of Food and Agriculture, 99*(12), 5499–5507.
40. Zhang, Y., Talalay, P., Cho, C. G., & Posner, G. H. (1992). A major inducer of anticarcinogenic protective enzymes from broccoli: Isolation and elucidation of structure. *Proceedings of the National Academy of Sciences, 89*(6), 2399–2403.
41. Hayes, J. D., Kelleher, M. O., & Eggleston, I. M. (2008). The cancer chemopreventive actions of phytochemicals derived from glucosinolates. *European Journal of Nutrition, 47*, 73–88.
42. Higdon, J. V., Delage, B., Williams, D. E., & Dashwood, R. H. (2007). Cruciferous vegetables and human cancer risk: Epidemiologic evidence and mechanistic basis. *Pharmacological Research, 55*(3), 224–236.
43. Sikorska-Zimny, K., & Beneduce, L. (2021). The glucosinolates and their bioactive derivatives in *Brassica*: A review on classification, biosynthesis and content in plant tissues, fate during and after processing, effect on the human organism and

interaction with the gut microbiota. *Critical Reviews in Food Science and Nutrition, 61*(15), 2544–2571.
44. Mehmood, A., & Zeb, A. (2020). Effects of different cooking techniques on bioactive contents of leafy vegetables. *International Journal of Gastronomy and Food Science, 22*, 100246.
45. Pilipczuk, T., Kusznierewicz, B., Chmiel, T., Przychodzeń, W., & Bartoszek, A. (2017). Simultaneous determination of individual isothiocyanates in plant samples by HPLC-DAD-MS following SPE and derivatization with *N*-acetyl-*l*-cysteine. *Food Chemistry, 214*, 587–596.
46. Hanschen, F. S., Herz, C., Schlotz, N., Kupke, F., Bartolomé Rodríguez, M. M., Schreiner, M., Rohn, S., & Lamy, E. (2015). The *Brassica* epithionitrile 1-cyano-2,3-epithiopropane triggers cell death in human liver cancer cells *in vitro. Molecular Nutrition & Food Research, 59*(11), 2178–2189.
47. Kim, Y. B., Li, X., Kim, S. J., Kim, H. H., Lee, J., Kim, H., & Park, S. U. (2013). MYB transcription factors regulate glucosinolate biosynthesis in different organs of Chinese cabbage (*Brassica rapa* ssp. *pekinensis*). *Molecules, 18*(7), 8682–8695.
48. Fenwick, G., Roger, H., Robert, K., Mullin, W. J., & VanEtten, C. H. (1982). Glucosinolates and their breakdown products in food and food plants. *CRC Critical Reviews in Food Science and Nutrition, 18*(2), 123–201.
49. Halkier, B. A., & Gershenzon, J. (2006). Biology and biochemistry of glucosinolates. *Annual Review of Plant Biology, 57*, 303–333.
50. Dickson, M. H., Lee, C. Y., & Blamble, A. E. (1988). Orange-curd high carotene cauliflower inbreds, NY 156, NY 163, and NY 165. *Hort Science, 23*, 778, 779.
51. Lee, H. W., Zhang, H., Liang, X., & Ong, C. N. (2020). Simultaneous determination of carotenoids, tocopherols and phylloquinone in 12 *Brassicaceae* vegetables. *LWT–Food Science and Technology, 130*, 109–125.
52. Grout, B. W. W. (n.d.). Cauliflower (*Brassica oleracea* var. *botrytis* L.). In Y. P. S. Bajaj (Ed.), *Crops II. Biotechnology in Agriculture and Forestry* (Vol. 6, pp. 211–225). Springer.
53. Crisp, P. (1982). The use of an evolutionary scheme for cauliflowers in screening of genetic resources. *Euphotic, 31*(3), 725–741.
54. Jongen, W. (2002). Fruit and vegetable processing—Improving quality. In *Woodhead Publishing Limited* (pp. 119–146). Cambridge.
55. Picchi, V., Migliori, C., Lo Scalzo, R., Campanelli, G., Ferrari, V., & Di Cesare, L. F. (2012). Phytochemical content in organic and conventionally grown Italian cauliflower. *Food Chemistry, 130*(3), 501–509.
56. Hooper, L., & Cassidy, A. (2006). A review of the health care potential of bioactive compounds. *Journal of the Science of Food and Agriculture, 86*(12), 1805–1813.
57. Quintero, C., & Bowers, M. D. (2012). Changes in plant chemical defenses and nutritional quality as a function of ontogeny in *Plantago lanceolata* (Plantaginaceae). *Oecologia, 168*, 471–481.
58. Freudenheim, J. L., Genco, R. J., LaMonte, M. J., Millen, A. E., Hovey, K. M., Mai, X., Nwizu, N., Andrews, C. A., & Wactawski-Wende, J. (2016). Periodontal disease and breast cancer: Prospective cohort study of postmenopausal women. *Cancer Epidemiology, Biomarkers & Prevention, 25*(1), 43–50.

59. Singh, J., Upadhyay, A. K., Prasad, K., Bahadur, A., & Rai, M. (2007). Variability of carotenes, vitamin C, E and phenolics in *Brassica* vegetables. *Journal of Food Composition and Analysis, 20*, 106–112.
60. Xia, N., Diaper, A., Förstermann, U., & Li, H. (2017). Antioxidant effects of resveratrol in the cardiovascular system. *British Journal of Pharmacology, 174*(12), 1633–1646.
61. Ramirez, D., Abellán-Victorio, A., Beretta, V., Camargo, A., & Moreno, D. A. (2020). Functional ingredients from *Brassicaceae* species: Overview and perspectives. *International Journal of Molecular Sciences, 21*, 1998–2019.
62. Ohara, M., Kimura, S., Tanaka, A., Ohnishi, K., Okayasu, R., & Kubota, N. (2011). Benzyl isothiocyanate sensitizes human pancreatic cancer cells to radiation by inducing apoptosis. *International Journal of Molecular Medicine, 28*, 1043–1047.
63. Vauzour, D., Rodriguez-Mateos, A., Corona, G., Oruna-Concha, M. J., & Spencer, J. P. (2010). Polyphenols and human health: Prevention of disease and mechanisms of action. *Nutrients, 2*(11), 1106–1131.
64. Kim, J. H., Lee, J., Choi, I. J., Kim, Y. I., Kwon, O., Kim, H., & Kim, J. (2018). Dietary carotenoids intake and the risk of gastric cancer: A case-control study in Korea. *Nutrients, 10*, 1031–1052.
65. Tsunoda, T., Krosse, S., & Van Dam, N. M. (2017). Root and shoot glucosinolate allocation patterns follow optimal defense allocation theory. *Journal of Ecology, 105*(5), 1256–1266.
66. Chiu, L. W., Zhou, X., Burke, S., Wu, X., Prior, R. L., & Li, L. (2010). The purple cauliflower arises from activation of a MYB transcription factor. *Plant Physiology, 154*(3), 1470–1480.
67. Kim, S. M. (2016). Cellular and molecular mechanisms of 3,3′-diindolylmethane in gastrointestinal cancer. *International Journal of Molecular Sciences, 17*, 1155–1178.
68. Thangam, R., Suresh, V., Rajkumar, M., Vincent, J. D., Gunasekaran, P., Anbazhagan, C., Kaveri, K., & Kannan, S. (2013). Antioxidant and *in vitro* anticancer effect of 2-pyrrolidinone rich fraction of *Brassica oleracea* var. *capitata* through induction of apoptosis in human cancer cells. *Phytotherapy Research, 27*, 1664–1670.
69. Saban, G. M. (2018). The benefits of *Brassica* vegetables on human health. *Journal of Human Health Research, 1*, 104–117.
70. Youn, H. S., Kim, Y. S., Park, Z. Y., Kim, S. Y., Choi, N. Y., Joung, S. M., Seo, J. A., Lim, K. M., Kwak, M. K., Hwang, D. H., et al. (2010). Sulforaphane suppresses oligomerization of TLR4 in a thiol-dependent manner. *Journal of Immunology, 184*, 411–419.
71. Kallifatidis, G., Labsch, S., Rausch, V., Mattern, J., Gladkich, J., Moldenhauer, G., Büchler, M. W., Salnikov, A. V., & Herr, I. (2011). Sulforaphane increases drug-mediated cytotoxicity toward cancer stem-like cells of pancreas and prostate. *Molecular Therapy: Journal of the American Society of Gene Therapy, 19*, 188–195.
72. Lamy, E., Hertrampf, A., Herz, C., Schüler, J., Erlacher, M., Bertele, D., Bakare, A., Wagner, M., Weiland, T., & Lauer, U. (2013). Preclinical evaluation of 4-methylthiobutyl isothiocyanate on liver cancer and cancer stem cells with different p53 status. *PLoS ONE, 8*, e7051.

CHAPTER 2

Cabbage (*Brassica oleracea* var. *capitata*)

TANVEER AHMAD,[1,*] MUHAMMAD ANWAR-UL-HAQ,[1] HAMMAD HUSSAIN,[1] MUHAMMAD TARIQ,[1] and MUHAMMAD AWAIS GHANI[2]

[1]*Department of Horticulture, MNS University of Agriculture, Multan, Punjab, Pakistan*

[2]*Institute of Horticultural Sciences, University of Agriculture, Faisalabad, Punjab, Pakistan*

*Corresponding author

ABSTRACT

Cabbage (*Brassica oleracea* var. *capitata*) is a globally significant cruciferous vegetable valued for its dense, leafy heads, which are used in various culinary preparations. This chapter reviews its historical origins, botanical characteristics, and traditional medicinal uses. A primary focus is its rich profile of bioactive compounds, including glucosinolates (e.g., glucoraphanin), their hydrolytic products (isothiocyanates and indoles like sulforaphane and 3,3'-diindolylmethane), and key antioxidants such as vitamins C and E, carotenoids (lutein, β-carotene), and phenolic compounds (notably anthocyanins in red cabbage). These phytochemicals underpin cabbage's potent health-promoting properties. The chapter details their mechanisms, including their chemopreventive effects against various cancers (stomach, lung, prostate, breast), their anti-inflammatory effects, and their ability to mitigate the risks of cardiovascular and neurodegenerative diseases such as Alzheimer's. The influence of cultivar, agronomic

Postharvest Quality Attributes of Fresh Vegetables and Grains.
Riadh Ilahy, Imen Tlili, Thouraya R'him, and Muhammad Azam (Eds.)
© 2027 Apple Academic Press, Inc. Co-published with CRC Press (Taylor & Francis)

DOI:10.1201/9781779642851-3

practices, and processing on phytochemical content is discussed. Finally, the chapter addresses the potential for sustainable management of cabbage waste and underscores the need for further research to optimize its cultivation for enhanced therapeutic benefits.

2.1 INTRODUCTION

Cabbage (*Brassica oleracea* var. capitata) is one of the most widely distributed vegetables in the world. It fits the genus Brassica of the Cruciferae family. Except for Antarctica, the Cruciferae group is known worldwide to include about 372 genera and 4060 accepted species that are widely distributed [1–3]. In addition to cabbage, the genus *Brassica* also includes leafy vegetables such as kale, cauliflower, Brussels sprouts, garden cress, mustard greens, and turnip. It originated in the Mediterranean and became widespread in Western Europe; later, it became an essential part of cuisines worldwide. The edible portion of this vegetable is the head, which consists of many leaves of distinct shapes, colors, textures, and sizes. Red cabbage, savoy cabbage, and white cabbage are the main varieties of cabbage. There are many ways to eat them, including stewing, chopping, steaming, soaking, fermenting, boiling, or eating raw [2].

2.1.1 HISTORY AND DISTRIBUTION

It is very complicated to indicate the exact background of cabbage because it has been cultivated since ancient times. Cabbage has been grown since 3000 BC. Some consider that the ancestors of cabbage grew in coastal regions of Europe approximately 8000 years ago [4]. Around 37 million years ago, Cruciferae was initiated as a tropical–subtropical family and was considered to have originated in the Iran–Turanian region. The review by Franzke et al. [5] comprehensively explains the age and early diversification of cruciferous plants. Some evidence suggests that modern cabbage varieties are descendants of wild kale grown across the Baltic Coast in the East Mediterranean [6]. Moreover, some wild alternates that develop naturally on the shores of the Adriatic Sea are termed stenoendemic [7].

2.2 BOTANICAL DESCRIPTION

Cabbage is a biennial green leaf vegetable usually grown annually, and its head consists of a multitude of leaves. It is directly linked with other same-genus plants like brussels sprouts, cauliflower, and broccoli [8, 9]. Dixon states that cabbage seedlings have heart-shaped cotyledons and a thin taproot system [10]. The first leaf is an ovate shape, and the petiole is lobed. The opening leaves form a rosette-type shape, and later, a new cabbage leaf develops a shorter petiole, and the leaves begin to form a head inside the cup. The cabbage weight range is usually 0.05–4.0 kg, and its color can be white, pale green, or dark green. Generally, the head of cabbage develops in the first year. Still, the emergence of inflorescences and the production of seeds usually occur in the second year of the plant's life cycle. The terminal part of the inflorescence is raceme type, and the height range of the inflorescence is 50 to 100 cm with white- or yellow-colored florets. Besides, each flower contains four sepals and petals, and six stamens arranged in a vertical pattern, with one containing a two-celled ovary, style, and stigma. The filaments are shorter than the six stamens. Its fruit is like a silicate. It opens when mature with round black or brown seeds. The cabbage crop is cross-pollinated, and generally, pollination occurs through insects (entomophilous). The root system of growing plants is shallow and fibrous [2, 3].

2.3 TRADITIONAL USES

In many countries, cabbage is vital in traditional dishes and cultures and has many uses in traditional medicines. Because of its low-cost availability in the home market, it has become a significant source of plant-based nutrients in human food. It can be kept for a long time, so it is available year-round. In addition, it is usually used fresh in salads and in many other ways, such as fried, boiled, or fermented. In many cultures (such as Chinese culture), on the occasion of the New Year, white cabbage is presented as a traditional dish associated with beliefs of good luck [2, 11]. Throughout history, everyone from Caesar to enslaved people has consumed cabbage for nutritional supplements and to support digestion. It can also prevent scurvy. Before drinking, Aristotle ate cabbage to prevent the alcohol from "fooling his cautious academic mind" [12]. Throughout history, it has also been referred to as the "poor people's medicine" and has been consumed to comprehensively improve health and treat various inflammatory conditions, support blood purification, or treat gastrointestinal problems [13].

2.4 HEALTH BENEFITS AND BIOLOGICAL ACTIVITIES

The cabbage vegetable occupies an important position in the global human diet, but it is very well-known in China, India, the United States of America, Japan, and European countries [14]. The cabbage crop is a highly nutritious vegetable in terms of nutritional status; various epidemiological analyses have announced that large amounts of leafy vegetables can mitigate the overall hazard of Alzheimer's, diabetes, heart disease, cancer, and many other chronic diseases. The possessive effect of this vegetable is primarily due to its phytochemical antioxidant agents, such as ascorbic acid (AA), carotenoids, flavonoids, and vitamin E as well. Insufficient consumption of leafy vegetables like cabbage and spinach can interrupt oxygen metabolism and lead to oxidative stress. This is itself a risk factor for cancer and other chronic disorders. Eating cabbage will eventually increase the antioxidant level in cells, tissues, and blood. As a result, they can stop free radical reactions and reduce the risk of long-term health problems to defend the body from oxidative damage [15, 16].

2.4.1 CANCER

At the beginning of the 20th century, it was first discovered that cruciferous plants had chemopreventive effects. According to studies and data, they have favorable benefits in preventing many types of cancers [17, 18]. There are a large number of reports that the anticancer properties of cabbage are produced by the biologically active sulfur-containing compound isothiocyanate, and isothiocyanate is a hydrolysate of thioglycollate. There has been an indication that these substances may suppress carcinogenesis, promote tumor cell death, and block active carcinogens from binding to DNA, as well as RNA, and other cellular proteins, thereby inhibiting the formation of cancerous cells [19, 20]. These anticancer substances have inhibitory effects on stomach cancer, lung cancer, prostate cancer, and breast cancer. Cabbage and cabbage-like vegetables are adequate to prevent other types of cancers, such as rectal, colon, pancreatic, and liver cancer [21, 22].

2.4.2 ANTI-INFLAMMATORY

Chronic inflammation is associated with significant health complications, including cancer, obesity, and heart disease. According to Penas et al. [23], lipopolysaccharides and other endotoxins activate macrophages, which release cytokines and nitric oxide. Targeting and blocking these promoters may reduce the chance of acquiring related disorders. Cabbage includes a high concentration of phytochemicals, which have been shown to decrease inflammation effectively in this approach. This primarily comprises anthocyanins, which are found at the highest concentration in red cabbage. It is claimed that anthocyanins may decrease inflammation by inhibiting the activity of pro-inflammatory enzymes (e.g., cyclooxygenase and inducing nitric oxide synthase). Mizgier et al. [24] revealed anthocyanins in red cabbage, including anthocyanins, p-coumarin, erucic acid, and ferulic acid. Red cabbage extract reduces the activity of the pro-inflammatory enzyme cyclooxygenase-1 by 38%. Zielinska et al. [25] also showed that red cabbage has an anti-inflammatory effect, and a mouse model of IBD observed significantly reduced levels of pro-inflammatory factors, tumor necrosis factor, and interleukin-1. In the same study, researchers discovered that red cabbage lowered inflammation and reduced the macroscopic damage score in a rat model of Crohn's disease (2.70 in the red cabbage group compared to 5.78 in the control group).

Cabbage contains anti-inflammatory phytochemicals such as indole-3-carbinol, 3,3-diindolylmethane, vitamin C, sulforaphane, and 3,3-diindolylmethane. Martinez-Villaluenga et al. [26] investigated the impact of cabbage fermentation on the anti-inflammatory effects of this vegetable. It was noted that fermentation caused a 12 times rise in the content of ascorbigen. The results showed that the anti-inflammatory activity of cabbage was significantly increased, and its nitric oxide inhibitory effect was increased by 2.6 times. Cho et al. [27] subsequently demonstrated the anti-inflammatory properties of an anti-inflammatory compound (3,3-diindolylmethane), which continuously inhibits lipopolysaccharide-induced macrophage pro-inflammatory mediator release while down-regulating AP-1 (activator protein) transcriptional activity and AP-1 DNA binding activity; in addition, it activates the phosphorylation of protein kinases. In addition, a study by Heiss and Gerhäuser [28] showed that sulforaphane exerts an anti-inflammatory effect in mouse macrophages because this compound inhibits the release of pro-inflammatory nitric oxide induced by lipopolysaccharide from these cells.

2.4.3 CARDIOVASCULAR DISEASE (CVD)

Recently, it was proven that, in blood, low-density lipoproteins (LDLs) increase the risk of cardiovascular disorders because they assist in the production of atherosclerotic lesions and foam cells. Cabbage has a wide range of antioxidants, which may scavenge free radicals and prevent phospholipid peroxidation, making it a powerful defense against CVD risk factors. Bacchetti et al. [29] observed 38 healthy members and asked them to supplement their daily diet with 300 grams of red and black cabbage over two weeks to conclude the effect of the cruciferous group of vegetables—effects of blood lipids and oxidative hypocholesterolemia. The results exhibited that several cabbage types had a substantial impact on the lipid distribution of the participants. LDL cholesterol was reduced by 0.03 mmol/L, and oxidized LDL and total cholesterol levels were significantly reduced. In general, this treatment has been shown to lower an individual's risk of cardiovascular disease by 5%. The addition of black and red cabbage to the diet increased plasma lutein content by 204% and β-carotene by 80%. These carotenoids were able to diminish the amount of lipoproteins in the body.

In recent research, Jiang et al. [30] found that the red cabbage diet reduced total fecal cholesterol by 32% in an obese mouse model, while also lowering liver triacylglycerol and cholesterol levels in the liver and regulating weight. Red cabbage's capacity to reduce cholesterol levels and the participation of lecithin-cholesterol acyltransferase enzymes in cholesterol esterification and transport, have been described. It was also revealed that increased consumption of fruits and vegetables such as white and Chinese cabbage, as well as other plant-based foods, had a positive impact on a sample of overweight people aged from 35 to 65. After 16 weeks of diet treatment, calorie use, waist circumference, body weight, and sagittal abdominal diameter decreased considerably, while serum α- and β-carotene concentrations, folic acid, and dietary fiber increased. As a result of these findings, long-term consumption of more vegetables, such as cabbage, might help lower the risk of coronary heart disease [31].

2.4.4 ALZHEIMER'S DISEASE

Phytochemicals found in cabbage may help prevent Alzheimer's disease, a neurodegenerative illness characterized by the insoluble beta-amyloid peptide (β), which is the primary component of senile plaques in the brain, which may induce neurodegeneration. Patients with Alzheimer's disease are more likely than the general population to have an increase in amyloid as a result of oxidative stress caused by free radicals in their brains [32].

Heo and Lee [33] revealed that red cabbage mainly had a protective impact on mitochondrial and amyloid-peptide-induced membrane and nerve damage because of its high antioxidant content. According to the author, the antioxidant properties of cabbage, particularly anthocyanins, may play a part in the vegetable's ability to defend against oxidative damage. A biologically active nutrient derived from cabbage and broccoli juice has been recently shown to guard against Alzheimer's disease development. To protect against cytotoxicity and apoptosis, the chemical inactivates anti-apoptotic signals, therefore increasing the activity of antioxidant enzymes and reducing the oxidative stress induced by neuronal damage [32, 34].

2.4.5 OTHER HEALTH BENEFITS

Cabbage also has various other health benefits. Patients with type 2 diabetes have also been shown to benefit from a more balanced diet that includes a variety of vegetables, including cabbage and its varieties. This is because blood sugar is more stable when α-Glucosidase is inhibited first in these vegetables rather than carbohydrates [35]. Due to the inhibition of α-Glucosidase, blood sugar is more stable when these vegetables are consumed before carbohydrates. Cabbage is a natural source of minerals, especially iodine (I_2). Iodine (I_2) is an essential trace element that cannot be synthesized by the human body and is an integral part of the human diet. The failure to produce thyroid hormones may lead to a variety of severe health issues, including hypothyroidism, goiter, intellectual disability, and nervous system malfunction if this substance is deficient. High consumption of cabbage may provide sufficient iodine and support the normal functioning of the thyroid, nervous system, and brain [36].

2.5 ANTIOXIDANT PHYTOCHEMICALS OF FRUIT, SEEDS, PEELS, WASTES, AND THEIR BYPRODUCTS

Cabbage is a rich source of many beneficial plant-based chemicals and several antioxidants, which have a positive effect on human health. The main phytochemicals and antioxidants are vitamins E and C, phenolic compounds, and carotenoids.

2.5.1 VITAMINS

As stated earlier, the main antioxidant vitamins in cabbage are carotenoids, vitamin C, and vitamin E. These vitamins can act alone or synergistically to scavenge free radicals, thereby decreasing the levels of active substances. Cabbage consists of both types of antioxidants, fat-soluble (vitamins A and E) and water-soluble (ascorbic acid). These antioxidants are very effective because they can scavenge lipids and free radicals in the water phase [16, 37].

2.5.2 CAROTENOIDS

There are several plant pigments known as carotenoids, which are principally responsible for the orange, yellow, and red colors that appear in many vegetables. These pigments are precursors of vitamin A, which implies that they are transformed in the human body into the physiologically active form of vitamin A. It is necessary to maintain normal development and eyesight. Carotenoids and lutein are the two most important types of carotenoids [38, 39]. There are two primary antioxidants in this vegetable: lutein and carotene. They both scavenge free radicals and block single oxygen molecules [16]. Beta-carotene is a member of the carotene family and is reddish-orange in color, as indicated by its name. Beta-carotene levels in red cabbage (0.0670 mg per 100 g), savoy cabbage (0.06 mg per 100 g), and white cabbage (0.0420 mg per 100 g) are the greatest among these three types of cabbage, as stated by the USDA's National Nutrition Database [40].

To be clear, these are conventional values, and there may be variations between comparable and dissimilar varieties of cabbage. Comparing the measured levels of Beta-carotene concentration in cabbage between the research studies. For instance, according to Chandra-Hioe et al. [39], Chinese white cabbage has 2.054 mg of Beta-carotene per 100 g of the vegetable. Singh et al. [41] investigated the carotenoid content of 18 cabbage varieties. They concluded that the savoy cabbage line contains the maximum levels of Beta-carotene (0.074 mg per 100 g). In contrast, red cabbage contains the lowest levels of Beta-carotene (0.04 mg per 100 g), which directly contradicts the United States Department of Agriculture's statement above.

Different growing seasons, storage conditions, processing, genotypes, and agronomic factors are thought to be clear explanations for these variations. According to Burton [42], this carotenoid may enhance the antioxidant activity of other cabbage antioxidants. Carotenoids were more helpful in breaking chains under low oxygen pressure, whereas vitamin E was more effective at breaking chains under high oxygen pressure in mammalian tissues. There are two types of xanthophylls: lutein and zeaxanthin, both of which are present in carotenoids, although lutein is more prevalent in green chlorophyll in cabbage and other leafy vegetables. It is obtained in Brassica plants like cabbage, as well as in the lens of the eye and macular pigment. Cataracts and macular degeneration, which are responsible for 56% of all blindness cases worldwide, are thought to be prevented by its antioxidant properties [41, 43]. The USDA National Nutrition Database states that white cabbage has 0.03 mg lutein per 100 g, whereas red and savoy cabbage have 0.07 and 0.32 mg lutein per 100 g, respectively [44]. These are the average values, which vary depending on cabbage variety.

2.5.3 VITAMIN C

Cabbage contains vitamin C, which is a water-soluble antioxidant vitamin. It is an essential part of the human diet because the human body can't synthesize it by itself. It is an antioxidant that acts as an electron donor or acceptor and a free radical scavenger, making it an essential ingredient that helps protect cell membranes from lipid peroxidation. It scavenges reactive oxygen species (ROS) and removes lipid peroxidation. Electrons are subtracted from hydrogen radicals to achieve this function [44, 45]. Vitamin C performs an imperative role in biological functions in the human body. This significant enzyme cofactor alpha-ketoglutarate (α)-dependent plus dioxygenase helps the catalysis of various collagen synthesis reactions.

The "hypoxia-inducible factor-1," a transcription factor for these genes, is a significant component in controlling cell apoptosis, tumor growth, and energy metabolism. In addition, reactive protein in the blood has been demonstrated to be decreased by this vitamin. It is a sign of infection, usually a sign of inflammation, and therefore the dual functions of anti-inflammatory compounds and antioxidants [16, 46]. The vitamin C content of cabbage varies by species. Vitamin C is usually obtained in cabbage at varying levels of 07.08–51.7 mg/100 g [18]. It is evident from the existing scientific literature that vitamin C is obtained in abundance in red cabbage. Chun et al. [18] showed that red cabbage (51.07 mg per 100 g) has the highest vitamin C content, followed by green cabbage (41.00 mg per 100 g), and the minimum is savoy cabbage (7.08 mg per 100 g). Singh et al. [41] also studied the antioxidant phytochemicals found in several cabbage varietals.

As previously reported, red cabbage had the highest vitamin C content (24.38 mg per 100 g fresh weight), while savoy cabbage had the highest amount of double vitamin C (14.49 mg per 100 g fresh weight). The findings were in accordance with earlier research on red cabbage, and the results exhibited that savoy cabbage had the maximum vitamin C content (24.38 mg per 100 g fresh weight), and double the amount of vitamin C. Because of its vulnerability to heat, air, and light, these conditions will cause this vitamin to decay via oxidation processes. As a result, it is reasonable to predict that the AA concentration of two different cabbages will vary. During the growing season, other elements, including plant maturity, fermentation, and storage, might have an impact.

2.5.4 VITAMIN E

Cabbage contains a fat-soluble antioxidant called vitamin E (tocopherol). Most of the time, it is found in the four forms of α, β, γ, and δ-tocopherol, and it is found in the form of alpha-tocopherol (α) in crucifers. Without hydrogen atom donation, it performs as a hydrogen peroxide free radical scavenger to avoid lipid peroxidation and preserve the cell membrane to bind vitamin A [41, 37]. In the case of lipid peroxidation, vitamin E provides a hydroxyl group that reacts with peroxy groups to form lipid hydrogen peroxide and tocopheroxyl radicals, which stop the chain reaction. At this point, vitamin C helps to replenish vitamin E by oxidizing tocopherol free radicals and converting them back to their reduced form [45]. In addition to preventing heart disease, vitamin E has been proven to reduce oxidized LDL in artery walls [37, 16].

Cabbage is an effective source of vitamin E, one of the few antioxidants available in relatively small concentrations. Research carried out by Singh et al. [41] exhibited that not all cabbage varieties contain α-tocopherol. The alpha-tocopherol was present in 14 of 18 cabbage varieties when examined. The range of alpha-tocopherol was 0.03 to 0.5 mg per 100 g.

According to studies, red cabbage has the highest concentration of this vitamin, while white cabbage contains the least.

Additionally, identical findings were made by another researcher, Podsedek [16]. Red cabbage had the greatest concentration of vitamin E (0.05 mg/100 g), whereas white cabbage had the lowest value (0.04 mg/100 g). Chun et al. [18] said that the difference in vitamin E content between different types of cabbage is caused by the weather, especially the amount of sunlight, the maturity of the vegetables, harvesting methods, and the uneven distribution of natural vitamin E.

2.5.5 PHENOLIC COMPOUNDS

Secondary metabolites found in plants are known as phenolic compounds. Benzoic acid, cinnamic acid, and flavonoid derivatives are examples of phenolic substances [47]. Cabbage has a high concentration of flavonoids, a class of phenolic compounds. This category of antioxidants is distinguished by the presence of two aromatic rings joined by a 3-C bridge in their structure. Cabbage contains flavonoids, which have been shown to have more antioxidant activity than vitamins and carotenoids [16]. It is well known that flavonoids are antioxidants, but they also have a positive impact on the capillaries by lowering oxidative stress [18, 48]. The capability to scavenge oxygen-free radicals has been proven to limit the oxidation of lipoproteins, which minimizes the threat of heart disease and delays the progression of atherosclerosis [18]. The concentration of total flavonoids and phenols in various cabbage genotypes is considerably different from the content of other phytochemicals.

According to Chun et al. [18], four different cabbage types were studied, and it was found that the total flavonoid content (03.9–108.1 mg per 100 g), and the total phenolic content (75.1 mg per 100 g) were significantly associated with antioxidant activity. Singh et al. [41] found that the phenol level of various varieties of cabbage differed as well. The phenol content in 18 red and white cabbage types ranged from 18.74–101.30 mg/100 g to 19.74–101.30 mg/100 g. It was also found by Kim et al. [49] that 10 green cabbage cultivars possess variable amounts of phenolic chemicals, ranging from 110.2 mg GAE per 100 g Genesee to 153.3 mg GAE per 100 g Lynx. Red cabbage has consistently been shown to possess the most significant levels of phenolic compounds, notably flavonoids, owing to the existence of substantial concentrations of anthocyanins, which are favorable for the red color of this cabbage species [18, 48]. The most prevalent flavonoids in cabbage are kaempferol, quercetin, and celery aglycone, all found at low to medium amounts of 0.1–10.5 mg per 100 g [18, 50]. Red cabbage anthocyanin is the most prominent cyanidin and the primary anthocyanin in red cabbage (209.08 mg per 100 g) [44, 48].

2.5.6 GLUCOSINOLATES

In cabbage, glucosinolates are among the plant's most distinctive traits [22]. Although chemically similar to the enzymes that produce them, glucosinolates lack biological activity. Cabbage's pungent, bitter flavor is due to sulfur compounds present in the plant. The distribution of glucosinolates in cabbage varies from variety to variety. According to the findings of Possenti et al. [51], white cabbage has the highest concentration. It was found that white cabbage had a higher concentration of glucosinolates (148 mg per 100 g) than red cabbage (88 mg per 100 g) and savoy cabbage (88 mg per 100 g) of fresh weight. According to Wennberg et al. [52] and Park et al. [53], it is the most plentiful glucosinolate recognized in green cabbage, accounting for roughly 65.6% of mustard oil and glucose protein. Gluconin (49.4%) and glucosamine predominate among red cabbage's glucosinolates. Two-hydroxyphenyl, 4-methoxy glucoside, and glucoside are all found in minute levels in different cabbages and may be detected by mass spectrometry and quantified by HPLC. Other studies have also shown that many red and green cabbages contain the same glucosinolates, consistent with their results. According to the results of this research, glucosinolates, Negril, and glucoside are the most prevalent in cabbage [54–56].

There are several health advantages associated with hydrolyzing these glucosinolates, but only if they are activated in the body first. Myrosinase, an enzyme found in plants, can do this. Plants produce the enzyme myrosinase when subjected to physical stress, such as cutting, chewing, washing, and so forth. Myrosinase cleaves glucosinolates, decomposing them into different isothiocyanates (ITCs) and indole as a result of physical stress. Based on the structure of the parent glucosinolate and the hydrolysis process, the hydrolysate may vary considerably [19, 57, 58].

2.5.7 ISOTHIOCYANATES

Isothiocyanate is most important hydrolysis product of glucosinolates and has significant anticancer effects. The hydrolysis products of isothiocyanate-glucoside and isothiocyanate-isothiocyanate are the most important ITCs in cabbage. Isothiocyanate helps the liver and epithelial cells with phase I & II metabolic enzymes by regulating the metabolic balance of primary and secondary carcinogenic compounds. At stage II, they deactivate and generate water-soluble components that may be expelled via the urine before returning to the body to perform the detoxification impact of these carcinogens [59, 60]. As reported in the scientific literature, minor levels of other ITCs, such as 2-phenylethyl isothiocyanate, phenyl isothiocyanate, and benzyl isothiocyanate, have been detected in it [22, 57, 60, 61]. Further research has demonstrated that isothiocyanate prevents cancer by suppressing mitosis, altering metabolic hormones, and inducing cell death [22, 61].

2.5.8 INDOLES

Indole-3-acetonitrile and its digestive products, glucobrassicin and 3,3′-diindolylmethane, are used to identify common indole compounds in cabbage (indole-3-methanol) [21, 62]. Only cancer cell proliferation, mitotic inhibition, and tumor initiation may be prevented by this drug [52]. These indoles, like ITCs, have been demonstrated to stimulate the first and second stages of the detoxification pathways, which are responsible for eliminating carcinogens from the body. Prostate cancer is thought to be less likely to occur in men who use antiandrogens such as 3-indole-3-methanol 3-diindolylmethane, which has been shown to lower estradiol levels in the blood. The indole of cabbage has also been determined to trigger cancer cell apoptosis and inhibit cell growth in cancer patients [21, 62, 63].

The scientific literature reviewed shows that glucosinolate, isothiocyanate, and indole levels in various cabbage cultivars vary significantly. Bellostas et al. [56] stated that the glucosinolate content of cabbage changes, especially during seed development and germination, indicating that germination duration influences the change in glucosinolate content in this vegetable. Rouzaud et al. [59] found that heat treatment affects isothiocyanate release and that the growing season impacts the glucosinolate profile of this vegetable. Choi et al. [55] revealed that when cabbage is grown in spring, there is a 57% lower concentration of idolethioglycosides than when it is grown in autumn, which implies that spring rains play a vital role in this differentiation. Some glucosinolate content may vanish if the vegetables are sliced or injured before eating or stored improperly after harvest [64–66].

2.6 CONCLUSION AND FUTURE PROSPECTS

Cabbage is a popular vegetable used worldwide. Cabbage's antioxidants and phytochemicals may help prevent cancer, heart disease, eye disease, Alzheimer's disease, and the effects of inflammatory hormones. Although it contains several vital amino acids, vitamins C and K, lutein, and beta-carotene, it's the glucosinolate and isothiocyanate compounds that are most notable. Other vegetables do not have them in significant amounts. Features like agroclimatic conditions, genetic variation, and biotic stresses all affect the production of various phytochemicals; therefore, these factors must be considered when growing cabbage for its therapeutic potential. We need further research on cabbage genotypes, breeding techniques, and climatic conditions to properly grasp their impact on phytochemical composition variations.

At last, although cabbage is commonly eaten, much waste is created during harvest, processing, and even sale. Various environmental solutions have been proposed to manage cabbage waste better, such as manufacturing coal, bioethanol, briquettes, and using it as a substrate for fungal biomass production. There is still a need for large-scale research to confirm the practicality of these ideas. Notably, it will help tackle issues like landfills, ozone depletion, and pollution in the future.

KEYWORDS

- **antioxidant properties**
- **cabbage**
- **cabbage waste management**
- **Cruciferae family**
- **glucosinolates**
- **phytochemicals**

REFERENCES

1. Maccagni, A., & Willi, Y. (2022). Niche breadth and elevational range size: A comparative study on Middle European *Brassicaceae* species. *Philosophical Transactions of the Royal Society B: Biological Sciences*, *377*(1846).

2. Liu, C., Chen, C., Jiang, A., Zhang, Y., Zhao, Q., & Hu, W. (2021). Effects of aqueous ozone treatment on microbial growth, quality, and pesticide residue of fresh-cut cabbage. *Food Science & Nutrition*, *9*(1), 52–61.
3. Al-Shehbaz, I. A., Beilstein, M. A., & Kellogg, E. A. (2006). Systematics and phylogeny of the *Brassicaceae* (Cruciferae): An overview. *Plant Systematics and Evolution*, *259*(2), 89–120.
4. Al-Shehbaz, I. A. (2011). *Brassicaceae* (Mustard Family). In *eLS*. Wiley, Chichester.
5. Franzke, A., Lysak, M. A., Al-Shehbaz, I. A., Koch, M. A., & Mummenhoff, K. (2011). Cabbage family affairs: The evolutionary history of *Brassicaceae*. *Trends in Plant Science*, *16*(2), 108–116.
6. Balkaya, A., Yanmaz, R., Apaydin, A., & Kar, H. (2005). Morphological characterization of white head cabbage (*Brassica oleracea* var. *capitata* subvar. *alba*) genotypes in Turkey. *New Zealand Journal of Crop and Horticultural Science*, *33*(4), 333–341.
7. Jasprica, N. (2015). *Brassica* L. In Nikolić, T., Milović, M., Bogdanović, S., & Jasprica, N. (Eds.), *Endemi u Hrvatskoj flori (Endems in Croatian flora)* (pp. 107–115). Alfa, Zagreb.
8. Moreb, N., Murphy, A., Jaiswal, S., & Jaiswal, A. K. (2020). Cabbage. In *Nutritional Composition and Antioxidant Properties of Fruits and Vegetables* (pp. 33–54).
9. Lazaro, A., & Aguinagalde, I. (1998). Genetic diversity in *Brassica oleracea* L. (Cruciferae) and wild relatives (2n = 18) using isozymes. *Annals of Botany*, *82*(6), 821–828.
10. Dixon, G. R. (2007). *Vegetable Brassicas and Related Crucifers. Crop Production Science in Horticulture*, 14. CAB International. ISBN 978-0-85199-395-9.
11. Hatfield, G. (2004). *Encyclopedia of Folk Medicine: Old World and New World Traditions*. ABC-CLIO.
12. Cavender, A. (2006). Folk medical uses of plant foods in southern Appalachia, United States. *Journal of Ethnopharmacology*, *108*(1), 74–84.
13. Passalacqua, N. G., Guarrera, P. M., & De Fine, G. (2007). Contribution to the knowledge of the folk plant medicine in Calabria region (Southern Italy). *Fitoterapia*, *78*(1), 52–68.
14. Izzo, L., Carrasco, Y. R., Pacifico, S., Castaldo, L., Narvaez, A., & Ritieni, A. (2020). Digestion of red cabbage extract chemically profiled through UHPLC-Q-Orbitrap HRMS. *Antioxidants*, *9*, 1–17.
15. Yahia, E. (2010). The contribution of fruit and vegetable consumption to human health. In De la Rosa, L. A., Alvarez-Parrilla, E., & González-Aguilar, G. A. (Eds.), *Fruit and Vegetable Phytochemicals: Chemistry, Nutritional Value and Stability* (pp. 353). Wiley Blackwell, Iowa.
16. Podsedek, A. (2007). Natural antioxidants and antioxidant capacity of *Brassica* vegetables: A review. *LWT – Food Science and Technology*, *40*(1), 1–11.
17. Lin, T., Zirpoli, G. R., McCann, S. E., Moysich, K. B., Ambrosone, C. B., & Tang, L. (2017). Trends in cruciferous vegetable consumption and associations with breast cancer risk: A case-control study. *Current Developments in Nutrition*, *1*(8), 000448.
18. Sasaki, K., Neyazaki, M., Shindo, K., Ogawa, T., & Momose, M. (2012). Quantitative profiling of glucosinolates by LC-MS analysis reveals several cultivars of cabbage and kale as promising sources of sulforaphane. *Journal of Chromatography B*, *903*, 171–176.

19. Keum, Y. S., Jeong, W. S., & Kong, A. (2004). Chemoprevention by isothiocyanates and their underlying molecular signaling mechanisms. *Mutation Research, 555*(2), 191–202.
20. Bianchini, F., & Vainio, H. (2004). Isothiocyanates in cancer prevention. *Drug Metabolism Reviews, 36*(3, 4), 655–667.
21. Li, Z., Wei, X., Li, L., Liu, Y., Fang, Z., & Yang, L. (2017). Development of a simple method for determination of anticancer component of indole-3-carbinol in cabbage and broccoli. *Journal of Food and Nutrition Research, 5*(9), 642–648.
22. Johnson, I. T. (2002). Glucosinolates: Bioavailability and importance to health. *International Journal for Vitamin and Nutrition Research, 72*(1), 26–31.
23. Peñas, E., Martinez-Villaluenga, C., & Frias, J. (2017). Sauerkraut: Production, composition, and health benefits. In *Fermented Foods in Health and Disease Prevention* (pp. 557–576).
24. Mizgier, P., Kucharska, A. Z., Sokół-Łętowska, A., Kolniak-Ostek, J., Kidoń, M., & Fecka, I. (2016). Characterization of phenolic compounds and antioxidant and anti-inflammatory properties of red cabbage and purple carrot extracts. *Journal of Functional Foods, 21*, 133–146.
25. Zielińska, M., Lewandowska, U., Podsędek, A., Cygankiewicz, A. I., Jacenik, D., Sałaga, M., & Fichna, J. (2015). Orally available extract from *Brassica oleracea* var. *capitata rubra* attenuates experimental colitis in mouse models of inflammatory bowel diseases. *Journal of Functional Foods, 17*, 587–599.
26. Martinez-Villaluenga, C., Peñas, E., Sidro, B., Ullate, M., Frías, J., & Vidal-Valverde, C. (2012). White cabbage fermentation improves ascorbigen content, antioxidant and nitric oxide production inhibitory activity in LPS-induced macrophages. *LWT–Food Science and Technology, 46*(1), 77–83.
27. Cho, H. J., Seon, M. R., Lee, Y. M., Kim, J., Kim, J. K., Kim, S. G., & Park, J. H. Y. (2008). 3,3′-Diindolylmethane suppresses the inflammatory response to lipopolysaccharide in murine macrophages. *Journal of Nutrition, 138*(1), 17–23.
28. Heiss, E., & Gerhäuser, C. (2005). Bioactive natural products as potential modulators of transcription factors NF-κB and AP-1. *Antioxidants & Redox Signaling, 7*(11, 12), 1553–1567.
29. Bacchetti, T., Tullii, D., Masciangelo, S., Gesuita, R., Skrami, E., & Bruge, F. (2014). Effect of black and red cabbage on plasma carotenoid levels, lipid profile and oxidized low-density lipoprotein. *Journal of Functional Foods, 8*, 128–137.
30. Jiang, X., Huang, H., Xiao, Z., Yu, L., Pham, Q., & Sun, J. (2016). Red cabbage microgreen lower circulating LDL, liver cholesterol and inflammatory cytokines in mice fed a high-fat diet. *Journal of Agricultural and Food Chemistry, 64*(48), 9161–9171.
31. Järvi, A., Karlström, B., Vessby, B., & Becker, W. (2016). Increased intake of fruits and vegetables in overweight subjects: Effects on body weight, body composition, metabolic risk factors and dietary intake. *British Journal of Nutrition, 115*(10), 1760–1768.
32. Çelik, E., & Şanlı-Er, N. (2017). Effects of nutrient and bioactive food components on Alzheimer's disease and epigenetic. *Critical Reviews in Food Science and Nutrition*, 112–118.

33. Heo, H., & Lee, C. (2006). Phenolic phytochemicals in cabbage inhibit amyloid β protein-induced neurotoxicity. *LWT–Journal of Food Science and Technology, 39*, 330–336.
34. Masci, A., Mattioli, R., Costantino, P., Baima, S., Morelli, G., & Punzi, P. (2015). Neuroprotective effect of *Brassica oleracea* sprouts crude juice in a cellular model of Alzheimer's disease. *Oxidative Medicine and Cellular Longevity*.
35. Da-Silva-Dias, J. C., & Imai, S. (2017). Vegetables consumption and its benefits on diabetes. *Journal of Nutritional Therapy*, *6*(1), 110–118.
36. Pandey, V., Chura, A., & Pandey, H. (2018). Cabbage–A storehouse of nutraceuticals. In Thangaraj, P. (Ed.), *Medicinal Plants: Promising Future for Health and New Drugs* (1st ed., pp. 115–123). CRC Press, Boca Raton, FL.
37. Traber, M. G., & Atkinson, J. (2007). Vitamin E, antioxidant and nothing more. *Free Radical Biology and Medicine*, *43*(1), 415–423.
38. Van Duyn, M. A., & Pivonka, E. (2001). Overview of the health benefits of fruit and vegetable consumption for the dietetics professional: Selected literature. *Journal of the American Dietetic Association*, *100*(12), 1511–1521.
39. Chandra-Hioe, M. V., Rahman, H. H., & Arcot, J. (2017). Lutein and β-carotene in selected Asian leafy vegetables. *Journal of Food Chemistry and Nanotechnology*, *3*(3), 93–97.
40. N/A. (2018). *USDA Food Composition Databases*. https://ndb.nal.usda.gov/ndb/nutrients/report?nutrient1=321&nutrient2=&nutrient3=&fg=11&max=25&subset=0&offset=75&sort=f&totCount=543&measureby=g (accessed on 1st January 2026).
41. Singh, J., Upadhyay, A., Bahadur, A., Singh, B., Singh, K., & Rai, M. (2006). Antioxidant phytochemicals in cabbage (*Brassica oleracea* L. var. *capitata*). *Scientia Horticulture, 108*(3), 233–237.
42. Burton, G. W. (1989). Antioxidant action of carotenoids. *Journal of Nutrition, 119*(1), 109–111.
43. Nwachukwu, I., Udenigwe, C., & Aluko, R. (2016). Lutein and zeaxanthin: Production technology, bioavailability, mechanisms of action, visual function, and health claim status. *Trends in Food Science & Technology, 49*, 74–84.
44. N/A. (2018). USDA National Nutrient Database for Standard Reference Legacy Release: 11112, cabbage, red, raw. United States Department of Agriculture Agricultural Research Service. https://ndb.nal.usda.gov/ndb/foods/show?ndbno5%2011112&fg5%2011&man5%20&lfacet%205%20&format%205%20Abridged&count5%20&max5%2025&offset5%2075&sort5%20f&qlookup5%20&rptfrm5nl&nutrient15%20338&nutrient25%20&nutrient35%20&subset5%200&totCount5%20522&measureby5%20g (accessed on 1st January 2026).
45. Padayatty, S. J., Katz, A., Wang, Y., Eck, P., & Kwon, O. (2003). Vitamin C as an antioxidant: Evaluation of its role in disease prevention. *Journal of the American College of Nutrition, 22*(1), 18–35.
46. Traber, M., & Stevens, J. (2011). Vitamins C and E: Beneficial effects from a mechanistic perspective. *Free Radical Biology and Medicine, 51*(5), 1000–1013.
47. Cartea, M., Francisco, M., Soengas, P., & Velasco, P. (2011). Phenolic compounds in *Brassica* vegetables. *Molecules, 16*, 251–280.

48. Manach, C., Scalbert, A., Morand, C., Remesy, C., & Jimenez, L. (2004). Polyphenols: Food sources and bioavailability. *American Journal of Clinical Nutrition, 79*(5), 727–747.
49. Kim, D. O., Padilla-Zakour, O. I., & Griffiths, P. D. (2004). Flavonoids and antioxidant capacity of various cabbage genotypes at juvenile stage. *Journal of Food Science, 69*(9), 685–689.
50. N/A. (2018). National Nutrient Database for Standard Reference Legacy Release: 11109, cabbage, raw. United States Department of Agriculture Agricultural Research Service. https://ndb.nal.usda.gov/ndb/foods/show/11109?n1%205%20%7BQv%3D1%7D&fgcd%205%20&man%205%20&lfacet%205%20&count%205%2025&max%205%2025&sort%205%20default&qlookup%205%20cabbage&offset%205%20&format%205%20Full&new%205%20&measureby%205%20&Qv%205%201&ds%205%20SR&qt%205%20&qp%205%20&qa%205%20&qn%205%20&q%205%20&ing%205 (accessed on 1st January 2026).
51. Possenti, M., Baima, S., Raffo, A., Durazzo, A., Giusti, A., & Natella, F. (2017). Glucosinolates in food. In J. M. Mérillon & K. Ramawat (Eds.), *Glucosinolates* (pp. 87–132). Springer International Publishing, Cham, Switzerland.
52. Wennberg, M., Ekvall, J., Olsson, K., & Nyman, M. (2006). Changes in carbohydrate and glucosinolate composition in white cabbage (*Brassica oleracea* var. *capitata*) during blanching and treatment with acetic acid. *Food Chemistry, 95*(2), 226–236.
53. Park, S., Arasu, M., Lee, M. K., Chun, J. H., Seo, J., & Lee, S. W. (2014). Quantification of glucosinolates, anthocyanins, free amino acids, and vitamin C in inbred lines of cabbage (*Brassica oleracea* L.) Suhyoung. *Food Chemistry, 145*, 77–85.
54. West, L., Meyer, K., Balch, B., Rossi, F., Schultz, M., & Haas, G. (2004). Glucoraphanin and 4-hydroxyglucobrassicin contents in seeds of 59 cultivars of broccoli, raab, kohlrabi, radish, cauliflower, brussels sprouts, kale, and cabbage. *Journal of Agricultural and Food Chemistry, 52*, 916–926.
55. Choi, S. H., Park, S., Lim, Y., Kim, S. J., Park, J. T., & An, G. (2014). Metabolite profiles of glucosinolates in cabbage varieties (*Brassica oleracea* var. *capitata*) by season, color, and tissue position. *Horticulture, Environment and Biotechnology, 55*(3), 237–247.
56. Bellostas, N., Kachlicki, P., Sørensen, J., & Sørensen, H. (2007). Glucosinolate profiling of seeds and sprouts of *Brassica oleracea* varieties used for food. *Scientia Horticulture, 114*(4), 234–242.
57. Schramm, K., Vassao, D. G., Reichelt, M., Gershenzon, J., & Wittstock, U. (2012). Metabolism of glucosinolate-derived isothiocyanates to glutathione conjugates in generalist lepidopteran herbivores. *Insect Biochemistry and Molecular Biology, 42*(3), 174–182.
58. Lynn, A., Collins, A., Fuller, Z., Hillman, K., & Ratcliffe, B. (2007). Cruciferous vegetables and colo-rectal cancer. *Proceedings of the Nutrition Society, 65*(1), 135–144.
59. Rouzaud, G., Young, S., & Duncan, A. (2004). Hydrolysis of glucosinolates to isothiocyanates after ingestion of raw or microwaved cabbage by human volunteers. *Cancer Epidemiology, Biomarkers & Prevention, 13*(1), 125–131.
60. Banerjee, A., Penna, S., & Variyar, P. (2015). Allyl isothiocyanate enhances the shelf life of minimally processed shredded cabbage. *Food Chemistry, 183*, 265–272.

61. Wu, Q. J., Yang, Y., Wang, J., Han, L. H., & Xiang, Y. (2013). Cruciferous vegetable consumption and gastric cancer risk: A meta-analysis of epidemiological studies. *Cancer Science, 104*(8), 1067–1073.
62. Watson, G. W., Beaver, L. M., Williams, D. E., Dashwood, R. H., & Ho, E. (2013). Phytochemicals from cruciferous vegetables, epigenetics, and prostate cancer prevention. *AAPS Journal, 15*(4), 951–961.
63. Ge, X., Yannai, S., Rennert, G., Gruener, N., & Fares, F. (1996). 3,3′-Diindolylmethane induces apoptosis in human cancer cells. *Biochemical and Biophysical Research Communications, 228*(1), 153–158.
64. Chepkemoi, J. (2017). *The World Leaders in Cabbage Production.* World Atlas. https://www.worldatlas.com/articles/the-world-leaders-in-cabbage-production.html (accessed on 1st January 2026).
65. Chun, O. K., Smith, N., Sakagawa, A., & Lee, C. Y. (2004). Antioxidant properties of raw and processed cabbages. *International Journal of Food Sciences and Nutrition, 55*(3), 191–199.
66. Kusznierewicz, B., Smiechowska, A., Bartoszek, A., & Namieśnik, J. (2008). The effect of heating and fermenting on antioxidant properties of white cabbage 'Barbara.' *Food Chemistry, 108*, 853–861.

CHAPTER 3

Onion (*Allium cepa* L.)

HELA CHIKH-ROUHOU,[1,*] KARIMA KACEM,[1] IMEN TLILI,[2] IMEN HENANE,[2] and RIADH ILAHY[2]

[1]*Regional Research Center on Horticulture and Organic Agriculture (CRRHAB), Production and Protection for a Sustainable Horticulture, University of Sousse, Chott Mariem, Sousse, Tunisia*

[2]*Laboratory of Horticulture, National Agricultural Research Institute of Tunisia (INRAT), University of Carthage, Ariana Governorate, Tunisia*

*Corresponding author

ABSTRACT

Allium cepa, commonly known as onion, represents a significant horticultural commodity with extensive global agricultural distribution, demonstrating remarkable adaptability across diverse ecological zones and agricultural systems worldwide. It has been identified as a food of great interest, given its nutritional and bioactive components and its high consumption rate all year round. The increasing demand for onions both for consumption and medicinal purposes has increased the need for breeding new cultivars with a high yield and quality. Great diversity in biochemical and nutritional composition was found in onion genotypes from different countries, highlighting the importance of onion consumption for human health. We report here a review of onion consumption benefits, their phytochemical composition that differs depending on the cultivation environments and onion part (peel, bulb or seed), and their antioxidative properties associated with health benefits and medicinal bioactivities, demonstrating the importance of this species as functional food.

Postharvest Quality Attributes of Fresh Vegetables and Grains.
Riadh Ilahy, Imen Tlili, Thouraya R'him, and Muhammad Azam (Eds.)
© 2027 Apple Academic Press, Inc. Co-published with CRC Press (Taylor & Francis)

DOI:10.1201/9781779642851-4

3.1 INTRODUCTION

Onion (*Allium cepa* L., $2n = 16$), a member of the Alliaceae family, is a significant commercial vegetable crop cultivated extensively across nearly all countries worldwide. According to the Food and Agriculture Organization (FAO) database, onion ranks as the second most cultivated vegetable globally, following potato, occupying approximately 5.4 million hectares (ha) and yielding 103 million tons (Mt) of production, predominantly in Europe and Asia, with China and India as the principal producing nations. Most commercial onions are cultivated from the thin, dark seeds of the plant. Onion is highly regarded and stored as pickles for its flavor and nutritional values [1]. Seed production constitutes the initial and critical stage in the onion supply chain. Importantly, seed cultivation can occur in the field using seeds, seedlings, or bulbs as starting materials [1]. Bulbs can differ in size, shape, color, pungency, single center, firmness, and tightness of outer dry skins and neck. Cultivated onions are of two major types, known as short-day and long-day onions. However, a third group can be expressed as an intermediate-day-length type. Onions that require 14 h or more to bulb formation are classified as long-day, whereas those requiring 11–12 h are considered short-day onions [2].

Since ancient times, onions have held a significant place in traditional folk medicine, valued for their therapeutic properties such as diuretic effects, anti-inflammatory capabilities, and their ability to relieve coughs and respiratory ailments. Recent scientific investigations have also unveiled additional noteworthy biological activities [3]. Onions boast a rich nutritional profile, containing essential vitamins (B2, B6, C, and B1), selenium, potassium, folic acid, magnesium, calcium, and phosphorus [1]. Furthermore, they harbor various phytochemicals, including organosulfur compounds, phenolic compounds, polysaccharides, and saponins. The main bioactive constituents of onions include phenolic compounds and sulfur-containing substances, such as onionin A, cysteine sulfoxides, quercetin, and quercetin glucosides [4]. Among these constituents are vitamins, fibers, minerals, fructooligosaccharides (FOS), inulin, polyphenols like quercetin, saponins, and numerous other secondary metabolites. Notably, onions are particularly abundant in flavonoids and alkenyl-cysteine sulfoxides (ACSO), which confer significant health benefits to consumers [5]. Onions are among the most essential and widely consumed vegetables in the human diet have a long history of use in both cuisine and medicine. They are cultivated and traded globally for their unique aroma and significant nutritional benefits. Currently, both onion production and the demand for processed onion products are on the rise. Pungency and total soluble solids (TSS) are two important traits of onion bulbs in terms of processing and storage. TSS, consisting mostly of fiber, starch, and sugars [6], shapes various important traits such as flavor, texture, and overall storability of onions. The unique pungency is mainly attributed to the sulfur-containing compound (allyl propyl disulfide) [7]. Over the last few years, there has been a growing consumer preference for milder onion varieties, often called "sweet onions," leading to the expansion of a distinct market segment [8].

Consumption of phytochemical-rich vegetable cultivars may provide health benefits [9]. In recent years, the onion has been receiving more attention from consumers, who have shown an increased interest in the nutraceuticals (functional compounds) that provide health benefits, including the prevention of diseases. Foods with high flavonoid content, in addition to their nutritive value, can have a protective effect on human health. Quercetin, the dominant flavonoid in onion bulbs, is known to possess anti-cholesterol, anticancer, and antioxidant properties [10–12]. This flavonoid accounts for over 95% of the total flavonoid content [13]. Research increasingly focuses on selecting food plants rich in nutraceuticals as a strategy for disease prevention rather than treatment [14, 15].

We report here a review of onion consumption benefits, their phytochemical composition that differs depending on the cultivation environments and onion part (peel, bulb, or seed), and their antioxidative properties associated with health benefits and medicinal bioactivities.

3.2 PHYTOCHEMICAL COMPOSITION

Onions are rich in biochemical components and exhibit significant antioxidant activity. Various studies have observed considerable diversity among genotypes, emphasizing that multiple factors influence the antioxidant activity of onions, including genetic background, horticultural practices, storage conditions, and different plant parts (Figures 3.1 and 3.2), extraction methods, processing technologies [9, 11, 15], and onion color [16]. It has been noted that red onions generally exhibit higher antioxidant activity compared to yellow and white onions [16].

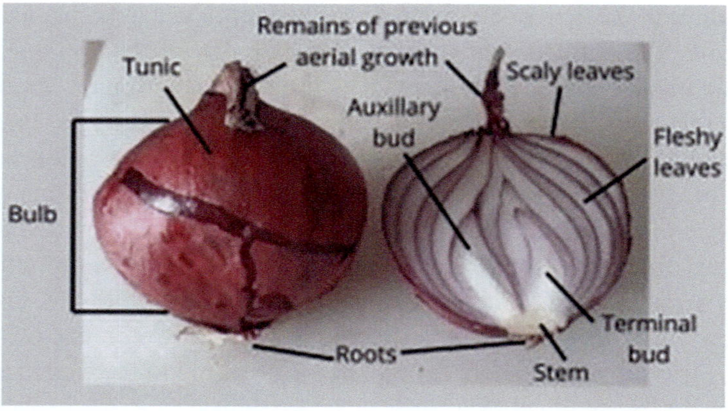

FIGURE 3.1 Parts of an onion bulb.

FIGURE 3.2 Onion seeds.

3.2.1 BULBS

Onions serve as an exceptional dietary source of polyphenols and essential nutrients. The composition of phytonutrients in onions is highly variable, influenced by factors such as geographical location, genetic background [17], maturity stage, environmental conditions, farming practices, storage duration, and bulb section [18]. Despite their high-water content, comprising

80–95% of their fresh weight, onions are low in calories. Approximately 65% of their dry weight consists of nonstructural carbohydrates (NSC), including glucose, fructose, sucrose, and FOS [19]. Regarding vitamins and minerals, onions are notably low in sodium while being high in vitamin B6, vitamin C, folic acid, calcium, magnesium, phosphorus, potassium, iron, selenium, and zinc. Their lipid content is minimal, with glutamic acid and arginine being the notable exceptions in the amino acid profile [20]. Among the FOS present in onion bulbs, kestose, nystose, and fructofuranosylnystose are predominant [21]. Furthermore, onions exhibit a significant amount of total dietary fiber (TDF) and a favorable soluble-to-insoluble dietary fiber ratio (SDF:IDF), which correlates with various metabolic and physiological effects [22]. Flavonoids constitute the primary phenolic compounds in onions, encompassing different subclasses such as flavones, flavanones, flavonols (e.g., quercetin and its derivatives), isoflavones, flavanonols, flavanols, chalcones, and anthocyanins. These flavonoids contribute to the coloration of onions, with variations in hues ranging from yellow to brown or red/purple [23].

The evaluation of morphological traits of Tunisian onion (landraces from the south of the country and breeding lines) was reported recently [24, 25]. The authors showed that the yellow genotype presented the smallest bulbs, with a mean weight of 185 g. The dry matter content ranged from 9.95% to 11.70%; the highest values were obtained in red onions and evidenced a great variability among these genotypes in terms of phenolic and flavonoid contents. Besides, Chalbi et al. [26] studied the bioactive compounds and activity in various Tunisian onion landraces (Figure 3.3) and showed that this genetic pool has great diversity in biochemical contents and composition.

FIGURE 3.3 Onion bulb diversity in Tunisia.

Flavonoids, one group of polyphenolic compounds, are secondary metabolites and important constituents of the human diet [27]. Plants of the *Allium* family are an important source of dietary flavonols. Foods with high flavonoid content, in addition to their nutritive value, can have a protective effect on human health [15].

Onion bulbs stand out as rich sources of dietary flavonoids, primarily comprising flavonols and anthocyanins [28]. Among onion cultivars, quercetin glycosides dominate as the primary flavonols, while anthocyanins are notably abundant in red onions, constituting nearly 10% of the flavonoid content expressed on a fresh weight basis [28]. Research shows that yellow onions have flavonol content ranging from 270 to 1187 mg/kg fresh weight (FW), while red onions contain between 415 and 1917 mg/kg FW [28]. Within all onion cultivars, quercetin derivatives, such as quercetin 4′-glucoside and quercetin 3,4′-diglucoside, are the predominant flavonols. The red coloration of onions is due to the presence of anthocyanins in the epidermal cells of the scale leaves, primarily cyanidin 3-glucoside [1, 23]. In red onions, quercetin is the major compound found in the skin, while quercetin-4-glucoside prevails in the bulb [4]. Notably, the quercetin content varies with skin color, with red and yellow bulbs containing more quercetin than white onions [29]. Compared with other foods such as apples (50 mg/kg), broccoli (100 mg/kg), and blueberries (40 mg/kg), onion has a significantly higher quercetin content, ranging from 5 to 10 times greater at 300 mg/kg [1]. Research suggests that quercetin exhibits various health benefits, including reducing the carcinogenic activity of cooked food mutagens, inhibiting activities associated with certain tumor cells, enhancing the antiproliferative activity of anticancer agents, and suppressing the growth of transformed tumorigenic cells [13, 30]. Approximately 65 to 80% of the dry matter in onion bulbs comprises nonstructural carbohydrates [31]. These carbohydrates primarily include glucose, fructose, sucrose, and FOS from the inulin and neoinulin series, with a degree of polymerization ranging from 3 to 20 [31, 32]. The dry matter content (DMC) and carbohydrate profile of onion bulbs exhibit significant genetic variability. Broadly, three distinct types of onion cultivars are identified based on their specific DMC and carbohydrate profiles: fresh market types (with DMC ranging from 5 to 9%), storage types (with DMC ranging from 8 to 12%), and dehydrator types (with DMC exceeding 18%) [33].

The fresh onion bulb contains fructans with a low degree of polymerization (10–40%), and heterogeneous polysaccharides, as well as flavonoids (mainly quercetin glucosides). In colored onions, 2.5–6.5% are saponosides (glycosides of furostanols), sterols (cholesterol, stigmasterol, β-sitosterol, etc.), vitamins (A, C, B1, and B2), pectin, anthocyanins, and sulfur compounds: trans-(+)-S-(1-propenyl)-L-cysteine (= isoalliin) sulfoxide and other cysteine derivatives (alkyl- and alkenyl cysteine and the corresponding sulfoxide derivatives). When the onion is cut or bruised, the sulfoxides are degraded by the alliinase contained in the vacuole of the cell, then there is a release of pyruvic acid and alkylthiosulfinates – simple compounds – unstable and quickly transformed into disulfides; the trans-(+)S-(1-propenyl) cysteine sulfoxide (precursor of tear factor), and for its part transformed, via 1-propenesulfenic acid, to S-oxide of Z-propenethial (=S-oxide of thiopropanal). Among these thiosulfinates: methyl methanethiosulfinate, alkyl (E)-1-propene thiosulfinate, (E,Z)-1-propenyl alkanethiosulfinates, propyl propanethiosulfinate, with the tear factor contribute to the taste and aroma of the onion, which undergo after a more or less short time, a decomposition into sulfides: di-, tri-, tetrasulfides, and other constituents responsible for the odor of onions, especially the methyl propyl disulfide, methyl propyl trisulfide, and dipropyl trisulfide [34]. Onion contains quercetin, fructose, quercetin-3-glucoside, isorhamnetin-4-glucoside, xylose, galactose, glucose, mannose, organo compounds, allylsulfides, flavonoids, flavonols, cycloalliin, selenium, thiosulfinates, and sulfur and seleno compounds [35].

Chalbi et al. [26], in their assessment of various Tunisian landraces using LC-ESI-MS identified five free phenolic acids and three flavonoids. These authors showed that the most abundant phenols were quercetin, quinic acid, and protocatechuic acid, whereas luteolin was the least abundant phenolic compound with low concentrations.

Fructans, polymers of fructose, constitute the primary reserve carbohydrates in onion bulbs, characterized by varying degrees of polymerization [36]. During the transition into dormancy, both fructans and simple sugars are known to decrease in onion bulbs [37]. Studies indicate that higher concentrations of fructans at harvest positively correlate with extended storage duration and delayed sprouting in onion bulbs [21, 38, 39]. Additionally, long-storing onion bulbs tend to exhibit higher fructan content compared to those with poor storage capabilities [21]. However, the precise mechanisms governing the mobilization and metabolism of fructans concerning onion dormancy remain unclear. Furthermore, the composition of sugars in onions plays a crucial role in determining their intended use. Petropoulos et al. [40] demonstrated that onions with dry matter content exceeding 15% and low reducing sugar content are conducive to prolonged storage. Moreover, they established a close relationship between onion sweetness, total soluble solids (TSS), and sugar compositions, which collectively contribute to sweetness.

The pungency of onions is assessed by pyruvic acid analysis, a compound produced stoichiometrically from thiosulfinates, providing a convenient method for measuring onion pungency [1]. Observations indicate correlations between pyruvic acid levels and flavor perception. The sweetness of onions, on the other hand, is influenced by both pungency and sugar concentrations [1]. High pungency can mask high sugar levels, preventing the onion from being perceived as sweet. Conversely, onions with low levels of both pungency and sugar may be perceived as bland. Ideally, a balance of high sugar content and low pungency results in a sweet onion [41]. The primary constituents of onion volatiles include dipropyl trisulfide, dipropyl disulfide, and propenyl disulfides, alongside various other compounds such as dipropenyl sulfide and dipropyl sulfide [1, 42–45].

As the demand for low-pungency onions has surged, the need to assess pungency levels in bulbs has become increasingly important [46]. Highly pungent onion varieties are typically utilized in cooking and in the production of sauces, canned soups, extracts, and dehydrated products [47]. Conversely, less pungent onion cultivars are now preferred for fresh consumption, particularly in regions like the USA, Japan, and many European countries [48]. Mild or low-pungency onions are characterized by a pyruvate concentration of less than 5 $\mu mol\ g^{-1}$ FW and often command a premium price [49]. Additionally, guidelines utilized by the sweet onion industry categorize onions based on their pungency levels, classifying them as low pungency/sweet (0–3 $\mu mol\ g^{-1}$ FW), medium pungency (3–7 $\mu mol\ g^{-1}$ FW), and high pungency (above 7 $\mu mol\ g^{-1}$ FW) [47].

3.2.2 SEEDS

Onion seeds serve as a valuable addition to the human diet, offering significant nutritional benefits [50]. They boast a notable oil content ranging from 22% to 26%, with a high proportion of linoleic acid (49–61%), alongside oleic and palmitic acids. Additionally, onion seeds contain proteins, comprising 16% to 26% of their composition, along with alcohols, acids, esters, and sulfur-containing compounds, which contribute to glycemic control and reduce oxidation phenomena [51]. Moreover, they contain steroidal saponins, which play roles in controlling cholesterol levels, diabetes, and potentially cancer [51]. Free amino acids and antioxidants further enhance their nutritional profile [52, 53]. Research by Dini et al. [53] highlights that onion seeds contain more than 20% of oil, 25% of raw proteins, and a noteworthy concentration of polyphenols. The accumulation of oils and their fatty acid composition in seeds is influenced by various factors such as environmental conditions, crop management practices, and genetic factors [50].

Yalcin and Kavuncuoglu [51] conducted a study revealing the presence of 48 compounds in the volatile fraction of onion seeds. The most prominent compound among them is 2-phenylethanol, an alcohol, with concentrations ranging from 0.42% to 41.17%. Additionally, 1-hexanol, with percentages varying from 5.85% to 31.61%, emerged as a major component in the volatile fraction. Various other alcoholic substances were also detected [51]. Aromatic hydrocarbons such as p-xylene (ranging from 0.13% to 0.90%) and mesitylene (0.10% to 0.85%) were identified. The monocyclic monoterpene limonene was found to vary from 0.75% to 3.17% in the samples. Furthermore, esters such as hexyl acetate, nonyl acetate, ethyl octanoate, and methyl hexanoate were among the predominant esters identified in onion seeds. The distinctive flavor profile of onion seeds, characteristic of the genus *Allium*, is predominantly influenced by various biologically active organosulfur compounds [54]. Notably, onion seeds contain carbondisulfide, 4-phenyl-2-thiazolethiol, methylsulfonylmethane, and dipropyl disulfide, which contribute to their characteristic aroma and flavor [54].

3.2.3 WASTES AND ITS BYPRODUCTS

Significant quantities of onion waste, including skin/peel, two external fleshy leaves, top, and bottom parts, are generated during industrial processing, and these byproducts are abundant in bioactive compounds and phytochemicals [55]. Studies have revealed the presence of various beneficial compounds in onion waste, such as phenolics, flavonoids, flavanols, anthocyanins, tannins, vanillic acid, and ferulic acid [56]. Onion waste is recognized as a valuable source of flavor compounds, fiber, nonstructural carbohydrates, and polyphenols [56]. This presents opportunities for its use as a source of functional food ingredients. Incorporating onion waste into novel products could enhance their antioxidant and prebiotic qualities, thereby enriching the nutritional value and health benefits of these products.

Onion peel stands out as a significant agricultural byproduct, abundant in bioactive compounds [57], making it a valuable candidate for utilization as a health-promoting ingredient, particularly in pharmacological and biomedical applications. Despite its nutritional potential, onion peel is often considered waste in the food industry, and improper disposal may lead to environmental pollution. Therefore, exploring ways to repurpose onion peel not only mitigates waste but also harnesses its health-enhancing properties, contributing to sustainable practices in both agriculture and industry.

The multilayered tissue of onion bulbs serves as a valuable reservoir of bioactive compounds, primarily falling into two main chemical groups: alkenyl cysteine sulfoxides and flavonoids. These compounds contribute significantly to the distinct aroma and taste of onions, thereby influencing the unique flavors found in different onion species and varieties [58]. Additionally, the red hue in certain onions is attributed to the presence of anthocyanins [59]. Notably, onion skins contain higher concentrations of quercetin aglycon compared to the flesh [60, 61]. The dried skin of onions serves as a rich source of flavonols, with a high percentage of aglycone forms, while the presence of anthocyanins is typically limited to red-colored varieties. Although most research has focused on the chemical and biological characterization of edible onion bulbs, less attention has been given to onion skins due to their inedible nature [57]. Despite this, the unique composition of onion skins holds promise for further exploration and utilization in various applications.

Benitez et al. [62] conducted a study focusing on the characterization of various compounds and the antioxidant activity found in brown and red onion skins from Spanish cultivars. Their findings highlighted the brown onion outer layer as a more promising waste source for valorization. Similarly, research on Californian onions [63] examined flavonoids in the outer layer of multiple onion varieties, revealing that Californian varieties possess higher flavonoid content compared to other Italian varieties assessed in the study. Additionally, Kumar et al. [57] emphasized the richness of bioactive compounds in onion peel, attributing several therapeutic benefits to this component. Specifically, the presence of biochemical compounds, including phytochemicals such as total flavonoids, total polyphenols, quercetin, and its derivatives, underscores the potential application of onion peel in biomedical and pharmaceutical fields [57].

Previous research has shown that onion peel extracts can demonstrate antimicrobial and neuroprotective properties, as well as protective effects against cancer (cervical, breast, and liver cancer cells), hyperglycemia, hypercholesterolemia, obesity, and erectile dysfunction [57]. Ifesan [64] revealed that onion peel contains a very high content of carbohydrate (88.56%) and low protein (0.88%). Onion peel extract yielded 98.52 μg ml^{-1} total flavonoid, 664.30 μg ml^{-1} total phenol, 1338.15 μg ml^{-1} total antioxidant, and 27.76 μg ml^{-1} scavenged DPPH radical. Moreover, several researchers have highlighted that onion peel contains a higher concentration of phytochemicals compared to the edible flesh, with quercetin being the predominant phenolic compound in the peel [65, 66]. The presence of all these bioactive compounds confers to onion peel its various therapeutic benefits in preventing cancer, obesity, diabetes, neurodegenerative disorders, cardiovascular disorders, microbial damage, and erectile dysfunction. Thus, onion peel in the form of powders or extracts can be employed in various industries as a therapeutic and pharmaceutical agent [57].

The waste biomass of onions primarily consists of outer scales, skin, roots, bulb tops, and deteriorated bulbs. Utilizing onion waste as a source of functional food ingredients represents an effective valorization approach, aligning closely with current food industry trends that focus on waste reuse and meeting the rising demand for functional foods [10]. Key functional components identified in onion waste include: (i) flavonols, such as quercetin and quercetin glucosides, recognized for their antioxidant properties; (ii) FOS, well-known prebiotics; (iii) cell-wall polysaccharides, including dietary fiber and pectin, which serve as food additives and potential sources of prebiotic oligomers; and (iv) organosulfur compounds. While the use of onion waste in food production is still being researched [10], promising results have been observed in studies that incorporate powdered onion waste concentrates and extracts into different food matrices. These innovative methods have produced products with improved antioxidant and prebiotic properties, highlighting the potential of valorizing onion waste to create value-added food products [10].

Furthermore, onion waste has been proposed as a promising and cost-effective source of pectin [67]. Innovative approaches to valorizing onion waste pectin include the production of pectic oligosaccharides with potential prebiotic benefits [68]. Additionally, the valorization of cellulose, the second-most-abundant cell wall polysaccharide in onion waste, could focus on its potential for fermentation and bioconversion [69]. A previous study by Benítez et al. [62] highlighted that different components of onion wastes are good sources of protein (8.3–15.6% DM), ash (4.4–8.6% DM), TDF (169–750 mg/g DM), minerals viz., potassium (11.1–15.9 mg/g), calcium (1.8–16.5 mg/g), magnesium (0.6–1.5 mg/g), iron (0.0196–0.8889 mg/g), zinc (0.0162–0.0538 mg/g), manganese (0.0065–0.0288 μg/g), selenium (0.00003–0.00093 μg/g), total phenolics (9.4–52.7 mg gallic acid equivalent (GAE)/g DM), flavonoids (7.0–43.1 mg quercetin equivalent (QE)/g DM), and flavonols (6.19–27 mg/g DM). Further studies by Campone et al. [66] showed that onion peel had a good amount of carbohydrates (88.56%) but had lower protein (0.88%), ash (0.39%), and crude fiber (0.15%; Table 3.2). On the other hand, research conducted by Michalak et al. [70] on onion skin powder indicated lower protein content (2.58–3.06%), crude fat (0.71–0.77%), while had good amount of ash (5.50–5.93%), TDF (7.78–62.09%), soluble dietary fiber (7.38%), and insoluble dietary fiber (54.71%). These findings highlight the diverse nutritional composition of onion waste components, emphasizing their potential as valuable sources of various nutrients and bioactive compounds. Table 3.1 summarizes the most important bioactive compounds in onion fractions reported in several investigations.

TABLE 3.1 Types of Bioactive Compounds in Onion (*Allium cepa* L.)

Bioactive Compounds	Part Used	Concentration	References
Organosulfur compounds: diallylmonosulfide (DMS), diallyldisulfide (DDS), diallyltrisulfide (DTS), diallyltetrasulfide (DTTS), PRENCSO, MCSO, alliin, asallicin, and thiopropanal S-oxide	Bulb	963.90–1393 mg/kg DW	[71]
Sulfur content	Skin	15.6 μmoles/g DW	[72]
Phenolic acids: protocatechuic acid, vanillic acid, ferulic acid, *p*-coumaric acid, cyanidin, peonidin, pelargonidin, isoeugenol, hydroxytyrosol, sylvopinol, *p*-octopamine, 4-methylcatechol	Bulb	72.12–551.74 μg/g DW	[73]
Flavonoids: quercetin, kaempferol, isorhamnetin, gallic acid, luteolin, rhamnazin, xanthomicrol, petunidin 3-galactoside	Bulb	0.46–48.27 μg/g DW	[73]
Quercetin, kaempferol, luteolin	Skin or peel	56.17–11,885.02 mg/kg DW	[74]

TABLE 3.1 *(Continued)*

Bioactive Compounds	Part Used	Concentration	References
Quercetin	Bulb	22.0–890.5 mg/kg FW	[75]
Anthocyanins (cyanidin, peonidin, pelargonidin, delphinidin, and petunidin)	Bulb	1555–250 mg/kg DW	[76]
Ascorbic acid	Bulb	35 mg/kg DW	[77]
	Leaves	390–5000 mg/kg FW	[78]
Organic acids	Bulb	0.10–0.99 μg/g DW	[73]
Alkaloids	Bulb	4.17–14.91 μg/g DW	[73]
Saponins: tropeoside A1	Bulb (red)	26.4 mg/kg FW	[79]
Ceposide A	Bulb (white)	9.8 mg/kg FW	[80]
Flavonoids	Waste	–	[81]
Linoleic acid, oleic acid, palmitic acid; proteins; alcohols, acids, and esters; sulfur-containing compounds	Seeds	2–26% linoleic acid, 49–61% oleic and palmitic acid; 16–26% proteins.	[51]

TABLE 3.2 Onion Compounds (*Allium cepa* L.)

Major Compounds	Minor Compounds	References
Water	–	[82]
Proteins	–	[82]
Carbohydrates	Inulin, fructooligosaccharides, isorhamnetin-4-glucoside, galactose, glucose, and mannose	[82]
Vegetal hormone	Glycoquinine	[83]
Steroids	Catechol, protocatechuic acid, thiocyanate, and thiopropionaldehyde	[84]
Vitamins	A, B complex, C, and E	[84]
Minerals	Selenium, phosphorus, iron, calcium, and chromium	[82]
Flavonoids	Quercetin, apigenin, rutin, myricetin, kaempferol, catechin, resveratrol, epigallocatechol-3-gallate, luteolin, and genistein	[84]
	Quercetin aglycone, quercetin diglucoside, quercetin 4-glucoside, and isorhamnetin monoglycoside or kaempferol monoglycoside	
Organosulfur compounds	Thiosulfinates, cepaenes, cysteine, S-methyl cysteine sulfoxide, diallyl disulfide, allyl methyl sulfide, allyl propyl disulfide, gamma-L-glutamyl-trans-S-1-propenyl-L-cysteine sulfoxide, S-propenyl cysteine sulfoxide, S-alk(en)yl cysteine sulfoxides, and S-allyl cysteine sulfoxide	[84]
Phenolic compounds	Phenolics, phenolic acids, anthocyanins, and hydroxycinnamic acid	[82]
Lipophilic antioxidants	Dialkyl disulfides, aglycones, anthocyanin, saponins, and fistulosin (octadecyl 3-hydroxyindole)	[82, 84]

3.3 HEALTH BENEFITS

Onions are a rich source of health-promoting phytochemicals—natural compounds that can provide various health benefits and help protect against diseases, including cancer [85, 86].

Organosulfur compounds found in onions exhibit a range of beneficial activities, including antimicrobial, antiallergenic, anti-inflammatory, and antithrombotic properties [85]. Additionally, flavonols present in onions, such as quercetin and kaempferol, play important roles in human health by offering antiviral, antimicrobial, anti-inflammatory, and anticancer effects, while also contributing to the protection of the heart and brain [86].

3.3.1 ANTIOXIDANT ACTIVITY

Numerous studies have highlighted variations in antioxidant activity among different onion cultivars or varieties, potentially attributed to their genetic backgrounds [4]. Furthermore, organic cultivation practices [87], the use of sulfur bentonite-organic-based fertilizers [88], and the presence of mycorrhizal fungi [89] have been shown to improve phytochemical concentration and the radical scavenging activity in onions.

A wealth of literature has demonstrated the antioxidant properties of onion flavonoids [90–92]. Reactive oxygen species (ROS) are known to cause damage to cells in various organs [93], and flavonoids have been observed to stabilize the free electrons generated from ROS *in vitro* [94]. The hydroxyl configuration of flavonoids is believed to play a critical role in stabilizing ROS through their antioxidant activity.

Flavonoids also have a metal-chelating quality that inhibits the formation of free radicals [95]. Quercetin, a flavonol of onion, plays a better role as an antioxidant compared to selenium and sulfur-containing amino acids [35, 92, 96–99]. Kaempferol is also considered a potent antioxidant because it induces the production of many antioxidant enzymes, such as superoxide, dismutase and catalase. Kaempferol has also been shown to prevent low-density lipoprotein (LDL) oxidation, giving protection in atherosclerotic conditions [90].

Prolonged exposure to oxidative stress can lead to cellular damage, affecting nucleic acids, proteins, and lipids, and contributing to the development of serious conditions such as cancer, diabetes mellitus, Alzheimer's disease, and cardiac and liver diseases. Antioxidants, also referred to as free radical scavengers [1], counteract the harmful effects of free radicals, thereby preventing or reducing cellular damage, DNA mutations [100], and subsequently lowering the risk of major diseases like cancer. Numerous studies have highlighted the antioxidant potential of onions, attributed to their high levels of organosulfur compounds, polyphenols, and flavonoids, all of which possess natural antioxidant properties [101].

Škerget et al. [97] investigated the ethanol and acetone extracts of skin and edible portions of red onion and found the highest antioxidant activity in skins extracted with 60% ethanol and using the β-carotene bleaching test. The antioxidant capacity of many plants is largely attributed to their phenolic compound content [98]. Additionally, the method used to extract polyphenols from plant materials plays a crucial role in determining their antioxidant activity [99].

Essential oils extracted from onions are rich sources of antioxidant compounds with a wide range of biological activities [102]. Onions are particularly abundant in phenolic compounds, especially flavonoids, which not only possess antioxidant properties but also demonstrate various pharmacological effects, including antibiotic, antidiabetic, anti-atherogenic, and anticancer activities [1]. The antioxidant capacity of onion essential oils can be evaluated using a variety of methods, such as the diphenyl-1-picrylhydrazyl (DPPH) assay, the β-carotene bleaching assay, the azinobis (3-ethylbenzothiazoline-6-sulfonic acid) (ABTS) assay, the oxygen radical absorbance capacity (ORAC) assay, and the Trolox equivalent antioxidant capacity (TEAC) assay. Research has consistently shown that different onion genotypes exhibit significant antioxidant activity, regardless of the assessment method used [103]. Furthermore, the antioxidant capacity of onions can vary based on their color. For example, red onion bulbs have been reported to exhibit higher antioxidant activity compared to yellow and white onions, a conclusion supported by numerous studies [25, 101, 104]. Additionally, the presence of other compounds, such as sulfur compounds and gallic acid, may also contribute to the antioxidant properties of onions.

3.3.2 ANTIBACTERIAL ACTIVITY

Onion extracts and their bioactive compounds, including thiosulfinate compounds, phenolic compounds, polysaccharides, and essential oils, are well-known for their strong antibacterial properties [4], antifungal activities [105], and antiviral effects [106]. Extensive research has highlighted the antimicrobial effects of phenolics and flavonoids [94]. Several mechanisms have been proposed to explain the inhibitory actions of onion-derived compounds. One key mechanism involves disrupting microbial growth by inhibiting the catalysis of microbial enzymes, adhesins, and transport proteins [107]. Additionally, bioactive compounds derived from *Allium* species have been found to reduce oxygen supply, which disrupts the fungal cell wall and inhibits cell growth [108].

Quercetin, one of the most extensively studied flavonoids, has demonstrated significant inhibitory effects on bacterial growth. It exhibits great potential in completely inhibiting the growth of *Staphylococcus aureus* [109]. Similarly, experiments have revealed that kaempferol serves as an inhibitor against *Helicobacter pylori* [95]. Furthermore, onion extracts have been found to exert a bactericidal effect against *Streptococcus mutans*, *S. sobrinus*, *Porphyromonas gingivalis*, and *Prevotella intermedia* (Gram-positive bacteria), which are implicated in dental caries and adult periodontitis [110]. Although sulfur compounds also possess antimicrobial activity, their utilization is restricted due to their pungent odor, strong flavor, and volatile nature [42].

Composite materials based on onion fiber, combined with isolated flavonoids from onion skins, have been reported to possess antibacterial activity against various strains such as *Staphylococcus aureus* and *E. coli* [111]. Additionally, studies have demonstrated the bactericidal effects of onion peel extract against *Bacillus cereus*, exhibiting a bacteriostatic effect as well, attributed to the presence of antimicrobial components like quercetin and quercetin oxidation products [57]. These compounds demonstrate antioxidant activity by disrupting the cell wall, enabling the entry of antimicrobial agents into the cell, which causes the leakage of cellular contents. Ultimately, this results in cell death through interactions with bacterial DNA [59]. Onion essential oil has also been shown to exhibit fungicidal or inhibitory effects on the growth of fungal species isolated from food, including *Aspergillus*, *Fusarium*, and *Penicillium* species [112]. Moreover, the biosynthesis of silver nanoparticles using an onion endophytic bacterium has demonstrated a potent antifungal effect against the rice blast pathogen, resulting in abnormal mycelial morphology and an 88% inhibition rate of mycelium diameter. Furthermore, ethanol extract of red onion has been found effective in preventing *tinea pedis* caused by the fungal *Trichophyton rubrum*. Therefore, onions exhibit promising potential as natural preservatives in the food industry [4].

Numerous studies have investigated the antiviral activity of flavonoids [113], revealing mechanisms that involve the destruction or inhibition of viral protein synthesis [114]. Notably, quercetin and kaempferol have been identified for their prominent inhibitory roles against various viruses, including the rabies virus, poliovirus, Mengo virus, herpes simplex type I virus, Sindbis virus, parainfluenza type 3 virus, pseudorabies, and various respiratory viruses [90].

In vitro evaluations of extracts from onion, shallot, garlic, and green onion have demonstrated significant antiviral activities against adenoviruses [115]. These observations are particularly noteworthy in onions, which contain the highest content of quercetin. Studies have indicated that quercetin is responsible for the observed antiviral action, and cell culture investigations have suggested its potential inhibition against various respiratory viruses [116].

3.3.3 ANTICANCER ACTIVITY

Allium vegetables have garnered significant attention due to their beneficial effects against numerous diseases, particularly in the treatment of cancer and mitigation of side effects associated with current anticancer therapies. These effects are attributed to their bioactive compounds, including sulfur compounds, flavonoids, and saponins [117]. Consumption of *Allium* vegetables has been inversely associated with the risk of various cancers, such as breast cancer, gastric cancer, and colorectal cancer (CRC) [4].

Various studies have reported that garlic and onion intake diminish the risk of sarcoma and carcinoma in various tissues and organs, such as the stomach, colon, esophagus, prostate, bladder, liver, lungs, breast, skin, and brain [118]. Murayyan et al. [119] found that onions exhibit effectiveness in eliminating breast cancer cells, demonstrating their ability to activate pathways that induce cancer cell death. Additionally, onions create an unfavorable environment for cancer cells and disrupt communication between them, thereby inhibiting their growth. Spiraeoside, derived from red onion skin, has shown promising anticancer effects against HeLa cells by promoting apoptosis [4].

3.3.4 ANTIDIABETIC ACTIVITY

Potent *in vitro* antidiabetic activity has been reported in previous studies. Extracts derived from onion skin or solid waste have demonstrated significant inhibitory activity against α-glucosidase and α-amylase enzymes [56]. Additionally, silver nanoparticles synthesized using onion-based green methods have shown excellent inhibitory activities against α-glucosidase and α-amylase enzymes. Furthermore, onion fiber concentrates exhibited potent activity in reducing starch digestibility and glucose biosynthesis by inhibiting α-amylase activity [120].

Studies have investigated the effects of onions and their bioactive constituents on diabetic complications. Gomes et al. [121] demonstrated that quercetin may be beneficial for managing diabetic nephropathy. In their study, this polyphenol, administered orally at a dose of 10 mg/kg/day for 28 days, led to improvements in renal function. Specifically, it reduced proteinuria and plasma concentrations of creatinine, uric acid, and urea.

In general, onions and their bioactive constituents can be considered as preventive or therapeutic agents for diabetes. They exert their effects through various mechanisms, including antioxidant properties, as well as inhibitory effects on α-glucosidase and α-amylase enzymes [122]. The prolonged absorption rate of natural flavonoids, such as quercetin, has been suggested to mitigate collagen glycation, a contributing factor to cardiovascular issues in diabetic patients [123–125]. Moreover, bioactive compounds found in *Allium*, such as methiin and SAC sulfoxide, have been shown to play a crucial role in stimulating insulin secretion from the pancreas [126].

Flavonoids also possess an antidiabetic property that is directly related to the increase of insulin production by controlling pancreatic cells [126]. Quercetin helps in suppressing diabetic symptoms [127]. Moreover, it was found that the consumption of quercetin may help to control type 2 diabetes [128].

3.3.5 ANTI-INFLAMMATORY ACTIVITY

Onion also demonstrates anti-inflammatory properties, offering protective effects against inflammation-related diseases such as neuroinflammation, allergic inflammation, lung inflammation, colitis, and paw edema [4]. Ali et al. [129] revealed that onion extract can reduce the onset and progression of tumors, along with exhibiting antiasthmatic effects. Furthermore, allicin was found to inhibit the production of pro-inflammatory cytokine messengers in a study involving IBD [130].

3.4 CONCLUSION

Onion is rich in bioactive compounds and antioxidant properties, providing considerable medicinal, health, and economic benefits if properly consumed or utilized in food products and/or as value-added products. Onions possess a wide range of useful medicinal properties, which can be further exploited, confirmed, and proposed as a functional food. Future research should further focus on the phytochemical profile of different onion cultivars, as well as on their antioxidant activity in relation to the pungency.

KEYWORDS

- ***Allium cepa***
- **antioxidant properties**
- **flavonoids**
- **functional food**
- **nutraceuticals**
- **phytochemicals**

REFERENCES

1. Chakraborty, A. J., Uddin, T. M., Redwan Matin Zidan, B. M., Mitra, S., Rajib, D., Nainu, F., Dhama, K., Roy, A., Md. Jamal, H., Ameer, K., & Talha, B. A. (2022). *Allium cepa*: A treasure of bioactive phytochemicals with prospective health benefits. *Evidence-Based Complementary and Alternative Medicine, 2022*, 4586318.
2. Gökçe, A. F., Basar, N., Candar, A., Kaderlioglu, E., & Akbudak, N. (2012). Onion breeding program in Turkey. *Acta Horticulture, 969*, 93–98.
3. Biswas, S. K., Khair, A., Sarker, P., & Alom, M. S. (2010). Yield and storability of onion (*Allium cepa* L.) as affected by varying levels of irrigation. *Bangladesh Journal of Agricultural Research, 35*, 247–255.
4. Zhao, X. X., Lin, F. J., Li, H., Li, H. B., Wu, D. T., Geng, F., Ma, W., Wang, Y., Miao, B. H., & Gan, R. Y. (2021). Recent advances in bioactive compounds, health functions, and safety concerns of onion (*Allium cepa* L.). *Frontiers in Nutrition, 8*, 669805.
5. Griffiths, G., Trueman, L., Crowther, T., Thomas, B., & Smith, B. (2002). Onions–A global benefit to health. *Phytotherapy Research, 16*, 603–615.
6. Lin, M., Watson, J. F., & Baggett, J. R. (1995). Inheritance of soluble solids and pyruvic acid content of bulb onions. *Journal of the American Society for Horticultural Science, 120*, 119–122.

7. Wall, M. M., & Corigan, J. N. (1992). Relationship between pyruvate analysis and flavor perception for onion pungency determination. *Hort Science, 27*, 1029–1030.
8. Phaff, V. (2007). The European sweet onion market. In *Proceedings of the 2nd World Onion Congress*, Dronten, 22–24.
9. Biswas, S. K., Khair, A., & Sarker, P. K. (2010). Yield of onion and leaf purple blotch incidence as influenced by different levels of irrigation. *Agricultura Tropica et Subtropica, 43*(2), 221–230.
10. Aleksandra, V. C., Danijela, Š., Ana, M., Arijana, M., & Draženka, K. (2020). Onion solid waste as a potential source of functional food ingredients. *Engineering Power, 15*(3), 123–128.
11. Block, E. (2005). Biological activity of *Allium* compounds: Recent results. In Liu, G. (Ed.), *Proceedings of the Fourth International Symposium on Edible Alliaceae. Acta Horticulture, 688*, 41–57.
12. Omidiji, O., & Ehimidu, J. (1991). Changes in the content of antibacterial isorhamnetin 3-glucoside and quercetin 3′-glucoside following inoculation of onion (*Allium cepa* L. cv. Red Creole) with *Pseudomonas cepacia. Physiological and Molecular Plant Pathology, 37*, 281–292.
13. Leighton, T., Glinther, C., Fluss, L., Harte, W. K., Cansado, J., & Notario, V. (1992). Molecular characterization of quercetin and quercetin glycosides in *Allium* vegetables: Their effects on cell transformation. *ACS Symposium Series*, 221–238.
14. Hertog, M. G. L., Hollman, P. C. H., & Venema, D. P. (1992). Optimization of a quantitative HPLC determination of potentially anticarcinogenic flavonoids in vegetables and fruits. *Journal of Agricultural and Food Chemistry, 40*, 1591–1598.
15. Perez-Gregorio, M. R., Garcia-Falcon, M. S., Simal-Gandara, J., Rodrigues, A. S., & Almeida, D. P. F. (2010). Identification and quantification of flavonoids in traditional cultivars of red and white onions at harvest. *Journal of Food Composition and Analysis, 23*(6), 592–598. https://doi.org/10.1016/j.jfca.2009.08.013 (accessed on 1st January 2026).
16. Gökçe, A. F., Kaya, C., Serçe, S., & Özgen, M. (2010). Effect of scale color on the antioxidant capacity of onions. *Scientia Horticulture, 23*, 431–435.
17. Metrani, R., Singh, J., Acharya, P. K., Jayaprakasha, G. S., & Patil, B. (2020). Comparative metabolomics profiling of polyphenols, nutrients and antioxidant activities of two red onion (*Allium cepa* L.) cultivars. *Plants, 9*, 1077.
18. Downes, K., & Terry, L. A. (2010). A new acetonitrile-free mobile phase method for LC–ELSD quantification of fructooligosaccharides in onion (*Allium cepa* L.). *Talanta, 82*, 118–124.
19. Davis, F., Terry, L. A., Chope, G. A., & Faul, C. F. J. (2007). Effect on extraction procedure on measured sugar concentrations in onion (*Allium cepa* L.) bulbs. *Journal of Agricultural and Food Chemistry, 55*, 4299–4306.
20. Hannan, A., Humayun, T., Hussain, M. B., Yasir, M., & Sikandar, S. (2010). *In vitro* antibacterial activity of onion (*Allium cepa*) against clinical isolates of *Vibrio cholerae. Journal of Ayub Medical College, Abbottabad: JAMC, 22*(2), 160–163.
21. Jaime, L., Martín-Cabrejas, M. A., Mollá, E., López-Andréu, F. J., & Esteban, R. M. (2001). Effect of storage on fructan and fructooligosaccharide of onion (*Allium cepa* L.). *Journal of Agricultural and Food Chemistry, 49*(2), 982–988.

22. Jaime, L., Mollá, E., Fernández, A., Martín-Cabrejas, M. A., López-Andréu, F. J., & Esteban, R. M. (2002). Structural carbohydrate differences and potential source of dietary fiber of onion (*Allium cepa* L.) tissues. *Journal of Agricultural and Food Chemistry, 50*(1), 122–128.
23. Tedesco, I., Carbone, V., Spagnuolo, C., Minasi, P., & Russo, G. L. (2015). Identification and quantification of flavonoids from two Southern Italy cultivars of *Allium cepa* L. var. Tropea (red onion) and Montoro (copper onion) and their capacity to protect human erythrocytes from oxidative stress. *Journal of Agricultural and Food Chemistry, 63*(21), 5229–5238.
24. Chalbi, A., Chikh-Rouhou, H., Mezghani, N., Slim, A., Fayos, O., Bel-Kadhi, M. S., & Garcés-Claver, A. (2023). Genetic diversity analysis of onion (*Allium cepa* L.) from the arid region of Tunisia using phenotypic traits and molecular markers. *Horticulture, 9*(10), 1098.
25. Chikh-Rouhou, H., Chalbi, A., & Sta-Baba, R. (2018). Morphological and phytochemical characterization of onion new breeding lines (*Allium cepa* L.). In Direk, H., & Simsek, H. (Eds.), *Proceedings Book of the 5th International Conference on Sustainable Agriculture and Environment (ICSAE)*, Hammamet-Tunisia, 8–10 October 2018. *Proceedings ICSAE*, 17–20.
26. Chalbi, A., Chikh-Rouhou, H., Tlahig, S., Mallor, C., Garcés-Claver, A., Haddad, M., Sta-Baba, R., & Bel-Kadhi, M. S. (2023). Biochemical characterization of local onion genotypes (*Allium cepa* L.) in the arid regions of Tunisia. *Polish Journal of Environmental Studies, 32*(1), 15–26.
27. Lanzotti, V. (2006). The analysis of onion and garlic. *Journal of Chromatography A, 1112*, 3–12.
28. Slimestad, R., Fossen, T., & Vågen, I. M. (2007). Onions: A source of unique dietary flavonoids. *Journal of Agricultural and Food Chemistry, 55*, 10067–10080.
29. Patil, B. S., Pike, L. M., & Yoo, K. S. (1995). Variation in the quercetin content in different colored onions (*Allium cepa* L.). *Journal of the American Society for Horticultural Science, 120*(6), 909–913.
30. Kim, M. H., Jo, S. H., Jang, H. D., Lee, M. S., & Kwon, Y. I. (2010). Antioxidant activity and α-glucosidase inhibitory potential of onion (*Allium cepa* L.) extracts. *Food Science and Biotechnology, 19*(1), 159–164.
31. Benkeblia, N. (2013). Fructooligosaccharides and fructans analysis in plants and food crops. *Journal of Chromatography A, 1313*, 54–61.
32. Pöhnl, T., Böttcher, C., Schulz, H., Stürtz, M., Widder, S., Carle, R., & Schweiggert, R. M. (2017). Comparison of high-performance anion exchange chromatography with pulsed amperometric detection (HPAEC-PAD) and ultra-high performance liquid chromatography with evaporative light scattering (UHPLC-ELSD) for the analyses of fructooligosaccharides in onion (*Allium cepa* L.). *Journal of Food Composition and Analysis, 63*, 148–156.
33. Darbyshire, B., & Steer, B. T. (2020). Carbohydrate biochemistry. In *Onions and Allied Crops* (1st ed., p. 284). Boca Raton.
34. Boucif, L. (2020). L'évaluation des caractéristiques physicochimiques et organoleptiques sur l'oignon blanc d'Oulhaça menée en sec et en irriguée (Master's thesis).

35. Rodrigues, A. S., Domingos, P. F., Almeida, G., & Gregorio, M. R. P. (2017). Onions: A source of flavonoids. In *Flavonoids–From Biosynthesis to Human Health.*
36. Valluru, R., & Van Den Ende, W. (2008). Plant fructans in stress environments: Emerging concepts and future prospects. *Journal of Experimental Botany, 59*, 2905–2916.
37. Chope, G. A., Cools, K., Hammond, J. P., Thompson, A. J., & Terry, L. A. (2012). Physiological, biochemical and transcriptional analysis of onion bulbs during storage. *Annals of Botany, 109*, 819–831.
38. Suzuki, M., & Cutcliffe, J. A. (1989). Fructans in onion bulbs in relation to storage life. *Canadian Journal of Plant Science, 69*, 1327–1333.
39. Benkeblia, N., Ueno, K., Onodera, S., & Shiomi, S. (2006). Variation of fructooligosaccharides and their metabolizing enzymes in onion bulb (*Allium cepa* L. cv. Tenshin) during long-term storage. *Food Science, 70*, S208–S214.
40. Petropoulos, S. A., Fernandes, A., Barros, L., Ferreira, I., & Ntatsi, G. (2015). Morphological, nutritional and chemical description of "Vatikiotiko," an onion local landrace from Greece. *Food Chemistry, 182*, 156–163.
41. Vågen, I., & Slimestad, M. (2008). Amount of characteristic compounds in 15 cultivars of onion (*Allium cepa* L.) in controlled field trials. *Journal of the Science of Food and Agriculture, 88*(3), 404–411.
42. Rose, P., Whiteman, M., Moore, P. K., & Zhu, Y. Z. (2005). Bioactive S-alk(en)yl cysteine sulfoxide metabolites in the genus *Allium*: The chemistry of potential therapeutic agents. *Natural Product Reports, 22*(3), 351–368.
43. Munday, R., & Munday, C. M. (2004). Induction of phase II enzymes by aliphatic sulfides derived from garlic and onions: An overview. *Methods in Enzymology, 382*, 449–456.
44. Munday, R., & Munday, C. M. (2001). Relative activities of organosulfur compounds derived from onions and garlic in increasing tissue activities of quinone reductase and glutathione transferase in rat tissues. *Nutrition and Cancer, 40*(2), 205–210.
45. Yamazaki, Y., Iwasaki, K., Mikami, M., & Yagihashi, A. (2011). Distribution of eleven flavor precursors, S-alk(en)yl-L-cysteine derivatives, in seven *Allium* vegetables. *Food Science and Technology Research, 17*(1), 55–62.
46. Dhumal, K., Datir, S., & Rajpandey. (2007). Assessment of bulb pungency level in different Indian cultivars of onion (*Allium cepa* L.). *Food Chemistry, 100*, 1328–1330.
47. Galmarini, C. R., Goldman, I. L., & Havey, M. J. (2001). Genetic analyses of correlated solids, flavor, and health-enhancing traits in onion (*Allium cepa* L.). *Molecular Genetics and Genomics, 265*, 543–551.
48. Mallor, C., Balcells, M., Mallor, F., & Sales, E. (2011). Genetic variation for bulb size, soluble solids content and pungency in the Spanish sweet onion variety Fuentes de Ebro: Response to selection for low pungency. *Plant Breeding, 130*, 55–59.
49. Abayomi, L. A., & Terry, L. A. (2009). Implications of spatial and temporal changes in concentration of pyruvate and glucose in onion (*Allium cepa* L.) bulbs during controlled atmosphere storage. *Journal of the Science of Food and Agriculture, 89*, 683–687.
50. Dini, I. (2011). Red onion (*Allium cepa* L. var. tropeana) seeds: Nutritional and functional properties. In *Nuts and Seeds in Health and Disease Prevention* (pp. 981–990).

51. Yalcin, H., & Kavuncuoglu, H. (2014). Physical, chemical and bioactive properties of onion (*Allium cepa* L.) seed and seed oil. *Journal of Applied Botany and Food Quality, 87*, 87–92.
52. Parry, J., Zhou, K., Luther, M., & Yu, L. (2007). Phytochemical compositions and free radical scavenging capacities of selected cold-pressed edible seed oils. *ACS Symposium Series, 956*, 255–267.
53. Dini, I., Tenore, G. C., & Dini, A. (2008). Chemical composition, nutritional value and antioxidant properties of *Allium cepa* L. var. tropeana (red onion) seeds. *Food Chemistry, 107*, 613–621.
54. Gyawali, R., Seo, H. Y., Lee, H. J., Song, H. P., Kim, H. D., Byun, M. W., & Kim, K. S. (2006). Effect of γ-irradiation on volatile compounds of dried Welsh onion (*Allium fistulosum* L.). *Radiation Physics and Chemistry, 75*, 322–328.
55. Črnivec, I. O., Skrt, G., Šeremet, M., Sterniša, D., Farčnik, E., Štrumbelj, A., Poljanšek, N., Cebin, L., Pogačnik, S. S., Možina, M., Humar, D., & Komes, N. P. (2021). Waste streams in onion production: Bioactive compounds, quercetin and use of antimicrobial and antioxidative properties. *Waste Management, 26*, 476–486.
56. Sharma, K., Mahato, N., Nile, S. H., Lee, E. T., & Lee, Y. R. (2016). Economical and environment-friendly approaches for usage of onion (*Allium cepa* L.) waste. *Food & Function, 8*, 3354–3369
57. Kumar, M., Barbhai, M., Hasan, M., Punia, S., Dhumal, S. S., Radha, N., et al. (2022). Onion (*Allium cepa* L.) peels: A review on bioactive compounds and biomedical activities. *Biomedicine & Pharmacotherapy, 146*, 112498. https://doi.org/10.1016/j.biopha.2021.112498 (accessed on 1st January 2026).
58. Russo, M., Serra, D., Suraci, F., Di, R., & Fuda, S. (2014). The potential of e-nose aroma profiling for identifying the geographical origin of licorice (*Glycyrrhiza glabra* L.) roots. *Food Chemistry, 165*, 467–474.
59. Rice-Evans, C. A., Miller, N. J., Bolwell, P. G., Bramley, P. M., & Pridham, J. B. (1995). The relative antioxidant activities of plant-derived polyphenolic flavonoids. *Free Radical Research, 22*(4), 375–383.
60. Downes, K., Chope, G. A., & Terry, L. A. (2009). Effect of curing at different temperatures on biochemical composition of onion (*Allium cepa* L.) skin from three freshly cured and cold stored UK-grown onion cultivars. *Postharvest Biology and Technology, 54*, 80–86. https://doi.org/10.1016/j.postharvbio.2009.05.005 (accessed on 1st January 2026).
61. Lines, T. C., & Ono, M. (2006). FRS 1000, an extract of red onion peel, strongly inhibits phosphodiesterase 5A (PDE 5A). *Phytomedicine, 13*, 236–239.
62. Benítez, V., Mollá, E., Martín-Cabrejas, M. A., Aguilera, Y., López-Andréu, F. J., Cools, K., Terry, L. A., & Esteban, R. M. (2011). Characterization of industrial onion wastes (*Allium cepa* L.): Dietary fiber and bioactive compounds. *Plant Foods for Human Nutrition, 66*, 48–57.
63. Lee, J., & Mitchell, A. E. (2011). Quercetin and isorhamnetin glycosides in onion (*Allium cepa* L.): Varietal comparison, physical distribution, coproduct evaluation, and long-term storage stability. *Journal of Agricultural and Food Chemistry, 59*(3), 857–863.
64. Ifesan, B. O. T. (2017). Chemical composition of onion peel (*Allium cepa*) and its ability to serve as a preservative in cooked beef. *Human Journals, 7*(4), 25–34.

65. Marefati, N., Ghorani, V., Shakeri, F., Boskabady, M., Kianian, F., Rezaee, R., & Boskabady, M. H. (2021). A review of anti-inflammatory, antioxidant, and immunomodulatory effects of *Allium cepa* and its main constituents. *Pharmaceutical Biology, 59*, 287–302.
66. Campone, L., Celano, R., Piccinelli, A. L., Pagano, I., Carabetta, S., Sanzo, R. D., Russo, M., Ibañez, E., Cifuentes, A., & Rastrelli, L. (2018). Response surface methodology to optimize supercritical carbon dioxide/co-solvent extraction of brown onion skin byproduct as source of nutraceutical compounds. *Food Chemistry, 269*, 495–502.
67. Lecain, S., Ng, A., Parker, M. L., Smith, A. C., & Waldron, K. (1999). Modification of cell-wall polymers of onion waste–Part I. Effect of pressure-cooking. *Carbohydrate Polymers, 38*, 59–67.
68. Baldassare, S., Babbar, N., Van Roy, S., Dejonghe, W., Maesen, M., Sforza, S., & Elst, K. (2018). Continuous production of pectic oligosaccharides from onion skins with an enzyme membrane reactor. *Food Chemistry, 30*, 101–110.
69. Choi, I. S., Cho, E. J., Moon, J. H., & Bae, H. J. (2015). Onion skin waste as a valorization resource for the byproduct's quercetin and bio sugar. *Food Chemistry, 188*, 537–542.
70. Michalak-Majewska, M., Teterycz, D., Muszyński, S., Radzki, W., & Sykut-Domańska, E. (2020). Influence of onion skin powder on nutritional and quality attributes of wheat pasta. *PLoS ONE, 15*, 1.
71. Loredana, L., Giuseppina, A., Filomena, N., Florinda, F., & Donatella, A. (2019). Biochemical, antioxidant properties and antimicrobial activity of different onion varieties in the Mediterranean area. *Journal of Food Measurement and Characterization, 13*(2), 1232–1241.
72. Benítez, V., Mollá, E., Martin, A., Lopez, F., Downes, K., Terry, L., & Esteban, R. (2011). Study of bioactive compound content in different onion sections. *Plant Foods for Human Nutrition, 66*(1), 48–57.
73. Zhou, Y., Li, C., Feng, B., Chen, B., Jin, L., & Shen, Y. (2020). UPLC-ESI-MS/MS based identification and antioxidant, antibacterial, cytotoxic activities of aqueous extracts from story onion (*Allium cepa* L. var. *proliferum* Regel). *Food Research International, 130*, 108969.
74. Sagar, N. A., Pareek, S., & Gonzalez-Aguilar, G. A. (2020). Quantification of flavonoids, total phenols and antioxidant properties of onion skin: A comparative study of fifteen Indian cultivars. *Journal of Food Science and Technology, 57*, 2423–2432.
75. Kaur, C., Singh, M., Walia, S., Joshi, S., & Munshi, A. D. (2010). Variations in phenolics and antioxidants in Indian onions (*Allium cepa* L.): Genotype selection for breeding. *Nutrition & Food Science, 40*, 6–19.
76. Ren, F., Nian, Y., & Perussello, C. A. (2020). Effect of storage, food processing and novel extraction technologies on onions flavonoid content: A review. *Food Research International, 132*, 108953.
77. Sun-Waterhouse, D., Smith, B. G., O'Connor, C. J., & Melton, L. D. (2008). Effect of raw and cooked onion dietary fiber on the antioxidant activity of ascorbic acid and quercetin. *Food Chemistry, 111*(3), 580–585.
78. Lachman, J., Orsák, M., & Pivec, V. (1999). Flavonoid antioxidants and ascorbic acid in onion (*Allium cepa*). *Zahradnictví–UZPI, 26*(4), 125–135.

79. Corea, G., Fattorusso, E., Lanzotti, V., Capasso, R., & Izzo, A. A. (2005). Antispasmodic saponins from bulbs of red onion, *Allium cepa* L. var. *Tropea*. *Journal of Agricultural and Food Chemistry, 53*, 935–940.
80. Lanzotti, V., Romano, A., Lanzuise, S., Bonanomi, G., & Scala, F. (2012). Antifungal saponins from bulbs of white onion, *Allium cepa* L. *Phytochemistry, 74*, 133–139.
81. Manohar, C. M., Xue, J., Murayyan, A., Neethirajan, S., & Shi, J. (2017). Antioxidant activity of polyphenols from Ontario grown onion varieties using pressurized low polarity water technology. *Journal of Functional Foods, 31*, 52–62.
82. Nile, S. H., & Park, S. W. (2014). Total, soluble, and insoluble dietary fiber contents of wild growing edible mushrooms. *Czech Journal of Food Sciences, 32*, 302–307.
83. Corzo-Martínez, M., Corzo, N., & Villamiel, M. (2007). Biological properties of onions and garlic. *Trends in Food Science & Technology, 18*, 609–625.
84. Ashwini, M., & Sathishkumar, R. (2014). Onion (*Allium cepa*)–Ethnomedicinal and therapeutic properties. In *Handbook of Medicinal Plants and Their Bioactive Compounds* (pp. 27–34).
85. Block, E., Gulati, H., Putman, D., Sha, D., You, N., & Zhao, S. H. (1997). *Allium* chemistry: Synthesis of 1-(alk(en)ylsulfinyl) propyl alk(en)yl disulfides (cepaenes), antithrombotic flavorants from homogenates of onion (*Allium cepa*). *Journal of Agricultural and Food Chemistry, 45*, 4414–4422.
86. Ansari, M. A., Abdul, H. M., Joshi, G., Opii, W. O., & Butterfield, D. A. (2009). Protective effect of quercetin in primary neurons against Aβ (1–42): Relevance to Alzheimer's disease. *Journal of Nutritional Biochemistry, 20*, 269–275.
87. Ren, F. Y., Reilly, K., Gaffney, M., Kerry, J. P., Hossain, M., & Rai, D. K. (2017). Evaluation of polyphenolic content and antioxidant activity in two onion varieties grown under organic and conventional production systems. *Journal of the Science of Food and Agriculture, 97*, 2982–2990. https://doi.org/10.1002/jsfa.8138 (accessed on 1st January 2026).
88. Muscolo, A., Papalia, T., Settineri, G., Mallamaci, C., & Panuccio, M. R. (2020). Sulfur bentonite-organic-based fertilizers as tool for improving bio-compounds with antioxidant activities in red onion. *Journal of the Science of Food and Agriculture, 100*, 785–793. https://doi.org/10.1002/jsfa.10086 (accessed on 1st January 2026).
89. Mollavali, M., Bolandnazar, S. A., Schwarz, D., Rohn, S., Riehle, P., & Nahandi, F. Z. (2016). Flavanol glucoside and antioxidant enzyme biosynthesis affected by mycorrhizal fungi in various cultivars of onion (*Allium cepa* L.). *Journal of Agricultural and Food Chemistry, 64*, 71–77. https://doi.org/10.1021/acs.jafc.5b04791 (accessed on 1st January 2026).
90. Calderon-Montano, J. M., Burgos-Morón, E., Pérez-Guerrero, C., & López-Lázaro, M. (2011). A review on the dietary flavonoid kaempferol. *Mini Reviews in Medicinal Chemistry, 11*(4), 298–344.
91. Marrelli, M., Amodeo, V., Statti, G., & Conforti, F. (2019). Biological properties and bioactive components of *Allium cepa* L.: Focus on potential benefits in the treatment of obesity and related comorbidities. *Molecules, 24*, 119.
92. Santas, J., Almajano, M. P., & Carbó, R. (2010). Antimicrobial and antioxidant activity of crude onion (*Allium cepa* L.) extracts. *International Journal of Food Science & Technology, 45*(2), 403–409.

93. Kumar, S., & Pandey, A. K. (2013). Chemistry and biological activities of flavonoids: An overview. *The Scientific World Journal*, 1–16.
94. Pietta, P. G. (2000). Flavonoids as antioxidants. *Journal of Natural Products, 63*(7), 1035–1042.
95. Mishra, A., Sharma, A. K., Kumar, S., Saxena, A. K., & Pandey, A. K. (2013). *Bauhinia variegata* leaf extracts exhibit considerable antibacterial, antioxidant, and anticancer activities. *BioMed Research International*, 1–10.
96. Duan, Y., Jin, D. H., Kim, H. S., Seong, J. H., Lee, Y. G., Kim, D. S., & Jang, S. H. (2015). Analysis of total phenol, flavonoid content and antioxidant activity of various extraction solvents extracts from onion (*Allium cepa* L.) peels. *Journal of Korean Oil and Chemical Society, 32*(3), 418–426.
97. Škerget, M., Majhenič, L., Bezjak, M., & Knez, Ž. (2009). Antioxidant, radical scavenging and antimicrobial activities of red onion (*Allium cepa* L.) skin and edible part extracts. *Chemical and Biochemical Engineering Quarterly, 23*(4), 435–444.
98. Tan, L. H., Zhang, D., Wang, G., Yu, B., Zhao, S. P., Wang, J. W., & Cao, W. G. (2016). Comparative analyses of flavonoids compositions and antioxidant activities of hawk tea from six botanical origins. *Industrial Crops and Products, 80*, 123–130.
99. Santas, J., Carbó, R., Gordon, M. H., & Almajano, M. P. (2008). Comparison of the antioxidant activity of two Spanish onion varieties. *Food Chemistry, 107*(3), 1210–1216. https://doi.org/10.1016/j.foodchem.2007.09.056 (accessed on 1st January 2026).
100. Neha, K., Haider, M. R., Pathak, A., & Yar, M. S. (2019). Medicinal prospects of antioxidants: A review. *European Journal of Medicinal Chemistry, 178*, 687–704. https://doi.org/10.1016/j.ejmech.2019.06.010 (accessed on 1st January 2026).
101. González-de-Peredo, A. V., Vázquez-Espinosa, M., Espada-Bellido, E., Carrera, C., Ferreiro-González, M., Barbero, G. F., & Palma, M. (2021). Flavonol composition and antioxidant activity of onions (*Allium cepa* L.) based on the development of new analytical ultrasound-assisted extraction methods. *Antioxidants, 10*(2), 273. https://doi.org/10.3390/antiox10020273 (accessed on 1st January 2026).
102. Hasani-Ranjbar, S., Larijani, B., & Abdollahi, M. (2009). A systematic review of the potential herbal sources of future drugs effective in oxidant-related diseases. *Inflammation & Allergy - Drug Targets, 8*(1), 2–10.
103. Fredotović, Ž., Šprung, M., Soldo, B., Ljubenkov, I., Budić-Leto, I., Bilušić, T., Čikeš-Čulić, V., & Puizina, J. (2017). Chemical composition and biological activity of *Allium cepa* L. and *Allium × cornutum* (Clementi ex Visiani 1842) methanolic extracts. *Molecules, 22*(3), 448.
104. Cheng, A., Chen, X., Jin, Q., Wang, W., Shi, J., & Liu, Y. (2013). Comparison of phenolic content and antioxidant capacity of red and yellow onions. *Czech Journal of Food Sciences, 31*(5), 501–508.
105. Genatrika, E., Sundhani, E., & Oktaviana, M. I. (2020). Gel potential of red onion (*Allium cepa* L.) ethanol extract as antifungal cause of tinea pedis. *Journal of Pharmacy & Bio Allied Sciences, 12*, 733–736. https://doi.org/10.4103/jpbs.JPBS_256_19 (accessed on 1st January 2026).
106. Ahmadi, S., Rajabi, Z., & Vasfi-Marandi, M. (2018). Evaluation of the antiviral effects of aqueous extracts of red and yellow onions (*Allium cepa*) against avian

influenza virus subtype H9N2. *Iranian Journal of Veterinary Science and Technology, 10*, 23–27.

107. Mishra, A. K., Mishra, A., Kehri, H. K., Sharma, B., & Pandey, A. K. (2009). Inhibitory activity of Indian spice plant *Cinnamomum zeylanicum* extracts against *Alternaria solani* and *Curvularia lunata*, the pathogenic dematiaceous molds. *Annals of Clinical Microbiology and Antimicrobials, 8*(1), 9.
108. Gupta, N., & Porter, T. D. (2001). Garlic and garlic-derived compounds inhibit human squalene monooxygenase. *The Journal of Nutrition, 131*, 1662–1667.
109. Havsteen, B. (1983). Flavonoids, a class of natural products of high pharmacological potency. *Biochemical Pharmacology, 32*(7), 1141–1148. https://doi.org/10.1016/0006-2952(83)90262-9 (accessed on 1st January 2026).
110. Bakri, I. M., & Douglas, C. W. I. (2005). Inhibitory effect of garlic extracts on oral bacteria. *Archives of Oral Biology, 50*(7), 645–651.
111. Cheng, X. N., Song, R. Y., Qian, Y. F., & Wei, J. (2016). Preparation of onion-based composite and study on its antimicrobial activity. *Advanced Engineering Materials and Applied Mechanics, 15*, 529–534. https://doi.org/10.1201/b19268-93 (accessed on 1st January 2026).
112. Tanackov, S. K., Dimić, K., Mojović, L., Gvozdanović-Varga, J., Vuković, A. D., Tomović, V., Šojić, B., & Pejin, P. (2016). Antifungal activity of the onion (*Allium cepa* L.) essential oil against *Aspergillus*, *Fusarium* and *Penicillium* species isolated from food. *Journal of Food Processing and Preservation, 41*(4), e132522.
113. Arshad, M. S., Sohaib, M., Nadeem, M., Saeed, F., Imran, A., Javed, A., & Batool, S. M. (2017). Status and trends of nutraceuticals from onion and onion byproducts: A critical review. *Cogent Food & Agriculture, 3*(1), 1280254.
114. Zandi, K., Teoh, B. T., Sam, S. S., Wong, P. F., Mustafa, M. R., & AbuBakar, S. (2011). Antiviral activity of four types of bioflavonoid against dengue virus type-2. *Virology Journal, 8*, 560.
115. Chen, C. H., Chou, T. W., Cheng, L. H., & Ho, C. W. (2011). *In vitro* anti-adenoviral activity of five *Allium* plants. *Journal of the Taiwan Institute of Chemical Engineers, 42*, 228–232.
116. Chen, L., Li, J., Luo, C., Liu, H., Xu, W., Chen, G., & Jiang, H. (2006). Binding interaction of quercetin-3-β-galactoside and its synthetic derivatives with SARS-CoV 3CLpro: Structure–activity relationship studies reveal salient pharmacophore features. *Bioorganic & Medicinal Chemistry, 14*, 8295–8306.
117. Nohara, T., & Fujiwara, Y. (2017). Antitumor sulfur compounds from *Allium* species (onion, Welsh onion, and garlic). *Natural Product Communications, 12*, 1235–1236.
118. Le Marchand, L., Murphy, S. P., Hankin, J. H., Wilkens, L. R., & Kolonel, L. N. (2000). Intake of flavonoids and lung cancer. *Journal of the National Cancer Institute, 92*, 154–160.
119. Murayyan, A. I., Manohar, C. M., Hayward, G., & Neethirajan, S. (2017). Antiproliferative activity of Ontario grown onions against colorectal adenocarcinoma cells. *Food Research International, 96*, 12. https://doi.org/10.1016/j.foodres.2017.03.017 (accessed on 1st January 2026).
120. Benítez, V., Mollá, E., Martín-Cabrejas, M. A., Aguilera, Y., & Esteban, R. M. (2017). Physicochemical properties and *in vitro* antidiabetic potential of fiber concentrates from onion byproducts. *Journal of Functional Foods, 36*, 34–42.

121. Gomes, I. B. S., Porto, M. L., Santos, M. C. L. F. S., Campagnaro, B. P., Pereira, T. M. C., & Meyrelles, S. S. (2014). Reno protective, antioxidative and anti-apoptotic effects of oral low-dose quercetin in the C57BL/6J model of diabetic nephropathy. *Lipids in Health and Disease, 13,* Article 1.
122. Galavi, A., Hosseinzadeh, H., & Razavi, M. B. (2021). The effects of *Allium cepa* L. (onion) and its active constituents on metabolic syndrome: A review. *Iranian Journal of Basic Medical Sciences, 24*, 3–16. https://doi.org/10.22038/ijbms.2020.46956.10843 (accessed on 1st January 2026).
123. Kim, H. Y. (2007). Effects of onion (*Allium cepa*) skin extract on pancreatic lipase and body weight-related parameters. *Food Science and Biotechnology, 16*(3), 434–438.
124. Baragob, A. E., Al-Wabel, N. A., Ahmed, N. A., Babiker, M. F., Abdalkarim, A. S., & Elboshra, M. I. (2015). Study to investigate the pancreatic regeneration and evaluation of the antidiabetic and antihyperlipidemic potential of aerial parts of *Allium cepa. Biochemistry and Biotechnology Research, 3*, 19–29.
125. Urios, P., Grigorova-Borsos, A. M., & Sternberg, M. (2007). Flavonoids inhibit the formation of the cross-linking AGE pentosidine in collagen incubated with glucose, according to their structure. *European Journal of Nutrition, 46*(3), 139.
126. Srinivasan, K. (2005). Plant foods in the management of diabetes mellitus: Spices as beneficial antidiabetic food adjuncts. *International Journal of Food Sciences and Nutrition, 56*(6), 399–414.
127. Kobori, M., Masumoto, S., Akimoto, Y., & Takahashi, Y. (2009). Dietary quercetin alleviates diabetic symptoms and reduces streptozotocin-induced disturbance of hepatic gene expression in mice. *Molecular Nutrition & Food Research, 53*(7), 859–868.
128. Knekt, P., Kumpulainen, J., Järvinen, R., Rissanen, H., Heliövaara, M., Reunanen, A., Hakulinen, T., & Aromaa, A. (2002). Flavonoid intake and risk of chronic diseases. *American Journal of Clinical Nutrition, 76*(3), 560–568.
129. Ali, M., Thomson, M., & Afzal, M. (2000). Garlic and onions: Their effect on eicosanoid metabolism and its clinical relevance. *Prostaglandins, Leukotrienes and Essential Fatty Acids, 62*(2), 55–73.
130. Lang, A., Lahav, M., Sakhnini, E., Barshack, I., Fidder, H. H., Avidan, B., Bardan, E., Hershkoviz, R., Bar-Meir, S., & Chowers, Y. (2004). Allicin inhibits spontaneous and TNF-induced secretion of proinflammatory cytokines and chemokines from intestinal epithelial cells. *Clinical Nutrition, 23*(5), 1199–1208.

CHAPTER 4

Garlic (*Allium sativum* L.)

CHADHA AYED* and BOUTHAINA DRIDI

Agrobiodiversity and Ecotoxicology, High Agronomic Institute of Chott-Mariem, University of Sousse, Chott Mariem, Sousse, Tunisia

*Corresponding author

ABSTRACT

Garlic (*Allium sativum* L.) is a widely utilized horticultural crop with a long history of medicinal and culinary applications. This chapter reviews the bioactive compounds of garlic, focusing on organosulfur derivatives such as allicin, ajoenes, and S-allyl cysteine, as well as phenolic compounds, saponins, and polysaccharides. These constituents contribute to garlic's diverse pharmacological activities, including antioxidant, anti-inflammatory, immunomodulatory, cardiovascular protective, hepatoprotective, anticancer, antimicrobial, and antidiabetic effects. Special attention is given to black garlic, a processed form with enhanced bioactive content and improved sensory attributes, which exhibits potent biological functions. The chapter also discusses traditional uses, mechanisms of action, and safety profiles, highlighting garlic's potential as a functional food and therapeutic agent. Future research directions are suggested to further elucidate its chemical composition and health benefits across different consumption forms.

4.1 INTRODUCTION

Garlic (*Allium sativum* L.) is a miraculous vegetable used worldwide and is among the oldest known horticultural plants [1]. Its usage has been traced back to more than 6000 years in Ayurvedic medicine, 4500 years in

Postharvest Quality Attributes of Fresh Vegetables and Grains.
Riadh Ilahy, Imen Tlili, Thouraya R'him, and Muhammad Azam (Eds.)
© 2027 Apple Academic Press, Inc. Co-published with CRC Press (Taylor & Francis)

DOI:10.1201/9781779642851-5

Babylonia, and 3000 years in the Old Egyptian Codex Ebers and ancient Chinese medicine [1, 2]. The worldwide production of garlic is 28,164,055 t, with China as the largest producer at 22,160,465 t, followed by India at approximately 1,693,000 t [1]. The word garlic in English is derived from "garleac," literally meaning "spear-leek" (gar signifies spear, matching the spear-like shape of a clove, and leac is the Old English form of leek) [3]. *Allium* species, the most important genus in the *Alliaceae* family, are among the earliest cultivated vegetables. They have been used as ornamentals, spices, vegetables, and as medicines to cure various diseases. The *Allium* genus includes over 700 species, widely distributed around the world and appreciated for their flavor, ease of growing, and long storage time [4]. Several species of the genus *Allium* have been renowned for their medicinal properties, generally ascribed to the presence of organosulfur compounds (OSCs), mainly cysteine sulfoxide. Through enzymatic reactions, these OSCs form thiosulfinates, which are the principal flavor compounds in *Allium*; the types of thiosulfinates and their ratios vary by species [5]. Nowadays, garlic is a commonly used spice worldwide and one of the first herbal drugs with a long tradition of medicinal use for a wide range of human health complications [6]. It is regarded as the second-most widely used *Allium* species after onion (*Allium cepa* L.), which is used as a remedy for several common disorders, including cold, influenza, snake bites, and hypertension [7]. It is a significant part of the pharmaceutical, food, and spice industries due to its numerous health, flavor, and therapeutic benefits [1]. Since ancient times, humans have used garlic for its remarkable biological functions, including antioxidant, cardiovascular protective, anticancer, anti-inflammatory, immune-modulatory, anti-obesity, anti-bacterial [8], antifungal [9], anti-viral, anti-protozoal, anti-diabetic, reno-protective, anti-atherosclerotic, anti-hypertensive [7], hypo-lipidemic, hepato-protective, anti-coagulant [6], and neuro-protective properties, as well as protection against mitochondrial damage [2] and natural or chemical toxicities [6]. Furthermore, it is employed in conventional medicine to treat gastrointestinal, respiratory, and urinary tract infections, and exhibits carminative, antipyretic, sedative, aphrodisiac, and diuretic properties [7]. In addition, garlic's therapeutic uses include increasing stamina, aiding digestion to prevent diarrhea, fatigue, worm infestations, and arthritis [10], and treatment of lung disorders, whooping cough, stomach disorders, colds, and earaches [7]. These qualities result from the presence of various bioactive substances, including allicin (responsible for most of the possible pharmacological functions), organosulfur compounds, selenium, arginine (as a micronutrient), and phenolic and steroid compounds. Furthermore, garlic is rich in various chemical components, including essential oils, sucrose, various vitamins [1], sulfurous components such as alliin, allicin, sulfides, ajoenes, and vinyldithiins, flavonoids such as quercetin [3, 7], phenolic compounds, polysaccharides [8], alkaloids, steroidal saponins, and terpenes [11]. Each compound in garlic has biological properties that allow it to be used as a treatment or an adjunctive therapy for some diseases [12]. For example, saponins found in the *Allium* genus have been shown to reduce cholesterol levels [9]. At the same time, OSCs such as allicin and diallyl disulfide (DADS) are the primary chemicals responsible for the pungent effects and spicy aroma [7]. Many processed garlic products (dehydrated garlic, garlic oil, aged garlic extract) have additional medicinal benefits and are being widely used [10]. Among them, black garlic (BG) is commercially available and well known as the most rapidly growing and studied health food [13].

Based on the information provided above, the main objective of this chapter is to focus on the available scientific evidence regarding the bioactive compounds of garlic and its important biological functions, traditional uses, and pharmacological activities. We anticipate that this chapter will draw further attention to garlic as a natural food and medicine and encourage its use in the management and treatment of various diseases.

4.2 BLACK GARLIC (BC)

Despite its numerous health benefits, the strong taste and odor of raw garlic deter some consumers from using it [13]. Recently, BG has been considered one of the fastest-growing healthful foods in Asia [10]. Many scientists have stated that the real origin of black garlic remains unidentified. The literature revealed that BG has been consumed since antiquity in Thailand, Japan, and Korea [10, 13]. With consumer demand, it has garnered global attention [10]. BG is a type of aged garlic produced from raw garlic via the Maillard fermentation process of whole garlic bulbs under high temperature (60–90°C) and humidity (70–90%) for several days [13], resulting in black cloves with a sweet taste. BG has a sweet taste and lacks the pungent odor, making it suitable to eat as a fruit once the outer coat is removed [10]. This is due to the change of garlic components during heat processing. In fact, the polysaccharide degrades, and the reducing sugar content increases [8].

In addition, the contents of organosulfur compounds, fructose, polyphenol, volatile compounds, total flavonoids, 5-hydroxymethylfurfural, and water-soluble antioxidant molecules (S-allyl cysteine, S-allylmercaptocysteine) are enhanced in black garlic compared to fresh garlic after heat treatment [8, 13, 14]. Furthermore, the aging process not only modifies the nutritional and sensory aspects of BG but also boosts its bioactivity [13]. BG and its bioactive compounds (such as phenol and organosulfur) have multiple biological activities and pharmacological properties which preserve and show greater efficacy in preventing many illnesses. Most of these benefits are due to its antioxidation, anti-inflammation, anti-obesity, hepato-protection, hypo-lipidemia, anticancer, anti-allergy, immune-modulation, nephron-protection, cardiovascular protection, and neuroprotection [13]. Due to its numerous health benefits, the demand for black garlic is rapidly expanding, with its incorporation into a wide range of products, including candies, beverages, vinegar, sausages, white bread, tofu, ice cream, yogurt, jam, and alcohol, mainly in America and Europe. It is also used in non-food items such as shampoo, cosmetics, skin protector, facial creams, and soaps [10].

Over the past few decades, substantial research has been conducted on BG and its components for various prevalent human ailments to determine their therapeutic benefits [13]. However, the impacts of the processing methods like fermentation and heating on garlic require further investigation, as they could modify its biological properties and safety [8].

4.3 CHEMICAL COMPOSITION OF GARLIC

Garlic contains various bioactive compounds, including organosulfur compounds, polysaccharides, phenolic compounds, and saponins [8]. The intact garlic bulb is particularly notable for its unique sulfur amino acids, known as S-alk(en)yl-l-cysteine sulfoxides (ACSO). Among these, the three primary ACSOs are S-2-propenyl-l-cysteine sulfoxide (alliin), S-1-propenyl-l-cysteine sulfoxides (isoalliin), and S-methyl-l-cysteine sulfoxide (methiin) [15]. In general, organosulfur compounds found in raw garlic are easier to digest than in cooked garlic [8]. Slicing or crushing garlic activates the enzyme C-S lyase (alliin lyase, EC 4.4.1.4), which converts S-alk(en)yl-l-cysteine sulfoxides into thiosulfinates and other volatile sulfur compounds. These compounds are key contributors to garlic's distinctive flavor as well as its antimicrobial and other health-promoting properties

[15]. Allicin [S-(2-propenyl)-2-propene-1-sulfinothioate, diallyl thiosulfonate], a lipid-soluble sulfur molecule that makes up around 70% of total thiosulfinates, is the main biologically active sulfur-containing compound in garlic and is responsible for its taste and smell [7]. It can be easily degraded by cooking [16]. However, saponins were discovered to be more stable while cooking. Purple garlic contains nearly 40 times more saponins overall compared to white garlic and is the only known source of certain saponin compounds, including proto-desgalactotigonin, voghieroside D1, desgalactotigonin-rhamnose, sativoside R1, and sativoside B1-rhamnose. Moreover, garlic contains over 20 phenolic compounds, surpassing those of many common vegetables. The predominant phenolic compound is β-resorcylic acid, followed by gallic acid, quercetin, pyrogallol, rutin, and protocatechuic acid [8]. Quercetin, the main flavonoid found in garlic, has been noted to interact with certain medications like vitamins C and E [7]. Additionally, garlic polysaccharides primarily consist of 85% fructose, 14% glucose, and 1% galactose [8, 16].

Under different conditions, thio-sulfinates are degraded to create additional sulfur constituents such as vinyldithiins, methyl allyl, diallyl, (E)- and (Z)-ajoene, and diethyl mono-, di-, tri-, tetra-, penta-, and hexasulfides [9]. Mnayer et al. [4] reported that the two major compounds of the total garlic essential oil (EO) are diallyl disulfide (DADS) (37.90%) and diallyl trisulfide (DATS) (28.06%). Other compounds include allyl methyl trisulfide at 7.26%, diallyl sulfide at 6.59%, diallyl tetrasulfide at 4.14%, and allyl methyl disulfide at 3.69%. Also, according to them, DADS is mostly abundant in garlic and Chinese chive EO, whereas dipropyl disulfide and dipropyl trisulfide are the most typical chemicals found in the essential oils of onion, leek, shallot, and chive [4]. Furthermore, allicin or DADS combines with cysteine, aided by S-allylmercapto cysteine, to produce allyl mercaptan, the molecule responsible for the garlic odor [7]. Overall, according to Quesada et al. [2], garlic and its preparations include four different types of organosulfur compounds (OSCs):

- Sulfur amino acid derivatives such as S-allyl-cysteine (SAC) are stable and odorless compounds present in whole cloves of the garlic bulb and aged garlic extract (AGE).
- Thiosulfinates, such as allicin, are unstable and highly reactive compounds present in crushed cloves and give fresh garlic its typical aroma and flavor.

- Ajoene and vinyldithiines are formed when allicin is kept at a moderate temperature in a lipophilic medium. They are present in stir-fried garlic and garlic macerated oil.
- Allyl (poly) sulfides, including diallyl trisulfide (DATS), diallyl disulfide (DADS), and diallyl sulfide (DAS), are found in steam-distilled garlic oil which is often used as a condiment in dishes worldwide owing to its potent aroma [2].

4.4 PHYTOCHEMICALS DISTRIBUTION

Garlic (*Allium sativum*) is a vital dietary source of health benefits and is widely used in conventional medicine across the globe. Despite all of garlic's health benefits, some people avoid eating it fresh due to its pungent, offensive taste and smell, as well as gastrointestinal discomfort. To alleviate these discomforts, researchers are exploring the use of raw garlic in various forms, including garlic oil macerate, smoked garlic, supplement pills, garlic juice, and garlic powder [13].

The distribution and the amount of S-alk(en)yl-l-cysteine sulfoxides (specifically alliin, isoalliin, and methiin), important non-volatile precursors of flavor-active compounds, in four parts of garlic plants (leaves, false stems, bulbs, and roots) are determined throughout the whole vegetation by Hornickova et al. [15]. In plants collected in March, at the start of the growing season, the highest concentration of alliin was found in the leaves, while the lowest was in the bulbs of the semi-bolter morphotype. Over the following two months (April and May), the content of alliin in garlic plants of morphotypes F and S decreased. In the later stages of growth, a steady increase in alliin content was observed in the bulbs until the end of the vegetation period. In contrast, it decreased in leaves, false stems, and roots. Therefore, the highest alliin concentrations were found in the bulbs sampled in July, just before harvest. At the end of vegetation, elevated levels of alliin, along with the next target analyte methiin, were also found in the scape, spathe, flowers, and bulbils. The changes in methiin content in the other parts of the garlic plant mirrored those of alliin, with methiin levels decreasing in the false stems and leaves, while increasing in the bulbs. The highest methiin concentration in the bulbs was found in July, at the end of the vegetation period. Isoalliin concentrations were highest in the roots, particularly in those collected in March, but decreased thereafter, with the most significant drop occurring at the end of the garlic growth season. The target S-alk(en)yl-l-cysteine sulfoxides contents depend not only on the portion of garlic plants but also on the morphotype of the garlic (such as flowering plants, semi-bolters, and scape absent) and the stage of the vegetation period [15].

Thiosulfinates are commercially important for their health benefits. The quantity of three main thiosulfinates, specifically allicin, allyl trans-1-propenyl thiosulfinates (ATPThs), and allyl methyl thiosulfinates (AMThs) in leaves of distinct *Allium* species was recently examined by Singh et al. [5]. *A. sativum* had the greatest total thiosulfinates content, followed by *A. tuberosum*, *A. guttatum*, and Zimmu (an interspecific hybrid of *Allium sativum* L. and *Allium cepa* L.). Garlic might be considered a rich source of allicin (82% of the total thiosulfinates) and ATPThs on fresh (fw) and dry (dw) weight basis [5]. The allicin content in garlic leaves (0.013–2.941 mg/g dw [5]) was less than that in garlic bulbs (5.03–18.24 mg/g dw) [17]. Leaves of the most promising *Allium* accessions could be used as food, a as a source of medicines, and as a natural and non-toxic food preservative. Flat and non-waxy leaves contained more allicin and other allyl thiosulfinates compared to fistular and waxy leaves [5]. Likewise, leaf chemistry has been correlated to wax content and leaf morphology. Total thiosulfinates content was also considerably higher in semi-domesticated and cultivated *Allium* species. Thiosulfinate compounds, therefore, appear to be positively linked to the degree of domestication: the higher the thiosulfinate content, the greater the degree of domestication and, in turn, the greater the preference for cultivation. Species with higher amounts of thiosulfinate compounds are more commonly cultivated due to their health benefits and greater tolerance to biotic and abiotic stresses. Allicin and other allyl thiosulfinates have extensive antimicrobial and health-promoting properties, making them natural and safe alternatives to synthetic preservatives [5].

4.5 ANTIOXIDANT PROPERTIES

Increasing reports have noticed that garlic has excellent antioxidant properties [8]. Garlic extracts possess an antioxidant effect in light of the presence of compounds such as thiosulfinates, allicin, S-allylcysteine (SAC), ajoenes, phenolic compounds, S-allyl-mercaptocysteine, and polysaccharide along with its derivatives [12]. The antioxidant capacity of phenolic compounds stems from their redox activity. Phenolic compounds play a key part in absorbing/neutralizing the free radicals [10]. Different processing methods affected the garlic's antioxidant activity. As a rule, raw garlic showed a greater antioxidant effect than cooked garlic, as measured by DPPH, ABTS, and FRAP assays. Stir-fried garlic had stronger antioxidant capacities as demonstrated by β-carotene bleaching. Fermented garlic, particularly black garlic, exhibited higher antioxidant activity than crude garlic. In a separate study, the DPPH and oxygen radical absorption capacity (ORAC) tests showed that the ethanolic extract of garlic sprouts had greater antioxidant properties compared to the ethanolic extract of raw garlic. In addition, ABTS, DPPH, FRAP, H_2O_2 scavenging, and Fe^{2+} chelating assays revealed that aged garlic had stronger antioxidant activity than fresh garlic. In comparison to multi-clove garlic extract, single-clove garlic extract contained a higher quantity of phenolic components and showed stronger antioxidant activity. Furthermore, increased pressure enhanced the antioxidant activity of garlic paste [8].

Frequent garlic intake enhanced endogenous antioxidant function and mitigated oxidative damage by boosting the synthesis of endogenous antioxidants (glutathione and cellular antioxidant enzymes) or decreasing the production of oxidants, such as oxygen-free radical species, which are responsible for numerous illnesses. Several studies have indicated that aged garlic extract has strong radical-scavenging activity due to its richness in phenols, flavonoids, and diverse sulfur compounds, such as SAC. Additionally, allicin, DADS, and DATS are the principal antioxidative substances that exhibited an antioxidant effect in lower doses at the physiological level [7]. In the same case, phenolic levels are a good indicator of antioxidant activity, thanks to their high redox potential, which allows them to act as radical scavengers or hydrogen donors [4]. Garlic is one of the most phenolic-rich vegetables consumed by humans. The main polyphenols are tannins, flavonoids, and phenolic acids [10, 13]. Different factors (agronomic, environmental, and genomic) influence the qualitative and quantitative traits of garlic. During the conversion of fresh garlic into black garlic, the heating process significantly increases the phenolic content [10] and the antioxidant activity of black garlic, compared to fresh garlic [8]. Hydroxy-methyl-furfural is a key antioxidant component found in black garlic. Garlic's aging process converts odorous and unstable chemicals into S-allylcysteine (SAC), which enhances the antioxidant potential of the black garlic. In fact, the amount of SAC increases six times as compared to raw garlic. The presence of different vitamins (B and folic acid) and antioxidant phytochemicals slows the progression of atherosclerosis and oxidative stress [13].

4.6 HEALTH BENEFITS AND BIOLOGICAL ACTIVITIES

Garlic is a frequently consumed spice around the world [8]. It is one of the earliest cultivated plants, used as a vegetable, a distinctive pungent seasoning for foods, and for medicinal applications [15]. Since ancient times, humans have utilized garlic for its cardioprotective, anti-tumor, anti-diabetic, anti-inflammatory, anti-hypertension, and antimicrobial properties [12]. Vast evidence and studies confirm that garlic exerts numerous and diverse biological functions, such as antioxidant, anticancer, anti-diabetic, anti-obesity, hepatoprotective, digestive system protective, neuroprotective, immunomodulator, renal protective, mitochondrial protector, anti-aging properties, and activates the immune system to defend against infections, as supported by epidemiological data from human clinical studies [2, 8]. As well, it is used to treat colds, lung disorders, earache, whooping cough, and stomach issues [7]. Overall, garlic serves as an excellent natural source of bioactive sulfur-containing compounds, offering great potential for developing functional foods or nutraceuticals to prevent and manage specific diseases [8]. These properties are attributed to the presence of various bioactive substances, including sulfur-containing compounds, such as allicin (accountable for the majority of the possible pharmacological functions), alliin, ajoene, diallyl sulfide (DAS), diallyl disulfide (DADS), diallyl trisulfide (DATS), 2-vinyl-4H-1,3-dithiin (2-VD), S-allyl-cysteine, and other organosulfur, selenium, and arginine as micronutrients, and phenolic and steroid components in garlic [2, 8, 12]. Each garlic compound has its own biological properties and can thus be utilized as a treatment or adjunctive therapy for certain diseases [12]. For instance, allicin, the active ingredient in garlic, has shown an interesting potential in general healthcare. In fact, its antioxidant and immunomodulatory properties result from mechanisms that may define its potential to inhibit apoptosis, promote proliferation, and induce apoptosis in various human cancers. Nonetheless, it may stimulate the stomach, especially in high doses [3]. In addition, allicin and other allyl thiosulfinates have a wide range of antimicrobial and health-promoting properties, making them natural and safe alternatives to synthetic preservatives [5]. Saponins also have biological roles and medicinal properties, including antimicrobial, antiviral, insecticidal, anti-inflammatory, anticancer, cytotoxic, and molluscicidal action. Furthermore, saponins have been shown to lower cholesterol levels in both animals and humans [11].

Owing to its beneficial effects on human health, the market for black garlic is developing rapidly [10]. The bioactive components of BG carry out various physiological activities in the body, which could have therapeutic potential for treating different diseases [13]. It has been found that black garlic is effective in decreasing diabetic, hyperlipidemic, hypercholesterolemic atherosclerosis, inflammation, oxidative stress, hypertension, cancer, and different neurological risks [10]. Furthermore, it has various bioactivities, including anti-obesity, hepatoprotection, anticancer, antiallergy, immunomodulation, cardiovascular prevention, and neurodegenerative protection. The wellness benefits of consuming BG are mostly due to its bioactive substances like organosulfur and phenol [13].

4.6.1 ANTI-INFLAMMATORY

Garlic has a high potential for treating inflammatory diseases in humans due to its low or absent toxicity. Overall, both *in vivo* and *in vitro* tests suggest that garlic may restrain inflammation by blocking key inflammatory mediators like interleukin (IL)-1β, nitric oxide (NO), and tumor necrosis factor alpha (TNF-α) [8], additionally by decreasing the reactive oxygen species (ROS) production [2]. Garlic extracts and its related phytochemicals had anti-inflammatory activity [7]. Alliin protects against lipopolysaccharide (LPS)-induced pro-inflammatory responses, as shown by the upregulation of anti-inflammatory genes and the downregulation of pro-inflammatory genes, such as IL-6 and monocyte chemoattractant protein-1 [18]. The water-soluble garlic polysaccharide inhibited the expression of pro-inflammatory cytokines IL-6, IL-1β, and TNF-α and reduced the colonic tissue damage [19]. AGE inhibited the production of inflammatory mediators and improved the liver function of adenosine monophosphate-activated protein kinase (AMPK) [8]. The garlic sulfur compound thiacremonone and the garlic 14-kDa protein reduced the inflammatory mediators by repressing the transcription factor nuclear factor-kappa B (NF-κB) signaling pathway [7, 8]. Moreover, garlic oil is effective in reducing the inflammation responses [10]. Allicin can be utilized as a potential adjunctive treatment toward the schistosome infection-induced inflammatory response [8]. DAS reduced inflammation by suppressing CYP-2E1 hepatic enzyme [7], while DADS reduced it by suppressing glycogen-synthase kinase-3β and the nuclear localization of NF-κB [8]. Furthermore, BG has great potential in treating various inflammation-related diseases such as rheumatoid arthritis, endometriosis, lethal sepsis, and inflammatory bowel disease (IBD) [13]. Significant anti-inflammatory effects were exhibited by 2-linoleoylglycerol, pyruvate, and 5-hydroxymethylfurfural, three active chemicals extracted from aged black garlic [10]. They acted by reducing prostaglandin E2, cyclooxygenase-2 (COX-2) expression, and inducible nitric oxide synthase (iNOS), thereby preventing the formation of NO, IL-6, and TNF-α [7]. In this context, the anti-inflammatory activity of aged garlic extract is most likely to interrupt the expression of classical mitogen-activated protein kinases (MAPKs) [13]. Additionally, aged black garlic extract may suppress the activation of the toll-like receptor 4 (TLR4) signaling pathway in macrophages, reduce nuclear NF-κB levels, and enhance NF-κB and IκB cytosolic levels [7]. Hexane, chloroform, and methanolic extracts of aged black garlic have also shown anti-inflammatory properties [7, 13]. Aged black garlic displayed potent antioxidant activities, which may be responsible for its anti-inflammatory effect [7]. Commonly, anti-inflammatory activity is directly proportional to antioxidant activity, yet this isn't always the case in aged black garlic. To investigate the antioxidant and anti-inflammatory properties of fresh and black garlic, a comparative study was conducted using pro-oxidant (hydrogen peroxide) and pro-inflammatory (lipopoly-saccharide) stressors. The results indicated that combining fresh garlic with aged black garlic, which has a higher sugar content, helps reduce its anti-inflammatory effect. One of the possible explanations for the aged black garlic's poorer anti-inflammatory efficacy than fresh garlic could be due in part to its high sugar content [10].

4.6.2 IMMUNOMODULATORY ACTIVITY

The primary purpose of traditional medicine is to develop immune-stimulators that are safer, completely biodegradable, and cost-effective in order to prevent diseases. In general, activities of innate immune cells, such as antibody-dependent cell-mediated cytotoxicity and phagocytosis, are crucial strategies for inhibiting and eradicating microbes and tumor cells. In addition, activated macrophages release cytokines like TNF-α, interleukin (IL)-6, and IL-12, as well as chemical mediators like prosta-glandins, NO, and ROS, which promote the proliferation of nearby innate and adaptive immune system cells [18]. Many herbal plants have immuno-modulatory properties, such as modulating cytokine secretion, promoting phagocytosis, activating macrophages, and producing immunoglobulins to help with allergic reactions. Among them, garlic exerts immunostimulatory and immunomodulatory effects, boosting host immune responses and improving the body's immune defense against various pathogens. The immune-stimulating and anti-infection role of garlic extract was attributed to the enhancement and activation of natural killer cells (NK), T-lymphocytes, and other immune functions. It's known that the adaptive immune system role in the body defense relies on lymphocytes, which come in three main forms: B cells, $CD8^+$ T cells, and $CD4^+$ T cells [20]. Garlic oil promotes a type 2 T helper (Th2)-type response at high levels, whereas it stimulates a Th1 immune response at low doses [18]. Garlic extract promotes the proliferation of CD4 T cells which belong to the TH1 lineage, are required to activate macrophages, and play an imperative role in the host defense mechanism against the intracellular and extracellular pathogens that survive phagocytosis by macrophages. Garlic extract increases the number of gamma delta (γδ) T cells which stimulate the rapid activation and proliferation of other lymphoid cells as well as the transportable anti-pathogen effect in αβ T-cell-deficient mice. Furthermore, γδT cells induce immunoglobulin A (IgA) responses in the mucosal immunity. This immunoglobulin is fast effective against mucosal pathogens. Aged garlic extract (AGE) has numerous pharmaco-immunological impacts, including tumor cell growth inhibition, chemopreventative effects, immunomodulation, and anti-allergic actions [20]. It also increases the NK and cytotoxic T cell activation in mice and enhances the immune response in humans by increasing NK and Tγδ cell proliferation [21]. Fermented garlic extract outperforms fresh garlic extract as an immune booster and anti-tumor agent. This is because the type of microbial strain and fermentation procedure converts organosulfur compounds in raw garlic into S-allyl cysteine and S-allyl mercaptocysteine, and enhances antioxidant activity and immune-stimulating potency by boosting levels of total polyphenols, vitamins, tetrahydrocarboline derivatives, and flavonoids. Its function is mediated by the stimulation of macrophage actions (by releasing immunostimulatory cytokines like TNF-α, IL-6, and IL-12), the production of NO, and the lytic activity of NK cells induced by interferon gamma (INF-γ) [18]. Garlic polysaccharides regulate the expressions of TNF-α, IL-10, IL-6, and interferon-γ, which provide a main immunomodulatory effect in RAW 264.7 macrophages [8]. Garlic consumption can boost the number of immune cells and the bone marrow cell proliferation, as well as improve intestinal health by increasing gut immunity through cytokine production and immune cell stimulation, thereby strengthening and sustaining a strong immune system [18]. Melguizo-Rodríguez et al. [21] claimed that the immunomodulatory action of garlic, especially its organosulfur compounds, influences various cell populations and supports its antiparasitic, antiapoptotic, and anticarcinogenic properties, helping to maintain immune system stability. This could explain its effectiveness in treating infections and cancer [21].

Garlic possesses immunostimulatory properties due to its high levels of antioxidants and bioactive substances, including vitamins, phenolic acid, and flavonoids. Allicin enhances macrophage phagocytic activity and peripheral blood cell immunological response of peripheral blood cells [18]. Garlic's protein fraction has been proven to stimulate macrophages, lymphocytes, and natural killer (NK) cells [21]. Polysaccharides seem to be the primary immune-modulating components in garlic [8]. They promoted phagocytosis, modified the expressions of immune-related cytokines (including TNF-α, IL-6, IL-10, and IFNγ), and enhanced NO release [19]. Fresh garlic polysaccharides are more potent in immunomodulation than black garlic polysaccharides. This is most likely due to degradation of the fructan constituent during processing, and therefore, no black garlic phagocytic activity was observed [8, 19].

4.6.3 CARDIOVASCULAR PROTECTION

CVD causes major morbidity and mortality, but it can be prevented by eating natural products and exercising regularly [22]. Since the 1960s, garlic's effects on CVD have been thoroughly examined [23]. In fact, consuming garlic powder effectively lowers total cholesterol, blood pressure, low-density lipoprotein (LDL), and other risk factors associated with CVD [7, 23]. Garlic and its chemical contents possess several benefits on cardiovascular health, including antihypertensive, anti-atherosclerotic, and antilipidemic activities [16, 22]. It can alleviate hyperlipidemia, platelet aggregation, and heart disease [7]. Garlic-based preparations have cardioprotective and vasculoprotective effects due to their natural ability to suppress cholesterol biosynthesis and reduce oxidation of LDL, thus improving blood lipid profile and blood pressure, inhibiting platelet aggregation, and optimizing fibrinolytic activity [16].

Furthermore, BG also has positive cardioprotective effects. A 12-week regimen of aged garlic extract combined with regular exercise reduced cardiovascular risk factors in post-menopausal women by lowering body mass index, body weight, total cholesterol, homocysteine, low-density lipoprotein-cholesterol (LDL-C), and malondialdehyde levels. Furthermore, a year-long daily supplementation of aged garlic extract (2400 mg) reduced the IL-6 levels in female patients with a low-risk profile of cardiovascular illnesses [13]. Garlic's mechanisms of action could be mostly related to the reduction of lipid peroxidation, inhibition of angiotensin-converting enzymes (ACEs), increase of H_2S and NO production, and decrease of oxidative stress [8]. The volatile compounds ajoenes have powerful bioactive properties and are responsible for platelet aggregation inhibition, promoting the dissolution of clots and protecting the organism from the risk of CVDs [23].

4.6.3.1 HEART PROTECTION

Garlic can protect the heart [23]. Garlic extract therapy improved heart tissue in a gentamicin-induced chronic renal failure rat model, reduced oxidative stress, and regulated Na^+/K^+-ATPase activity and Ca^{2+} levels. It increased the level of Na^+/K^+ -ATPase protein and decreased cardiac hypertrophy and remodeling in rats induced by isoproterenol [8, 23]. However, SAC with atenolol was more efficacious in preventing isoproterenol-induced myocardial dysfunction in rats [8]. Garlic extract activated the sirtuin-3-manganese superoxide dismutase pathway via acetylation of manganese superoxide dismutase, protecting the heart role in streptomycin-induced diabetic rats. It reduced heart rate variability leading to improved cardiac function in obese insulin-resistant rats. The addition of garlic and fenugreek to the rats' diet resulted in a pathological improvement of their heart tissues. In the presence of thiols, allicin was rapidly metabolized to organic diallylpolysulfides, which could effectively deliver H_2S to protect the heart [8, 23]. Garlic can up-regulate AMPK, which ameliorates cardiac damage and improves its function by activating the AKT pathway and inhibition of apoptotic signaling cascades leading to an increase in hydrogen sulfide concentration in the myocardium and reduction of oxidative stress through *Nrf2*-dependent pathway [23]. Short-term treatment with garlic powder (2 tablets daily for 3 months) improved endothelial function and enhanced brachial flow-mediated dilation in patients with coronary artery disease. Among various dietary supplements, aged garlic extract (AGE) is the most studied product associated with health benefits [2]. In rat hearts, polyphenols from aged black garlic extract relaxed coronary arteries and improved myocardial contractility before and after ischemia-reperfusion (I/R) [8].

4.6.3.2 GARLIC AND PLATELET AGGREGATION

Platelet activation is fundamental to many critical physiological and pathological processes, including various arterial phenomena such as strokes and myocardial infarction. Platelets activate when they bind to exposed collagen, laminin, and von Willebrand factor in the damaged vessel wall. The activation can also be triggered by adenosine-5-diphosphate (ADP) and thrombin. Once activated, platelets undergo shape changes, release granules, shed pseudopodia, and stick to other platelets in order to initiate the platelet aggregation process. Aggregate formation is also promoted by platelet-activating factor, a cytokine generated by monocytes, neutrophils, and platelets. The antiplatelet aggregation effect of garlic may be attributed to its antioxidant compounds and antioxidant activity [8]. The most stable compound in garlic, 2-vinyl-4H-1,3-dithiin (2VD), inhibits platelet aggregation and hypolipidemic activity associated with CVD and inflammation via a mechanism related to the sulfhydryl signaling pathway [2]. Fermented garlic is also effective in controlling hypercholesterolemia and platelet aggregation. In fact, daily oral intake of fermented garlic at a dose of 300 mg/kg, along with a hypercholesterolemic diet for one month, restrained collagen- and ADP-induced platelet aggregation and adenosine triphosphate (ATP) release in male Sprague-Dawley rats. Furthermore, BG can significantly lessen the platelet's ability to aggregate. Aged garlic extract inhibited platelet aggregation, adhesion, and platelet functions selectively. It also significantly reduced the platelet aggregation rate [13].

4.6.3.3 GARLIC AND ATHEROSCLEROSIS

Atherosclerosis-caused cardiovascular disorders contribute significantly to the global disease burden and are a major cause of death worldwide [16]. Atherosclerosis could result from chemical interactions within the cellular components of the arterial wall, such as cholesterol and lipids. These deposits are called plaques. Over time, these plaques can narrow or completely block arteries, leading to systemic complications [13]. The onset and progression of atherosclerosis are closely associated with mild systemic inflammation, a key mechanism of atherogenesis. The atherosclerotic process is accelerated not only by the release of inflammatory chemokines and cytokines, but also by the production of ROS, which contribute to the proliferation of vascular smooth muscle cell and endothelial dysfunction. Therefore, current therapeutics for atherosclerotic vascular disease should have pleiotropic anti-inflammatory and antioxidant effects [2]. Garlic OSCs have shown beneficial effects on numerous clinically relevant risk factors, as well as atherosclerosis, and the treatment of patients with associated illnesses [24]. Consuming garlic containing 2VD would help to reduce atherogenesis and inflammation. Nonetheless, additional research is required to comprehend the 2VD action mechanisms [2]. Women's risk of CVD and atherosclerosis progression increases significantly after menopause, possibly due to reduced estrogen levels. An isoflavonoid-rich herbal preparation, comprising tannins from garlic powder, was discovered to reduce atherosclerotic plaques progression and even suppress the formation of new ones in post-menopausal women after 12 months of treatment [25]. Furthermore, BG's effectiveness in treating and preventing atherosclerosis has been shown and confirmed in several preclinical and clinical studies [13]. To prevent atherosclerosis, AGE suppressed vascular inflammation and lipid deposition by lowering serum levels of C-reactive protein (CRP), thromboxane B-2, TNF-α and IL-1 receptor-associated kinase 4 protein, and raising AMPK activity in the liver, during early-stage atherosclerosis in apolipoprotein E-knockout mice [8]. It suppressed, also, the severity of coronary artery calcification and the metabolic risk in humans by repressing the adipose tissue volume progression rate [8, 13]. After taking 2400 mg/day of AGE for a year, 104 patients who were predisposed to cardiovascular events showed markedly lower levels of IL-6, blood pressure, and glucose. Furthermore, a one-year treatment of 60 asymptomatic patients revealed that AGE with supplements reduced the levels of epicardial, periaortic, pericardial, and subcutaneous adipose tissue [13].

4.6.4 ANTI-HYPERTENSIVE ACTIVITY

Hypertension is a common cause of cardiovascular disease and is increasing in prevalence worldwide. According to published reports, 24% and 80% of the populations in developing and developed nations, respectively, are affected by hypertension [10]. Since garlic is known to dramatically lower both systolic and diastolic blood pressure, Varshney and Budo [24] detailed its crucial role within the management of cardiovascular risk factors. Garlic-based treatments have been widely utilized to relieve and inhibit cardiovascular disorders such as hyperlipidemia, atherosclerosis, hypertension, thrombosis, and arrhythmia. Garlic extracts and their derived bioactive molecules have shown an antihypertensive impact in several experimental and human investigations [7]. The regular garlic diet can help lower high blood pressure, as delineated by clinical studies [10]. The antihypertensive mechanism of garlic extracts is due to its richness in numerous active sulfur molecules effective to promote endothelium-constricting and -relaxing factors, which lower blood pressure. Garlic has moreover been proven to promote the production of hydrogen sulfide (H_2S) and nitric oxide (NO) that eventually causes vasodilation [7]. Additionally, it can relieve hypertension by reducing oxidative stress and inhibiting the ACE [8]. The garlic's gamma-glutamylcysteine was reported to diminish the blood pressure by repressing the ACE [7]. S-1-propylenecysteine was demonstrated to be the key antihypertensive composite in the AGE. It reduced the systolic blood pressure and improved peripheral blood circulation in spontaneous hypertensive rats, while having no impact on the systolic blood pressure of control rats [8]. In rats with dexamethasone-induced hypertension, allicin significantly improved body weight and food intake and reversed the systolic blood pressure. Moreover, garlic has been shown to play a noteworthy role in inhibiting thrombosis and platelet aggregation or adhesion in humans [7]. In a placebo-controlled study involving 44 hypertensive patients, the consumption of enzymatic browning-processed garlic impressively reduced both diastolic and systolic blood pressures [8]. Hence, garlic is adopted as a medicinal plant for managing blood pressure globally [7].

On the other hand, several studies proved the potential impacts of black garlic (BG) on inhibiting ACE and decreasing arterial hypertension. Aged garlic extract consumption for 12 weeks enhanced gut microbial profile and reduced inflammation, blood pressure, and arterial stiffness in 49 participants with uncontrolled arterial hypertension [26]. BG extract had the lowest IC50 value (0.04 mg/mL) and demonstrated greater ACE inhibitory activity than raw garlic extract. Two Amadori chemicals in BG extract, notably N-(1-deoxy-D-fructos-1-yl)-l-methionine and N-(1-deoxy-D-fructos-1-yl)-l-arginine, were most likely attributed to ACE inhibitor activity [27]. According to Jang et al. [28], the BG extract had greater ACE inhibitor effects (88.8%) than conventional garlic extract (52.7%). Furthermore, a recent assessment revealed that elevated oxygen free radical levels in the hypothalamic paraventricular nucleus (PVN) may potentiate the cardiac sympathetic afferent reflex and the vasoconstrictor angiotensin II, all of which lead to hypertension [29]. As reported by Miao et al. [30], bioactive components of BG may be transported from SHRs across the blood–brain barrier into PVN to scavenge excess oxygen-free radicals. This process lowers blood pressure by blocking angiotensin II and cardiac sympathetic afferent reflex potentiation. In any case, there's still not sufficient data to recommend BG as a standard clinical treatment for hypertensive patients. More properly designed and analyzed experiments are required for a firm conclusion [13].

4.6.5 ANTI-HYPERLIPIDEMIC AND ANTI-HYPERCHOLESTEROLEMIA ACTIVITIES

Hypercholesterolemia and hyperlipidemia are the major factors that contribute to atherosclerosis. Garlic oil reduces triglyceride levels, whereas aged garlic extract and powder lower total serum cholesterol levels. Garlic extract can substantially lower triglyceride levels and serum total cholesterol in various experiments exploring its impact on patients with hypercholesterolemia and hyperlipidemia [10]. According to other studies, garlic can reduce blood lipids in both humans and animals. It can lower triglyceride, LDL cholesterol, and total cholesterol levels [8]. Daily eating of half to one clove of garlic can reduce cholesterol by up to 9% [22]. Furthermore, the intake of 300 mg of garlic daily for eight weeks decreased cholesterol and LDL levels while increasing high-density lipoprotein levels. However, in patients with diabetic dyslipidemia, garlic had no effect on triglyceride levels [8]. Extensive assays have been conducted to examine the impact of garlic in comparison to black garlic. The therapy with aged black garlic considerably reduced hydroperoxide concentration of serum lipids, plasma, and urinary F2-isoprostanes. The supplementation with aged garlic decreased oxidative stress. In another study, incorporating 1.5% black garlic extract into a high-fat diet for male Sprague-Dawley rats could effectively modulate lipids and cholesterol metabolism while diminishing the total levels of triglyceride, blood lipids, and cholesterol. This effect may be due to the suppression of mRNA expression of sterol regulatory element binding protein-1c. Moreover, a 13-week dose of aged garlic decreased the concentrations of F2-isoprostanes in plasma and urine, as well as the activities of lipid hydroperoxide and myeloperoxidase in serum in 41 patients with hypercholesterolemia. Overall, aged garlic presented better effects than raw garlic [8]. Garlic plays a vital role as an anti-atherosclerosis agent [10].

Interestingly, various evidence supports the pivotal role of garlic preparations and their phytochemicals in managing hypercholesterolemia by inhibiting cholesterol biosynthesis in the liver and impeding the oxidation of low-density lipoproteins (LDL and HDL). Moreover, garlic lowers cholesterol levels by stimulating the excretion of acidic and neutral steroids or by reducing the lipogenic and cholesterogenic effects of 3-hydroxy-3-methyl-glutaryl-CoA reductase, malic enzyme, glucose-6-phosphate dehydrogenase, and fatty acid synthase in hepatocytes [7].

4.6.6 EFFECT ON DYSLIPIDEMIA

Dyslipidemia is the leading cause of myocardial infarction and CVDs. It is defined by a low HDL level and high levels of LDL, total cholesterol, and triglyceride. Garlic has a significant impact on dyslipidemia by moderately elevating HDL cholesterol and significantly decreasing serum total cholesterol, triglycerides, and LDL cholesterol levels. Different garlic preparations were used in a variety of clinical and experimental trials on humans and animals. The inconsistent results were attributed to the difference in garlic preparation composition, quantity of active sulfur compounds present in each preparation, and mode of action of each compound [7]. For instance, the ethanolic garlic extract uptake decreased the cholesterol level and body weight in albino rabbits [31]. Furthermore, in rabbits fed with cholesterol, AGE substantially precluded the formation of thickened, lipid-laden lesions in neointima induced by balloon-catheter damage of the right carotid artery [32]. A study involving 65 intermediate-risk patients showed that a supply of aged garlic extract capsules containing L-arginine, vitamin B12, vitamin B6, and folic acid increased high-density lipoprotein as well as lipoprotein, while decreasing homocysteine, LDL-C, immunoglobulin G, and triglycerides [13]. In clinical trials, giving patients 300 and 60 mg of garlic daily for 12 months and 12 weeks, respectively, increased HDL while decreasing total cholesterol, LDL, and triglyceride. Moreover, giving diabetic patients with dyslipidemia 600 mg of garlic pills daily for 12 weeks raises HDL and lowers LDL and total cholesterol levels [7].

4.6.7 ANTICANCER ACTIVITY

Cancer is recognized as a major cause of death worldwide [8]. The cure for cancer varies and frequently involves radiation therapy, surgery, and chemotherapy. The application of natural ingredients is a promising research area in cancer therapy [13]. Recent studies have shown that garlic and its active compounds can help protect against diverse cancers, including gastric, colorectal, lung, and bladder cancers [8, 10]. Various Chinese studies have shown that consuming garlic and onions greatly reduces the risk of cancer [10]. Shang et al. [8] reported that garlic extracts' anticancer mechanisms were attributed to the stimulation of apoptosis, the inhibition of invasion and migration, the inhibition of cell growth and proliferation, the regulation of carcinogen metabolism, the prevention of angiogenesis, as well as the reduction of the negative effects of anticancer therapies. Iciek et al. [33] have detailed the anti-tumor functions of organo-sulfur compounds (OSC) counting alliin, allicin, DAS, DADS, S-allyl cysteine (SAC), and allyl mercaptan, extracted from garlic. DATS successfully prevented the angiogenesis of human breast cancer cells [8]. Allicin repressed colorectal cancer (CRC) metastasis through bolstering the immune system, hindering the formation of tumor vessels, and increasing survivin gene expression to boost the cancer cell's apoptosis. It can improve pancreatic cancer treatment likewise by restraining cancer cell proliferation and inverting gene silencing [7]. It has shown an interesting potential in general healthcare; in fact, its antioxidant and immunomodulatory effects arise from mechanisms that may define their potential in inhibiting apoptosis, proliferation, and induction in different human tumors [3]. Polyphenols also have a great antioxidant potential and protect humans against cancer and numerous other illnesses. Higher phenolic content contributes to health promotion [10]. Moreover, Z-ajoene exhibited anti-proliferative properties against several cancer types, and it suppressed the growth of glioblastoma multiform cancer stem cells and human breast cancer cells. It induced apoptosis in human leukemic cells by increasing peroxide production and activating caspase-8 and caspase-3-like activities. The administration of garlic powders to rats inhibited the DNA harm caused by N-nitrosodimethylamine in the liver by 35%, and this impact was attributed to the high alliin concentration (up to 60%) in the samples [7]. Fleischauer and Arab [34] proclaimed that regular garlic intake may decline different kinds of cancer propagation, including those of the prostate, lung, stomach, colon, and breast. Ethanol-based garlic extract increased stress on the endoplasmic reticulum, which inhibited the growth of mammary tumor cells *in vivo* as well as the growth of multiple myeloma and prostate cancer cells *in vitro* [8]. When tested against various cancer cells out of 33 raw vegetable extracts, raw garlic extract was the most effective and extremely specific anticancer drug without harming the non-cancerous cells [7]. Numerous studies have also proved BG's anti-tumor activities through annihilation of cell proliferation against gastric and colon cancers. Based on *in vitro* trials, BG is masterly capable of suppressing cancer cell growth and proliferation in many cancer cell lines. One of the BG compounds, S-allylmercaptocysteine, has recently been demonstrated to have anticancer properties against malignancies of the liver, prostate, ovaries, and bladder. The main mechanisms of S-allylmercaptocysteine were that it could: (i) effectively inhibit prostate cancer cells by modifying prostate biomarker expression and using testosterone to restore the expression of E-cadherin; (ii) facilitate MAPK inhibitor-induced apoptosis by triggering the transforming growth factor-betaTGF-β signaling pathway; (iii) restrain the invasion, survival, and migration of bladder cancer cells by blocking the inhibitor of differentiation-1 pathway; and (iv) inhibit epithelial ovarian cancer cells' proliferation and dissemination [13].

4.6.8 HEPATOPROTECTIVE ACTIVITY

Many substances, including drugs, dietary components, alcohols, and pollutants, can result in chronic and/or acute liver maladies like non-alcoholic steatohepatitis, liver fibrosis, non-alcoholic liver disease, as well as cirrhosis [13]. Numerous natural products, including garlic, possess hepatoprotective benefits [8].

Garlic oil improved the activities of hepatic antioxidant enzymes, reduced apoptosis, and blocked metabolic activation of 1,3-dichloro-2-propanol in the liver, which induce a protective effect against liver harm in rats. Several garlic's active ingredients, including DAS, DADS, and S-methyl-l-cysteine, may help prevent and treat liver damage, such as acute and chronic ethanol-induced liver impairment. DADS in garlic essential oil attenuated nonalcoholic fatty liver disease in rats induced by a high-fat diet. DADS markedly diminished the liver's production of pro-inflammatory cytokines and raised antioxidant function through minimizing the cytochrome P450 2E1 expression [8]. S-allyl-cysteine and polysaccharides enriched-BG extract inhibited CCl_4-induced hepatic damage by reducing inflammation and lipid peroxidation. The aged black garlic extract protected the liver treated with CCl_4 and D-galactosamine by reducing the alanine transaminase (ALT) and aspartate transaminase (AST) levels in the liver [13]. Single clove garlic exhibited a better protective effect than multi-clove garlic on CCl_4-induced acute liver damage in male rabbits. Fermented garlic extract by *Lactobacillus plantarum* BL2 (LAFGE) performed better than unfermented garlic extract in boosting hepatic steatosis and reducing liver lipid levels in mice [8]. The hepatoprotective benefits of fermented garlic extract were validated by regulation of oxidative stress, glycometabolism, inflammation, and lipometabolism. Furthermore, by increasing GSH-Px, superoxide dismutase, CAT, glutathione reductase, as well as glutathione peroxidase, and decreasing the levels of alkaline phosphatase (ALP), AST, ALT, total cholesterol, LDL-C, and malondialdehyde (MDA), black garlic showed a promise preventative measure against oxidative stress and cholesterol-linked liver diseases [13]. Likewise, fermented garlic extract by *Lactobacillus plantarum* BL2 reduced liver cell apoptosis partially by down-regulating p53 and eliminating MAPK phosphorylation, which prevented the liver from acetaminophen-induced damage [8]. Garlic protected against gentamycin- and acetaminophen-induced hepatotoxicity by controlling oxidative stress and enhancing antioxidant status [7]. However, a garlic overdose has detrimental impacts on the liver's morphology. Indeed, the weight of the liver increased, and bleeding and nodular swelling appeared on the external surfaces of the Wistar albinism rats' liver after 30 days of injecting 500 mg and 1000 mg/kg fresh garlic extract via their gastric tubes. Although garlic effectively alleviated chronic or acute liver injury, it's important to take into account the negative effects of its excessive consumption and determine the safe dose and duration of its usage [8].

4.6.9 NEPHRO-PROTECTION

Garlic has an intriguing ability to alleviate symptoms associated with nephrotoxicity and different genitourinary diseases [13]. Aged garlic extract suppressed gentamicin-induced nephrotoxicity through lessening oxidative stress and preserving glutathione reductase, glutathione peroxidase, and manganese superoxide dismutase activities [35]. Aged garlic extract increased urinary urea nitrogen level while decreased urinary albumin and blood urea nitrogen levels in diabetic nephropathy rats. It could protect against diabetic nephropathy through its hypolipidemic and anti-glycation qualities [36]. Aged garlic extract markedly mitigated the elevation of serum triglycerides and total cholesterol and reduced lipid peroxidation in kidney and liver tissues. The protective impact of the aged black extract on chronic kidney illness was that it rescued ethephon-induced kidney harm by restricting inflammation and apoptotic response and activating nuclear factor erythroid 2–related factor 2. According to a different study, aged black garlic extract halted the renal function deterioration as measured by standardized biomarkers including the blood urea nitrogen and creatinine levels in serum. The aged black garlic had lowered levels of oxidative stress markers, namely cyclooxygenase-2 (COX-2), NF-κβ, inducible nitric oxide (NO) synthase, 8-hydroxydeoxyguanosine, transforming growth factor-beta 1 (TGF-β1), and malondialdehyde [13]. Furthermore, garlic aqueous extract enhanced the renal plasma biochemical factors caused by alloxan in Wistar rats. As well, DATS activated the *Nrf2*-ARE pathway, which protects the kidney from arsenic-induced oxidative stress harm in rats [8].

4.6.10 DIGESTIVE SYSTEM PROTECTION

Several studies reported the therapeutic efficacy of garlic against gastric tissue damage. Black garlic extract acts by promoting gastrointestinal emptying, facilitating defecation, and stimulating gastrointestinal peristalsis. *In vitro*, the water fraction of black garlic outperformed the n-butanol and ethyl acetate fractions on improving gastrointestinal functions within the small intestine. Moreover, the AGE oral administration reduced the overall microbial content in the stomach and healed indomethacin-induced gastric mucosal harm in male rats [8]. BG is a long-term medicinal and dietary supplement effective to regulate the gastrointestinal system and promote digestion [13]. Garlic's bioactive compounds had a crucial role in the digestive system protection. Allicin relieved ulcerative colitis in mice by blocking the activation of AP-1/NF-κB/signal transducer and activator of transcription-1 (STAT-1) through the alleviation of JNK, p38, and the extracellular signal-regulated kinase 1/2 (ERK1/2)-regulated peroxisome proliferator-activated receptor (PPAR)-γ. In interferon-γ-stimulated intestinal cells, DADS decreased interferon-inducible protein-10, IL-6, and DAS reduced NO, as well as the STAT-1 expression. Likewise, DAS and DADS enhanced the dinitrobenzenesulfonic acid-induced colitis in mice [8]. A recent study reported that BG could treat inflammation-based gastroesophageal reflux disease, such as reflux esophagitis [13]. To assess SAC-enriched BG's preventive efficacy against reflux esophagitis, Kim et al. [37] conducted an essay by administering Sprague-Dawley rats an oral BG dosage of 100 mg/kg body weight two hours prior to the reflux esophagitis induction and comparing its effects to those of raw garlic. The results revealed that BG significantly reduced histological alterations induced by reflux esophagitis, compared with raw garlic. Supplementing with BG significantly increased catalase inhibition; however, the esophagus's superoxide dismutase levels tended to rise. These findings suggested that BG therapy can reduce the incidence of esophagitis by controlling NF-κβ-mediated inflammation. Nevertheless, raw garlic intake decreased *Helicobacter pylori* levels and bacterial urease activity in the stomach of 15 patients [8]. Furthermore, BG shown efficacy in the gastric ulcers treatment. Indeed, aged garlic extract could prevent indomethacin-induced ulcers via the reduction of oxidative stress and the elevation of NO, prostaglandin E-2, and glutathione levels in gastric tissue [38]. Additionally, it prevented intestinal damage by preserving cellular integrity and suppressing the increased plasma lactate and malondialdehyde levels in tissue [13]. Generally speaking, garlic's bioactive compounds can enhance gastrointestinal function and alleviate gastric ulcers, colitis, and other gastrointestinal disorders by decreasing *Helicobacter pylori*, inhibiting inflammation, and reducing oxidative stress [8].

4.6.11 ANTI-DIABETIC ACTIVITY

The World Health Organization defined diabetes as a chronic metabolic illness characterized by high blood glucose (or blood sugar) levels that may seriously harm the heart, eyes, blood vessels, nerves, and kidneys over time [6]. Chronic hyperglycemia and impaired protein, lipid, and carbohydrate metabolism are hallmarks of diabetes mellitus, a complex metabolic disorder caused by abnormalities in either insulin secretion, action, or both [13]. A worldwide agreement has been reached to stop the increase in obesity and diabetes by 2025. Natural products of plant origin have been used as alternative antidiabetic agents to synthetic drugs. Garlic is a rich natural source of bioactive sulfur-containing components with potential antidiabetic effects [6]. Garlic and its bioactive compounds, including organic sulfides (alliin, diallyl sulfide, ajoene, diallyl trisulfide, diallyl disulfide, and S-allyl cysteine), are potent antidiabetic agents, based on studies conducted in different animal models and human cell lines. Bioactive compounds in garlic and their therapeutic doses can furnish antidiabetic effects (i.e., glycemic control and the amelioration of diabetic complications) through two recognized mechanisms: insulin secretagogue and/or insulin sensitizer [6]. Published research has shown that garlic and its components have a promising role in the treatment of metabolic syndrome due to their hypolipidemic, antidiabetic, and hypotensive properties. Along this line, several studies reported the garlic efficiency in lowering blood glucose in both humans and animal models of type 1 and type 2 diabetes. Researchers hypothesized that the sulfur-containing compounds in garlic juice are primarily responsible for garlic's hypoglycemic effects. In this respect, an aqueous extract of garlic, garlic oil, garlic powder, garlic leaf extract, homogenized garlic, garlic bulb, aged garlic extract, crushed garlic, garlic juice, and garlic ethanol extract exhibited antidiabetic effects [6]. A meta-analysis involving 768 patients showed that garlic supplements are effective in the management of type 2 diabetes mellitus [8]. Garlic administration pills at 900 mg/day in patients with hyperlipidemia and type II diabetes decreased the serum lipids, cholesterol, and fasting blood sugar. Besides, allicin, allyl propyl disulfide, S-allyl cysteine sulfoxide, and cysteine sulfoxide reduced blood glucose levels by inhibiting hepatic insulin activation, boosting insulin secretion from pancreatic beta cells, preventing insulin from binding to receptors, and increasing insulin sensitivity in cells [7]. Garlic and its bioactive constituents may therefore be useful tools for treating diabetes and its sequelae [8].

Black garlic also has protective effects against diabetes and obesity. It contains abundant organosulfur compounds produced during the fresh garlic process into black garlic, which activate beta cells to boost insulin secretion. Garlic produced by microbial fermentation reduces insulin resistance and lowers total cholesterol and triglyceride levels [10]. In streptozotocin-induced diabetic rats, aged black garlic extract considerably reduced blood glucose and glycated hemoglobin, while significantly raised serum insulin levels, indicating its ameliorative action against glycometabolic biomarkers [13].

4.6.12 ANTI-OBESITY ACTIVITY

Obesity is the most prevalent health problem that may contribute to countless ailments like dyslipidemia, hypertension, insulin resistance, cardiovascular disorders, hyperglycemia, metabolic syndrome, liver damage, and other complicated issues [2, 7, 10, 13]. Raw garlic homogenate may reduce some abnormalities of metabolic syndrome, such as increased body weight [2]. The whole garlic, its oil, and several important natural composites, including 3-gallate, rutin, o-coumaric acid, gallic acid, and epigallocatechin, are considered very significant in the treatment of obesity [10]. The treatment of obese rats with garlic oil can reduce body weight and protect the liver from damage; interestingly, acupuncture-mediated garlic administration was more effective than oral administration. Through the KLF15 signal cascade, allicin potentially prevented obesity and related metabolic problems by boosting the brown adipocyte-specific genes expression, such as UCP-1 [2]. Ajoene isolated from garlic extracts could promote apoptosis, lessen the lipid accumulation in 3T3-L1 adipocytes, and drastically reduce the mice's body weight gain without affecting their food intake amount [7]. In addition, DATS, in combination with the synthetic anti-obesity drug Orlistat reduced body weight, ameliorated lipid profile, and restored liver function parameters in high-fat diet-induced obese rats [2]. Black garlic consumption proved a crucial role in preventing obesity. A 1.5% black garlic diet can ameliorate obesity and reduce epididymal fat through downregulating transcription and enzymes [10]. Shang et al. [8] proposed that fermented garlic products can help reduce obesity by modulating lipid metabolism and inhibiting lipogenesis. AGE can be used to overwhelm chronic inflammation in obese subjects [2]. The administration of 3 g of aged garlic extract daily for 6 weeks prevented increases in serum TNF-α and IL-6 and reduced LDL cholesterol levels in adults with obesity [13]. Moreover, significant improvements in adipokines and leptin were found in overweight women and men who undertook diet and exercise. Similarly, an increase in adiponectin was observed in the group that additionally took a nutritional supplement containing *Citrus aurantium*, garlic, caffeine, raspberry ketone, and capsaicin (4 capsules/day for 8 weeks), indicating better regulation of metabolic status [2].

4.6.13 ROLE IN OSTEOARTHRITIS

Osteoarthritis is a chronic degenerative illness that damages cartilage, joints, bones, and muscles. It also restricts movements in weight-bearing joints, such as the knee and/or hip, and can cause physical disability [12]. Garlic and its extract might be a novel remedy for osteoarthritis through regulating cytokine secretion, boosting immune cells, and reducing disease expressions [39]. S-allyl-mercaptocysteine, alliinase, alliin, ajoenes, allicine, and diallyl sulfide act as inflammatory alleviator factors like TNF-α, IL-1β, and IL-6, which cause the pain or even the cartilage destruction. TNF-α expression is induced by another inflammatory factor named resistin, produced in synovial fluids fibroblasts in osteoarthritis. Resistin upregulates IL-1β, TNF-α, and IL-6 production in the cartilage, which leads to chondrocytes and fibroblasts degeneration, synovial inflammation, and cartilage damage. A topical garlic formulation may be an effective way to lower resistin concentration. IL-6 can affect the immune system, bone, and cartilage. One theory of the garlic mechanisms in osteoarthritis therapy is the decline of IL-6 [12]. Hussein et al. [40] reported a substantial drop in IL-6 levels after 8 weeks of taking 900 mg/day garlic capsules with breakfast. Whereas, Tavokali et al. [12] found that garlic's healing effects are unlikely to be observed at a dosage of 1000 mg for 12 weeks.

4.6.14 ANTI-ALZHEIMER'S DISEASE ACTIVITY

Among the elderly with cerebrovascular and neurodegenerative ailments, Alzheimer's disease is the primary cause of dementia [7]. Several *in vivo* and *in vitro* trials have revealed that garlic has potent neuroprotective properties, acting primarily on the hippocampus. Garlic ethanol extract improved memory, and organic sulfur compounds exhibited a major role in neuroprotection [8]. AGE protected the brain against neurodegenerative disease by preventing ischemia-induced brain damage, sparing neurons from apoptosis, and reducing β-amyloid-induced oxidative stress. Additionally, AGE or SAC promoted learning and memory retention, inhibited frontal lobe degeneration, and prolonged lifespan [7]. Through its antioxidant qualities, AGE enhanced cognitive dysfunction by reducing malondialdehyde levels and increasing glutathione peroxidase and superoxide dismutase activities [13]. The prolonged garlic uptake promoted memory performance by influencing serotonin levels. Furthermore, garlic oil and extracts prevented free radical-induced oxidative damage and inhibited the acetylcholinesterase enzyme, which improved memory and delayed cognitive decline and dementia. Allicin has a protective impact against ischemic or traumatic neuronal injury, regulated by apoptosis and oxidative stress pathways. Nowadays, the most common Alzheimer's disease treatment is the combination therapy involving allicin and cholinesterase inhibitors such as rivastigmine, galantamine, and donepezil [7]. BG could also halt cognitive decline through shielding neurons from Aβ-induced neurotoxicity and apoptosis, thereby averting neuronal death following ischemia or reperfusion and enhancing memory retention and learning capacity [13].

4.6.15 FOOD SAFETY

Noted for its high nutritional value, garlic has been used as a food ingredient since ancient times [10]. The primary causes of food safety, quality, and shelf-life decline are lipid oxidation and microbial modification. In fact, many fatal illnesses may be brought on by food-borne pathogens such as Gram-positive *Listeria monocytogenes*, *Staphylococcus aureus*, and Gram-negative *Campylobacter jejuni*, *Salmonella Typhimurium*, and *Escherichia coli*. The essential oils of garlic, Chinese chive, and onion possessed the highest antibacterial activity towards the five aforementioned bacteria, whereas shallot and leek EOs were the strongest antioxidants. In order to satisfy the consumer demands by using fewer synthetic antioxidants and more natural components, EOs could be beneficial as food additives for the food industry and for preservation, and as a viable, healthy alternative to unpopular synthetic antioxidants [4]. Furthermore, ethanolic garlic extract had a higher inhibitory impact against *Escherichia coli* and *Salmonella Typhimurium* than the aqueous extract, which exhibited little to no inhibitory impact [7]. Use of *Allium* extracts and compounds as natural antimicrobial agents and food preservatives is widespread, and some products have even been patented [5]. The strong flavor and odor of raw garlic, as well as any gastrointestinal distress, are lessened by aged garlic preparations. The consumption of 100 mg/kg of raw crushed garlic, administered twice daily for one month, improved components of metabolic syndrome. In addition, after consuming a single acute dose of sofrito sauce (240 g/70 kg), containing tomato, olive oil, onion, and garlic, the inflammatory biomarkers hsCRP and TNF-α decreased significantly [2].

4.6.16 ANTIMICROBIAL ACTIVITY

Garlic possesses a wide range of antifungal and antibacterial qualities [8]. Phenolic contents and allicin derivative products (diallyl trisulfide, diallyl disulfide, diallyl sulfide) extracted from garlic essential oils have proved antioxidant and antimicrobial qualities [4]. Thiosulfinate compounds had a broad spectrum of applications owing to their antimicrobial activities, including both Gram-positive and Gram-negative bacteria, and extend to viruses, fungi, and protozoa as well [5]. Allicin exerts an anti-pathogen defense mechanism by stimulating immune cell activity and influencing signaling pathways in these cells [7].

4.6.16.1 ANTIBACTERIAL ACTIVITY

Allium-derived products (mainly propylpropane-thiosulfinate and propylpropanethiosulfonate) are effective even against strains of bacteria that have developed antibiotic resistance [41]. Different garlic extracts (aqueous, chloroform, ethanolic, and methanolic extracts) inhibited the growth of several pathogenic bacteria, including Gram-positive (*Bacillus subtilis* and *S. aureus*) as well as Gram-negative (*Kl. pneumoniae* and *E. coli*) strains, to varied degrees of susceptibility [7]. Garlic EO is the main antibacterial ingredient and highly potent against *Bacillus subtilis*, *E. coli*, *S. aureus*, and *Listeria monocytogenes* [4, 8]. It destroyed the bacterial cell's structure and metabolism [8]. Due to their hydrophobic nature, essential oils can penetrate to the cell's phospholipid membrane, making it more permeable. As a result, the bacteria died when they lost their cellular contents [4]. Raw garlic inhibited *Helicobacter pylori*, while AGE effectively inhibited *Burkholderia cepaciae* [8]. The majority of garlic's antimicrobial properties are ascribed to its allicin content which combated a wide range of microorganisms encompassing antibiotic-resistant, Gram-negative, and Gram-positive bacteria like *Pseudomonas aeruginosa, Shigella, Proteus vulgaris, Streptococcus mutans, Staphylococcus aureus, Streptococcus pyogenes, Streptococcus faecalis, Salmonella enterica, Klebsiella aerogenes, E. coli, Vibrio, Enterococcus faecalis*, and *Mycobacteria* [7]. Dental caries is mainly caused by the gram-positive bacterium *Streptococcus mutans*. Using garlic extract mouthwash significantly decreased the number of saliva *Streptococcus mutans* colonies compared to the control group [3]. Although the exact mechanism of action is yet unknown, the primary antibacterial activity of thiosulfinates, such as allicin, could be attributed to (i) their ability to permeate phospholipid membranes effectively; (ii) their chemical reactions with thiol groups of key enzymes (e.g., acetate kinase and phosphotransacetyl-CoA synthetase) which inhibit the formation of lipids and block acetate incorporation into fatty acids; (iii) the inhibition of RNA polymerase and synthesis [41]; and their chemical reactions with thiol-containing enzymes, which are essential for bacterial nutrition, such as thioredoxin reductase and alcohol dehydrogenase [3]. Black garlic had stronger antibacterial qualities than fresh garlic. Gram-positive bacteria (*L. monocytogenes*, *E. faecalis*, and *S. aureus*) were more susceptible to the effects of black garlic extract than gram-negative bacteria (*Salmonella*, *E. coli*, *P. aeruginosa*, *A. baumanni*, and *K. pneumonia*). An efficient method to increase the antibacterial activity of raw garlic is to process it (humidity, temperature, and fermentation) to obtain black garlic [10].

4.6.16.2 ANTIFUNGAL ACTIVITY

By inhibiting the germination and growth of a wide range of fungi, including *Aspergillus*, *Candida, Torulopsis, Ascosphaera, Cryptococcus, Trichophyton, Botrytis, Trichosporon, Alternaria, Meyerozyma, Rhodotorula, Microsporum, Rhizopus, Neofabraea, Penicillium,* and *Epidermophyton* species, garlic extracts exhibited a broad-spectrum fungicidal impact [7, 8]. Methanolic, aqueous, ethanolic, and petroleum ether extracts of garlic have antifungal activity against various human pathogenic fungi [7]. The mechanism involved altering the fungal cell wall and resulting in permanent changes to the fungal cells' ultrastructure that weakened their structural integrity and reduced their germination ability. These cytoplasmic content changes lead to cell organelles and nucleus damage that eventually results in cell death [8]. Moreover, garlic oil, allicin, DADS, DATS, and saponins also exhibited antifungal activities [7].

4.6.16.3 ANTIVIRAL ACTIVITY

Allium sativum was reported as an antiviral agent against viruses such as Herpes simplex Virus type 1 & 2, Parainfluenza virus type 3, Human Rhinovirus type 2, Influenza virus, and Respiratory Syncytial virus [42]. Several phytochemicals, such as allicin, ajoene, methyl allyl thiosulfinate, and allyl methyl thiosulfinate, present in garlic were responsible for its ability to enhance the production of neutralizing antibodies when administered. Ajoene's antiviral activity was derived from blocking leukocyte fusion and adhesive interaction, whereas allicin acted by avoiding many thiol enzymes. DATS acted by boosting NK-cell activity that destroys virus-infected cells, and it was also efficient against human cytomegalovirus replication and viral immediate-early gene expression [7]. The large consumption of garlic might be the likely reason for the current low incidences and mortality rate of COVID-19 [42]. SARS-CoV-2 is a new virus within the viral family Coronaviridae. Garlic oil's allicin has the ability to inhibit the virus's replication by blocking the SARS-CoV-2 main protease (M^{opar}), which is the key enzyme involved in SARS-CoV-2 replication. Its suppression could be helpful in managing the COVID-19 pandemic, though clinical trials are still in the early stages of testing [43].

4.7 CONCLUSION AND FUTURE PROSPECTS

In order to emphasize the value of garlic (*A. sativum*), we reviewed its primary bioactive components and biological roles, paying particular attention to the pertinent mechanisms of action. Throughout history, garlic has been used as an herbal remedy for a variety of ailments. Extracts and isolated compounds from garlic are reported to possess numerous and different biological functions, including anticarcinogenic, antioxidant, antidiabetic, reno-protective, anti-atherosclerotic, cardiovascular protective, hepatoprotective, digestive system protective, antibacterial, antifungal, antiprotozoal, anti-obesity, neuroprotective, renal protective, mitochondrial protector, antihypertensive, immunomodulatory, and anti-inflammatory activities, which lead to great health advantages. Furthermore, it often has a low toxicity or is non-toxic. Consequently, garlic and its bioactive components hold promise as nutraceuticals and functional foods for preventing and treating diseases. It contains a variety of bioactive compounds, including phenolic compounds, saponins, organic sulfides, and polysaccharides. The main bioactive ingredients in garlic are the organic sulfides, including S-allyl-cysteine, allicin, diallyl sulfide, alliin, diallyl disulfide, ajoene, diallyl trisulfide, and vinyldithiins. There is also an emerging global interest in consuming black garlic owing to its aromatic profile, gelatinous texture, and sweet taste, as the major challenge in its production is to lessen the pungent odor and unpleasant taste of the fresh garlic. With this chapter, we hope to raise awareness of garlic as a natural pharmaceutical and therapeutic food, and to promote its application in the management and treatment of many disorders. Further researches are needed to separate more chemical compositions of garlic and determine its biological roles, particularly in light of the various forms of garlic consumption (such as raw, aged, allicin, oil, macerated, powder, aqueous, and alcoholic extracts).

KEYWORDS

- ***Allium sativum***
- **bioactive compounds**
- **black garlic**
- **functional food**
- **organosulfur compounds**
- **pharmacological activities**

REFERENCES

1. Yilmaz, N., & Fidan, H. (2021). Collection of, and serological and molecular viral disease detection in local garlic (*Allium sativum* L.) accessions collected from Northern Cyprus. *Fresenius Environmental Bulletin, 30*(1), 569–576.
2. Quesada, I., de Paola, M., Torres-Palazzolo, C., Camargo, A., Ferder, L., Manucha, W., & Castro, C. (2020). Effect of garlic's active constituents in inflammation, obesity and cardiovascular disease. *Current Hypertension Reports*. https://doi.org/10.1007/s11906-019-1009-9 (accessed on 1st January 2026).
3. Torbati, M., Emamverdizadeh, P., Torbati, M., Maghalian, M., & Mirghafourvand, M. (2021). Effect of garlic (*Allium sativum*) extract on salivary *Streptococcus mutans*: A systematic review and meta-analysis. *Pharmaceutical Sciences*. https://doi.org/10.34172/PS.2021.37 (accessed on 1st January 2026).
4. Mnayer, D., Fabiano-Tixier, A., Petitcolas, E., Hamieh, T., Nehme, N., Ferrant, C., Fernandez, X., & Chemat, F. (2014). Chemical composition, antibacterial and antioxidant activities of six essential oils from the Alliaceae family. *Molecules*. https://doi.org/10.3390/molecules191220034 (accessed on 1st January 2026).
5. Singh, P., Mahajan, V., Shabeer, A., Banerjee, K., Jadhav, M. R., Kumar, D., & Gopal, J. (2020). Comparative evaluation of different *Allium* accessions for allicin and other allyl thiosulfinates. *Industrial Crops & Products*. https://doi.org/10.1016/j.indcrop.2020.112215 (accessed on 1st January 2026).
6. Saikat, A., Hossain, R., Mina, F. B., Das, S., Khan, I. N., Mubarak, M. S., & Islam, M. T. (2021). Antidiabetic effect of garlic. *Revista Brasileira de Pharmacognosy*. https://doi.org/10.1007/s43450-021-00193-y (accessed on 1st January 2026).
7. Batiha, G. E., Beshbishy, A. M., Wasef, L. G., Elewa, Y. H. A., Al-Sagan, A. A., Abd El-Hack, M. E., Taha, A. E., Abd-Elhakim, Y. M., & Devkota, H. P. (2020). Chemical constituents and pharmacological activities of garlic (*Allium sativum* L.): A review. *Nutrients, 12*(3), 872. https://doi.org/10.3390/nu12030872 (accessed on 1st January 2026).
8. Shang, A., Cao, S., Xu, X., Gan, R., Tang, G., Corke, H., Mavumengwana, V., & Li, H. (2019). Bioactive compounds and biological functions of garlic (*Allium sativum* L.). *Foods, 8*(7), 246. https://doi.org/10.3390/foods8070246 (accessed on 1st January 2026).

9. Buitrago Diaz, A., Rojas Vera, J., Rojas Fermin, L., Morales Mendez, A., Aparicio Zambrano, R., & Rodriguez Contreras, L. (2011). Composition of the essential oil of leaves and roots of *Allium schoenoprasum* L. (Alliaceae). *Boletín Latinoamericano y del Caribe de Plantas Medicinales y Aromáticas, 10*(3), 218–221.
10. Afzaal, M., Saeed, F., Rasheed, R., Hussain, M., Aamir, M., Hussain, S., Mohamed, A. A., Alamri, M. S., & Anjum, F. M. (2021). Nutritional, biological, and therapeutic properties of black garlic: A critical review. *International Journal of Food Properties*. https://doi.org/10.1080/10942912.2021.1967386 (accessed on 1st January 2026).
11. El Aziz, A. M. M., Ashour, A. S., & Melad, A. S. G. (2019). A review on saponins from medicinal plants: Chemistry, isolation, and determination. *Journal of Nanomedicine Research, 7*(1). https://doi.org/10.15406/jnmr.2019.07.00199 (accessed on 1st January 2026).
12. Tavakoli-Far, F., Amiri-Ardekani, E., & Tehrany, A. (2021). *Allium sativum* L. (garlic) role in osteoarthritis: A systematic review of clinical trials. *Bio Interface Research in Applied Chemistry*. https://doi.org/10.33263/BRIAC114.1210412119 (accessed on 1st January 2026).
13. Ahmed, T., & Wang, C. (2021). Black garlic and its bioactive compounds on human health diseases: A review. *Molecules, 26*(16), 5028. https://doi.org/10.3390/molecules26165028 (accessed on 1st January 2026).
14. Kang, M., Kang, J., Woo, M. S., Kang, D., & Shin, J. (2022). Alterations in the physicochemical properties and antioxidant activity during aging of stored raw garlic. *Foods, 11*(10), 1390. https://doi.org/10.3390/foods11101390 (accessed on 1st January 2026).
15. Horníčková, J., Velíšek, J., Ovesná, J., & Stavělíková, H. (2009). Distribution of S-alk(en)yl-L-cysteine sulfoxides in garlic (*Allium sativum* L.). *Czech Journal of Food Sciences*. https://doi.org/10.17221/610-CJFS (accessed on 1st January 2026).
16. Yuristo, E. (2022). Garlic and cardiovascular disorders: A current review of literature. *Eureka Herba Indonesia, 4*(1). https://doi.org/10.37275/EHI.v4i1.52 (accessed on 1st January 2026).
17. Khar, A., Banerjee, K., Jadhav, M. R., & Lawande, K. E. (2011). Evaluation of garlic ecotypes for allicin and other allyl thiosulfates. *Food Chemistry*. https://doi.org/10.1016/j.foodchem.2011.04.004 (accessed on 1st January 2026).
18. Gurung, P., Lim, J., Magar, T. B. T., Shrestha, R., & Kim, Y. (2022). Modulation of local and systemic immune responses by fermented garlic extract. *International Journal of Functional Nutrition*. https://doi.org/10.3892/ijfn.2022.29 (accessed on 1st January 2026).
19. Kumari, N., Kumar, M., Radha, R., Lorenzo, J. M., Sharma, D., Puri, S., et al. (2022). Onion and garlic polysaccharides: A review on extraction, characterization, bioactivity, and modifications. *International Journal of Biological Macromolecules*. https://doi.org/10.1016/j.ijbiomac.2022.07.163 (accessed on 1st January 2026).
20. Khalil, A. M., & Maher, Z. M. (2022). Garlic extract enhances the peripheral blood cellular immune responses in male ICR mice. *Scholars International Journal of Traditional and Complementary Medicine, 5*(8). https://doi.org/10.36348/sijtcm.2022.v05i08.00X (accessed on 1st January 2026).
21. Melguizo-Rodríguez, L., García-Recio, E., Ruiz, C., De Luna-Bertos, E., Illescas-Montes, R., & Costela-Ruiz, V. J. (2022). Biological properties and therapeutic

applications of garlic and its components. *Food & Function*. https://doi.org/10.1039/d1fo03180e (accessed on 1st January 2026).

22. Bazaraliyeva, A., Moldashov, D., Turgumbayeva, A., Kartbayeva, E., Kalykova, A., Sarsenova, L., & Issayeva, R. (2022). Chemical and biological properties of bioactive compounds from garlic (*Allium sativum*). *Pharmacia*. https://doi.org/10.3897/pharmacia.69.e93604 (accessed on 1st January 2026).
23. Ribeiro, M., Alvarenga, L., Cardozo, L. M. F. M., Chermut, T. R., Sequeira, J., Moreira, L. S. G., et al. (2021). From the distinctive smell to therapeutic effects: Garlic for cardiovascular, hepatic, gut, diabetes and chronic kidney disease. *Clinical Nutrition*. https://doi.org/10.1016/j.clnu.2021.03.005 (accessed on 1st January 2026).
24. Varshney, R., & Budoff, M. J. (2016). Garlic and heart disease. *The Journal of Nutrition*. https://doi.org/10.3945/jn.114.202333 (accessed on 1st January 2026).
25. Arent, S. M., Walker, A. J., Pellegrino, J. K., Sanders, D. J., McFadden, B. A., Ziegenfuss, T. N., & Lopez, H. L. (2018). The combined effects of exercise, diet, and a multi-ingredient dietary supplement on body composition and adipokine changes in overweight adults. *Journal of the American College of Nutrition*. https://doi.org/10.1080/07315724.2017.1368039 (accessed on 1st January 2026).
26. Ried, K., Travica, N., & Sali, A. (2018). The effect of Kyolic aged garlic extract on gut microbiota, inflammation, and cardiovascular markers in hypertensives: The GarGIC Trial. *Frontiers in Nutrition*, *5*, 122. https://doi.org/10.3389/fnut.2018.00122 (accessed on 1st January 2026).
27. Yu, J., Shan, Y., Li, S., & Zhang, L. (2020). Potential contribution of Amadori compounds to antioxidant and angiotensin I converting enzyme inhibitory activities of raw and black garlic. *LWT–Food Science and Technology*. https://doi.org/10.1016/j.lwt.2020.109553 (accessed on 1st January 2026).
28. Jang, E. K., Seo, J. H., & Lee, S. P. (2008). Physiological activity and antioxidative effects of aged black garlic (*Allium sativum* L.) extract. *Korean Journal of Food Science and Technology*, *40*, 443–448.
29. Han, Y., Fan, Z. D., Yuan, N., Xie, G. Q., Gao, J., De, W., Gao, X. Y., & Zhu, G. Q. (2011). Superoxide anions in the paraventricular nucleus mediate the enhanced cardiac sympathetic afferent reflex and sympathetic activity in renovascular hypertensive rats. *Journal of Applied Physiology*. https://doi.org/10.1152/japplphysiol.00908.2010 (accessed on 1st January 2026).
30. Miao, Y., Chen, J., Zhou, G., Xu, X., Zhang, Q., & Wang, J. (2014). The antihypertensive effect of black garlic (*Allium sativum*) in spontaneously hypertensive rats via scavenging of free radicals. *Research in Health and Nutrition*, *2*, 5–12.
31. Iweala, E. E., Akubugwo, E. I., & Okeke, C. U. (2005). Effect of ethanolic extracts of *Allium sativum* Linn. Liliaceae on serum cholesterol and blood sugar levels of albino rabbits. *Plant Products Research Journal*. https://doi.org/10.4314/pprj.v9i1.35240 (accessed on 1st January 2026).
32. Campbell, J. H., Efendy, J. L., Smith, N. J., & Campbell, G. R. (2001). Molecular basis by which garlic suppresses atherosclerosis. *The Journal of Nutrition*, *131*(3), 1006S–1009S. https://doi.org/10.1093/jn/131.3.1006S (accessed on 1st January 2026).
33. Iciek, M., Kwiecień, I., Chwatko, G., Sokołowska-Jeżewicz, M., Kowalczyk-Pachel, D., & Rokita, H. (2012). The effects of garlic-derived sulfur compounds on cell

proliferation, caspase-3 activity, thiol levels and anaerobic sulfur metabolism in human hepatoblastoma HepG2 cells. *Cell Biochemistry and Function.* https://doi.org/10.1002/cbf.1835 (accessed on 1st January 2026).

34. Fleischauer, A. T., & Arab, L. (2001). Garlic and cancer: A critical review of the epidemiologic literature. *The Journal of Nutrition, 131*(3), 1032S–1040S. https://doi.org/10.1093/jn/131.3.1032S (accessed on 1st January 2026).
35. Maldonado, P. D., Barrera, D., Medina-Campos, O. N., Hernández-Pando, R., Ibarra-Rubio, M. E., & Pedraza-Chaverri, J. (2003). Aged garlic extract attenuates gentamicin-induced renal damage and oxidative stress in rats. *Life Sciences.* https://doi.org/10.1016/S0024-3205(03)00609-X (accessed on 1st January 2026).
36. Shiju, T. M., Rajesh, N. G., & Viswanathan, P. (2013). Reno protective effect of aged garlic extract in streptozotocin-induced diabetic rats. *Indian Journal of Pharmacology.* https://doi.org/10.4103/0253-7613.106429 (accessed on 1st January 2026).
37. Kim, K. J., Kim, S. H., Shin, M. R., Kim, Y. J., Park, H. J., & Roh, S. S. (2019). Protective effect of S-allyl cysteine-enriched black garlic on reflux esophagitis in rats via NF-κB signaling pathway. *Journal of Functional Foods.* https://doi.org/10.1016/j.jff.2019.04.040 (accessed on 1st January 2026).
38. El-Ashmawy, N. E., Khedr, E. G., El-Bahrawy, H. A., & Selim, H. M. (2016). Gastroprotective effect of garlic in indomethacin-induced gastric ulcer in rats. *Nutrition.* https://doi.org/10.1016/j.nut.2016.01.010 (accessed on 1st January 2026).
39. Arreola, R., Quintero-Fabián, S., López-Roa, R. I., Flores-Gutiérrez, E. O., Reyes-Grajeda, J. P., Carrera-Quintanar, L., & Ortuño-Sahagún, D. (2015). Immunomodulation and anti-inflammatory effects of garlic compounds. *Journal of Immunology Research, 2015,* 401630. https://doi.org/10.1155/2015/401630 (accessed on 1st January 2026).
40. Hussein, N. A., & Sharara, G. M. (2007). Poster 9: Effect of combined garlic therapy and comprehensive rehabilitation program versus comprehensive rehabilitation program alone on control of clinical manifestations and quality of life of knee osteoarthritis patients. *Archives of Physical Medicine and Rehabilitation.* https://doi.org/10.1016/j.apmr.2007.06.584 (accessed on 1st January 2026).
41. Sorlozano-Puerto, A., Albertuz-Crespo, M., Lopez-Machado, I., Ariza-Romero, J. J., Baños-Arjona, A., Exposito-Ruiz, M., & Gutierrez-Fernandez, J. (2018). *In vitro* antibacterial activity of propyl-propane-thiosulfate and propyl-propane-thiosulfate derived from *Allium* spp. against Gram-negative and Gram-positive multidrug-resistant bacteria isolated from human samples. *BioMed Research International.* https://doi.org/10.1155/2018/7861207 (accessed on 1st January 2026).
42. Raimi, I. O., Musyoki, A. M., Olatunji, O. A., Jimoh, M. O., Dube, W. V., & Olowoyo, J. O. (2022). Potential medicinal, nutritive and antiviral food plants: Africa's plausible answer to the low COVID-19 mortality. *Journal of Herbmed Pharmacology.* https://doi.org/10.34172/jhp.2022.03 (accessed on 1st January 2026).
43. Ezeorba, T. P. C., Chukwudozie, K. I., Ezema, C. A., Anaduaka, E. G., Nweze, E. J., & Okeke, E. S. (2022). Potentials for health and therapeutic benefits of garlic essential oils: Recent findings and future prospects. *Pharmacological Research: Modern Chinese Medicine.* https://doi.org/10.1016/j.prmcm.2022.100075 (accessed on 1st January 2026).

PART II
Leafy Vegetables

CHAPTER 5

Lettuce (*Lactuca sativa* L.)

ANISSA RIAHI,[1,2,*] IMEN HENANE,[1,*] AMANI ROMDHANE,[1] NAJLA MEZGHANI,[3,4] HAYET BEN HAJ KOUBAIER,[5] and CHAFIK HDIDER[1]

[1]*Laboratory of Horticulture, National Agricultural Research Institute of Tunisia (INRAT), University of Carthage, Ariana Governorate, Tunisia*

[2]*Higher School of Agriculture of Kef, University of Jendouba, Kef Governorate, Tunisia*

[3]*National Gene Bank of Tunisia, Tunis, Tunis Governorate, Tunisia*

[4]*Research Laboratory of Horticultural Crop, High Institute of Agronomy of Chott Mariem, University of Sousse, Chott Mariem, Sousse, Tunisia*

[5]*Laboratory of Innovation and Valorization for a Sustainable Food Industry, Higher School of Food Industries of Tunis, University of Carthage, Tunis, Tunisia*

*Corresponding author

ABSTRACT

Lettuce is a leafy vegetable that is often eaten in salads and is known for its crunchy texture and refreshing taste. It also serves as a rich provider of essential vitamins and minerals, such as vitamin A, vitamin K, folate, and iron. Additionally, lettuce is rich in antioxidants, which are compounds that can aid in shielding the body from damage caused by free radicals. Eating lettuce may provide various health benefits, such as promoting heart health, aiding in digestion, and supporting the immune system while also potentially benefiting those with type 2 diabetes. Lettuce waste and

Postharvest Quality Attributes of Fresh Vegetables and Grains.
Riadh Ilahy, Imen Tlili, Thouraya R'him, and Muhammad Azam (Eds.)
© 2027 Apple Academic Press, Inc. Co-published with CRC Press (Taylor & Francis)

DOI:10.1201/9781779642851-7

its byproducts can be used in various ways, providing additional benefits beyond their nutritional value. Repurposing lettuce waste and its byproducts can provide a range of benefits, from reducing waste to promoting sustainability and creating economic opportunities. By finding innovative ways to use these byproducts, the food industry can help reduce its environmental impact while also creating value for consumers and businesses alike.

5.1 INTRODUCTION

Lettuce, with its scientific name as *Lactuca sativa* L., is a leafy green vegetable that is widely grown and consumed around the world [1]. It is a significant crop in many countries, with global production exceeding 27 million tons in 2021 [2]. China is the world's largest producer of lettuce, primarily cultivating a type known as stem lettuce (*Lactuca sativa* L. var. angustana), which is more popular in parts of Asia than in the US or Western Europe [3]. The earliest evidence of cultivated lettuce dates back to primitive writings in 2680 BCE, where it was used as a medicinal herb. Lettuce has been domesticated in the Mediterranean region from the wild species *Lactuca serriola* L., with Southwest Asia thought to be its center of origin [4]. In order to cultivate lettuce, different agricultural systems have been used, such as hydroponic cultivation in a controlled environment, conventional and organic farming in open fields, and soil cultivation in greenhouses [5].

Lettuce is a frequent ingredient in salads and sandwiches. It is also utilized as a decorative garnish and in a diverse array of dishes, including soups, stews, and stir-fries. In many countries, lettuce is increasingly becoming an important part of the diet. While lettuce is primarily used for fresh consumption, some recent cultivars have been bred for dual purposes, including fresh consumption and processing. This has helped to increase the versatility of lettuce as a crop and has expanded its potential applications in the food industry.

Lettuce exhibits a wide variety of colors, shapes, surface textures, and leaf margins, ranging from green to purple [3]. These attributes, along with their nutrient content, play an important role in consumer purchasing decisions [6]. Despite being a popular vegetable, lettuce is primarily composed of water (94–95%) and contains minimal calories. Nevertheless, it serves as an excellent reservoir of vitamins, minerals, and bioactive compounds, including polyphenols, carotenoids, vitamin C, and chlorophyll, which provide a multitude of benefits for health [7–9].

The nutrient composition and antioxidant profile of lettuce vary among different varieties, particularly the green and red types [8]. Red lettuce, for example, is rich in antioxidants and is a valuable addition to a daily diet [10]. Research studies have indicated that lettuce not only comprises a range of bioactive compounds but also differing amounts of these compounds among different lettuce types, with a direct relationship between the intensity of the red color and the amount of these phytochemicals [11–14]. Additionally, the nutrient requirements of edible lettuce leaves can be high, depending on the growing conditions and metabolic factors [15, 16]. Thus, understanding these differences in bioactive compounds, their bioaccessibility, and bioavailability can help consumers choose a healthier lettuce diet [8].

5.2 HISTORY

Lettuce (*Lactuca sativa* L.) is a highly valued vegetable crop belonging to the Compositae family, also known as Asteraceae. It is enjoyed in many countries as a popular ingredient in salads, due to its fresh, crisp texture and mild flavor [17]. The cultivation of *Lactuca sativa* is thought to have its roots in ancient Egypt.

Lettuce boasts a lengthy and storied history, with its origins tracing back to ancient times. The plant was first depicted in wall paintings of Egyptian tombs around 2500 BC [18, 19]. The Egyptians initially grew the plant for its seeds, which were used to extract oil. However, over time, they began to selectively breed the plant, eventually cultivating a variety that was primarily grown for its edible leaves. This represents a significant milestone in the evolution of lettuce as a food crop.

Lactuca sativa held significant cultural and religious importance in ancient Egypt as a sacred plant of the god Min, associated with fertility and sexuality. During Min's festivals, lettuce was carried and placed near his images, believed to enhance the god's endurance during sexual activities [20]. The cultivated variety of lettuce grown by the Egyptians resembled a taller version of today's romaine lettuce, reaching a height of approximately 30 inches (76 cm). This unique variety of lettuce subsequently spread beyond Egypt, first to Greece and then to Rome [4]. Christopher Columbus later brought the plant to the Americas in the late

15th century, as noted by Subbarao and Koike [21]. From the late 16th century to the early 18th century, numerous varieties of lettuce were developed in Europe.

Based on historical and botanical evidence, it is widely acknowledged that the domestication of lettuce involved the ancestor *L. serriola*. However, there is no consensus regarding the exact center of lettuce domestication. Several hypotheses have been proposed, including Egypt, the Mediterranean region, the Middle East, and Southwest Asia [18, 22].

5.3 PRODUCTION

Lettuce (*Lactuca sativa* L.) has the distinction of being one of the most cultivated and consumed vegetables around the world. According to recent data released by the FAO [23], the global production of endive and lettuce has shown a steady increase from 26,375,002 tons in 2015 to 27,011,747.55 tons in 2021. This production is spread across a total area of 1,213,340 hectares. These figures are a testament to the growing popularity of these vegetables as a preferred food choice around the world. The dominance of China in the production of endive and lettuce is evident, with a production of 14,367,831.42 tons in 2021 [2]. This is almost four times the production of the second-largest producer, the United States, which produced 3,395,480 tons in the same year [2]. India and Spain, with a production of 1,129,374.25 tons and 1,066,010 tons, respectively, are also significant contributors to the global production of endive and lettuce [2]. These values indicate that endive and lettuce have become essential crops for several countries, and their production is a significant contributor to their economy. It is also noteworthy that the increasing demand for these vegetables has led to more extensive cultivation, and the production trend is expected to continue in the coming years.

5.4 BOTANICAL DESCRIPTION

The Asteraceae are thought to be the largest family of plants, consisting of approximately 23,000 to 30,000 species [1]. The genus *Lactuca L.* is part of the family Asteraceae (Compositae), the largest of the dicotyledonous families [24, 25]. The tribe Lactuceae of subfamily Cichorioideae, previously identified as the Cichorieae [26–28], is one of the best-known and most easily recognized tribes of the family. In the genus Lactuca, there are

100 species. Among them, four are wild species (*L. sativa, L. serriola, L. virosa*, and *L. saligna*) that can be crossed by classical methods [29, 30]. The genus Lactuca appears in a variety of forms, comprising glabrous or pubescent herbs with abundant latex, rarely shrubs, rhizomatous, sometimes with underground stolons, or with fusiform and/or tuberous roots.

Modern lettuce (*Lactuca sativa* L.) varieties are generally classified into six principal types based on morphological resemblances: crisphead, romaine, butterhead, cos, stalk or stem, and Latin lettuce [3, 17, 31]. Lettuce (*Lactuca sativa*) is the most important crop among leafy vegetables. It is a diploid species with a chromosome number of $2n = 2x = 18$ and is self-fertilized [32–34]. Lettuce is self-pollinated; however, there is up to 5% cross-pollination among lettuce varieties [35].

Lettuce has a deep horizontal lateral root system, which is densest near the soil surface in order to absorb water and nutrients. The stem is often shortened. The leaves are arranged in a thick rosette. *Lactuca sativa* exhibits great morphological diversity, including color, surface, margin, texture, and shape of the leaves [36]. The leaf surface can be wrinkled or smooth. The leaves feature a variety of colors, ranging from yellow to green to dark purple. Anthocyanins can be combined to encompass all or part of the leaves. The leaf margin is entire to setose dentate, frequently curly. Stem leaves are oblong-elliptic, with a cordate base. The inflorescence (capitulum, head) is composed of 7–15 [35] yellow ligules (florets). The heads form a corymbose, densely bracted panicle [31]. The involucre is cylindrical and measures approximately 10 to 15 mm in length. The involucral bracts are typically broadly to narrowly lanceolate, light green in color, with white margins, erect at fruit maturity stage [36]. The stem elongation is the signal for the end of the vegetative phase and the beginning of the reproductive phase. Typically, a single stem is produced to support flowering [36]. A dense scarlet complexion is composed of numerous capitula, each one comprising numerous flowers. The small flowers usually range in number, varying from 12 to 20, but their number can reach 35 [36]. The seeds have several colors, including white, brown, yellow, gray, and black [36]. The fruit (achene) measures between 6 and 8 mm in length, and its color varies between white, gray, cream, brown, or black. The development of these traits is a result of the selection and breeding processes involved in domesticating lettuce for agricultural use, which has resulted in the many diverse and valuable lettuce varieties we enjoy today.

5.5 ANTIOXIDANT PHYTOCHEMICALS

5.5.1 LETTUCE FLESH

Nowadays, interest in green eating is continuously growing, leading to a rise in the consumption of vegetables, particularly leafy vegetables. Leafy vegetables are known for their abundance of bioactive secondary metabolites, which offer various health benefits for humans [7, 37]. The advantages of consuming lettuce hinge on its composition, particularly its antioxidants, which can serve as valuable nutrients [8]. Lettuce (*Lactuca sativa* L.) is one of the most valuable leafy vegetables grown, essentially for its fresh leaves.

Lettuce boasts a high-water content of 94–95% and is low in calories. It serves as a good source of fiber, folate, minerals, and various vitamins such as vitamin E and vitamin C. In addition, lettuce serves as an exceptional source of various health-beneficial bioactive compounds like polyphenols, carotenoids, and chlorophyll, which offer various health benefits [7, 38]. Phenolic acids, notably caffeic acid and chlorogenic acid, along with their derivatives, as well as flavonoids like kaempferol and quercetin derivatives, anthocyanins, and the flavone luteolin, are recognized as the major phenolic compounds discovered in lettuce.

Phytochemical compositions and contents are distinct depending on the types. In comparison to green lettuce, some studies have indicated that red lettuce tends to have higher contents of phenolic compounds. Therefore, red lettuce can be regarded as a principal source of antioxidants in the daily diet [10]. Research has shown that lettuce is rich in vitamin C [9].

Lettuce is typically consumed raw; it retains more of its nutrients compared to vegetables that are cooked or processed, like potatoes. Moreover, the trend of consuming salads made from young leaves (cotyledons or baby leaf) has gained popularity in recent years [39].

Certain thermal processing methods, including boiling, cooking, and blanching, have the potential to decrease the levels of anti-nutrients in lettuce. Further research is also needed on a wide variety of agronomic methods aimed at reducing the content and mitigating the effects of anti-nutrients in green leafy vegetables like lettuce and enhancing their nutritional value [8].

5.5.2 PHENOLIC COMPOUNDS

Phenolic compounds constitute a valuable group of plant secondary metabolites that play a key role in the defense system of plants [40].

More than 8000 polyphenols have been identified in plants. Phenolic compounds are widespread in all plant organs [40] and exert an influence on the quality properties of plant-based foods [41]. Polyphenols have been characterized as having superior antioxidant properties when compared to vitamin C and E [42].

Phenolics are necessary for plant secondary metabolites, which have a crucial function in the plant's defense mechanisms [40]. More than 8000 polyphenols have been identified in plants. It is believed that phenolics can extend the shelf life of lettuce and enhance the plant's resistance to stress, helping to reduce postharvest losses [43]. This is characteristic because the reactive oxygen species (ROS) can be scavenged, which is thought to be implicated in leaf senescence as well as the plant's antioxidant defense mechanism [43]. Depending on their chemical structure, phenolics can be categorized into various subgroups [41]. Phenolic subgroups associated with lettuce are flavonoids and phenolic acids. The most prevalent phenolic acids in lettuce are caffeic acid and chlorogenic acid, along with their derivatives [44]. Phenolic acids represent 94% of the total phenolic content of some green varieties, while they represent only 45% for red lettuce [13]. Chlorogenic acid is the predominant composition found among the free phenolics. Generally, romaine lettuces had the highest chlorogenic and caffeic acids, but with abundant variability among cultivars [43].

The total phenolic acid content, quantified as the content of caffeic acid derivatives, was between 0.2 and 2.0 mg g^{-1} FW (4–40 mg g^{-1} DW). Red leaf lettuce was found to have the highest concentration of phenolic acids, while crisphead lettuce had the lowest content [13]. Phenolic acids represent 70–94% of the total phenolic content in green romaine and crisphead (iceberg) lettuce and vary between 35 and 45% of the same content in the red leaf lettuces [13]. The qualitative and quantitative phenolic content of lettuce is largely dependent on the variety, which is determined by its genetic information. Red lettuce became popular in the diet as a salad mix constituent due to its higher phenolic content and anthocyanin levels, which are believed to offer greater health benefits than green lettuce [45]. Tiveron [45] reported that the phenolic composition of legumes and vegetables consumed in Brazil, including lettuce, attained the highest result of the content of phenolic compounds for lettuce with 1690 mgEAGg^{-1}. Higher phenolic content in red lettuce in comparison to green lettuce categories is due to the fact that phenolic acids and flavonols are the main phenolic compounds in red lettuce, while phenolic acids were the main group of phenolic compounds in green lettuce [13], the study also incorporated genetic analysis of antioxidant properties and bioactive compounds of the lettuce.

Among the factors that can also influence the levels of phenolic compounds are the cultivation systems used for lettuce. Negrão et al. [46] found that among the three cultivation systems under study, the organic cultivation system yielded lettuce with the highest total phenolic content (1845.69 mgEAGg^{-1}) and flavonoids (1199.37 mgEQg^{-1}). On the other hand, the conventional cultivation system produced lettuce with the highest tannin content (1.17 mgECg^{-1}).

5.5.3 FLAVONOIDS

Flavonoids represent a type of plant compound found in numerous vegetables and fruits. The composition of flavonoids and flavonols of lettuce includes quercetin and kaempferol derivatives, as well as anthocyanins, and flavone luteolin [13]. Moreover, quercetin was followed by isorhamnetin, an important (O-methylated flavonol) flavonoid in certain lettuce types. Anthocyanins are one of the primary distinctions between red and green lettuce, as these flavonoids are responsible for the red, blue, or purple color pigments found in certain vegetables and fruits, though not in all [6].

Llorach et al. [13] analyzed five lettuce cultivars, with three green leaf (romaine, iceberg, and green leaf subtype continental) and two red leaf (lollo rosso and subtypes oak leaf), for their flavonoid composition and flavonols. They identified quercetin and kaempferol derivatives, anthocyanins, and flavone luteolin in some but not all cultivars. Additionally, the content of these compounds varied depending on the type of lettuce. Total flavonoid values were greater in red leaf subtype lollo rosso (2.8 mg g^{-1} FW or 56 mg g^{-1} DW) and red oak leaf (1.1 mg g^{-1} FW or 22 mg g^{-1} DW), succeeded by green lettuces continental (0.2 mg g^{-1} FW or 4 mg g^{-1} DW) and romaine (0.17 mg g^{-1} FW or 3.4 mg g^{-1} DW), with the lowest content in green iceberg (0.01 mg g^{-1} FW or 0.2 mg g^{-1} DW). Generally, field-grown lettuces had significantly higher flavonol content than high tunnel-grown lettuces due to the higher solar radiation level [44].

There was an association between leaf color and lettuce type. Generally, the qualitative and quantitative phenolic content of lettuce relies on its variety, with different genetic factors. Red lettuce has gained popularity in salads due to its anthocyanin content, which is believed to offer greater health benefits compared to green lettuce [47].

However, several agronomic or environmental factors, for instance, light, weather, and postharvest conditions, and their tissue type, can affect the phenolic content found in lettuce [48]. Negrão et al. [46] reported that the organic production system had the highest levels of flavonoids (1199.37 $mgEQIIg^{-1}$) among the three cultivation systems studied. The hydroponic system followed with the second-highest levels.

5.5.4 CAROTENOIDS

Carotenoids are a kind of lipid-soluble pigment normally observed in yellow-orange-colored fruits and vegetables and dark-colored leafy vegetables [49]. Lettuce leaves can be regarded as a reservoir of phytonutrients with elevated carotenoid content, such as lutein, β-carotene, and lactucaxanthin, while lactucaxanthin is not frequently found in other plants [13].

In all the lettuce cultivars analyzed, β-carotene content was discovered to be the most abundant, constituting half of the total carotenoids present. Lutein made up 20% of the total carotenoids, while lactucaxanthin, violaxanthin, and neoxanthin made up 13%, 11%, and 6%, respectively [43]. The content of β-carotene and lutein varied from 9.0 to 24.0 mg per gram of fresh weight (180–480 mg per gram of dry weight) and from 12.5 to 31.0 mg per gram of fresh weight (250–620 mg per gram of dry weight), respectively [43]. López [43] found that romaine lettuce (including the mini romaine subtype) contained β-carotene and lutein contents ranging from 20.1 to 41.8 mg g^{-1} FW (402–836 mg g^{-1} DW) and 7.7 to 14.9 mg g^{-1} FW (154–298 mg g^{-1} DW), respectively.

Nevertheless, carotenoid content differs among lettuce types. Mampholo et al. [6] observed a moderate association between β-carotene content and leaf color value L*, indicating that β-carotene content increases with leaf brightness. Catherine et al. [50] reported that red oak leaf lettuce generally contains higher amounts of β-carotene and lutein than green lettuce varieties like butterhead and Batavia, indicating a positive correlation between carotenoid concentration and green chlorophyll content. However, other studies have given contradictory results. For instance, Mou [51] identified higher levels of carotenoids in green leaf lettuce compared to red lettuce, and similar β-carotene content in both red and green leaf lettuce, indicating that carotenoid content is not solely related to leaf pigmentation. Thus, the carotenoid levels are not entirely dependent on leaf pigmentation.

Besides, previous reports have shown that variety and other conditions could have an impact on the carotenoid content of lettuce. Carotenoid levels in the outer leaves tend to be higher than those in the inner leaves owing to the positive impact of light level on the biosynthesis of carotenoids in the outer leaves [6].

5.5.5 VITAMIN C

Vitamin content, especially vitamin C, is a notable indicator of the nutritional quality of fruits and vegetables. Vitamin C exists in two forms that can be converted to one another: ascorbic acid (AA), which acts as an antioxidant, and dehydroascorbic acid (DHAA), its oxidized form [52]. Wild species and traditional varieties had higher amounts of total ascorbic acid (TAA) compared to commercial varieties, with increases of 21% and 8%, respectively [9].

Vitamin C is essential for normal metabolism, maintaining the immune system, and it also acts as an antioxidant, helping to protect cells from damage caused by free radicals [53]. Regular dietary intake is crucial for vitamin C because of its water solubility, which makes it easy to be excreted from the body [4]. AA accounts for 24.5% of the total radical scavenging activity in lettuce, which indicates that vitamin C is a crucial antioxidant [50].

López et al. [43] conducted a study on seven cultivars of romaine lettuce and two cultivars of mini romaine lettuce, finding that the vitamin C content ranged from 61–109 mg g^{-1} FW (1220–2180 mg g^{-1} DW). Other studies found varying levels of vitamin C in different types of lettuce. Crisphead (iceberg), green romaine, and green continental lettuce contained vitamin C contents of 42, 28, and 195 mg g^{-1} FW (840, 560, and 3900 mg g^{-1} DW), respectively. Red leaf varieties, such as red oak leaf, had a vitamin C content of 146 mg g^{-1} FW (2920 mg g^{-1} DW), while pink lollo had a content of 117 mg g^{-1} FW (2340 mg g^{-1} DW). The vitamin C content of head (iceberg), butterhead, romaine, green, and red lettuce varied from 28 to 92 mg g^{-1} FW (560–1840 mg g-1 DW), with green having the highest content and head (iceberg) the lowest [13]. Mampholo et al. [6] found that the content of AA in red varieties of lettuce, such as Red Oak and Red Curly, ranged from 25.00 to 30.61 mg 100 g^{-1} FW, which was comparable to the levels reported for red leaf lettuce by Martínez-Sánchez et al. [54].

Generally, red lettuce had a higher phenolic content compared to green lettuce, while green leaf lettuce had the highest vitamin C content.

Additionally, baby leaf lettuce, particularly baby green romaine, may offer a plentiful source of vitamin C, with a higher content than mature romaine lettuce [38].

5.6 LETTUCE SEEDS

Lettuce (*Lactuca sativa* L.) belongs to the Asteraceae family, which is commonly used as salad. However, over the last decade, particular focus has been placed on lettuce byproducts, such as seeds, for reuse to support human health. A previous analysis of *L. sativa* seeds has revealed that they are a rich source of unsaturated fatty acids, specifically oleic acid (61.5%), stearic acid (20.4%), palmitic acid (9.7%), myristic acid (2.8%), cis-palmitoleic acid (1.2%), behenic acid (0.5%), lignoseric acids (0.3%), α-linolenic acid (8.5%), and palmitoleic acid (0.2%). In addition, *L. sativa* seeds also contain saturated fatty acids, including palmitic acid (13.6%) and stearic acid (1.3%) [55]. In 2008, researchers collected seeds of *L. sativa* from the Miquan Vegetable Planting Base in Xinjiang, China, and investigated their chemical constituents using 95% and 50% ethanol extracts. Their analysis led to the discovery of four compounds, including a new flavonol glycoside called lactucasativoside A, which has a rare structure type. The other three compounds identified were japonicin, isoquercitrin, and caffeic acid, which are already known compounds [56]. Based on a study conducted by Sayyah et al. [57], *Lactuca sativa* seeds were found to contain triterpenoids and phenolic carboxylic acids through phytochemical screening. These phytochemicals contribute to the potential health benefits of lettuce seeds.

Lettuce seeds have been shown to reduce insomnia symptoms during pregnancy, suggesting that they could be a safe and natural cure for the treatment of insomnia disorder in pregnancy [58]. According to Ranjbar et al. [59], the topical application of lettuce seed oil on the forehead and temporal areas of kids can be considered a safe remedy for sleeplessness. The essential oils and crude extracts derived from Omani lettuce represent promising candidates for use as antioxidant, antifungal, or antimicrobial agents in the preparation of new drugs for therapy of infectious diseases [60].

Main volatile constituents of the seed oil were recognized: linoleic acid (38.8%), oleic acid (17.9%), palmitic acid (4.9%), 3-methyl heptane (3.2%), 1-decene (2.8%), 1-hexyl-3-methyl cyclopentane (2.3%), 2,4-decadienal (E,E) (2.2%), limonene (2.1%), and carvone (2.0%) [61]. The essential oils from the seeds of the plant cultivated in Isfahan principally consist of a blend of saturated and unsaturated fatty acids [61].

5.7 WASTES AND THEIR BYPRODUCTS

The byproducts and processing waste from fruits and vegetables are abundant sources of bioactive compounds, and their utilization can contribute to reducing waste and enhancing the sustainability of the food processing industry [62]. Nowadays, there is an increasing demand from consumers for health-promoting products containing these bioactive compounds, and consequently, there is also a growing trend in different industries towards the development of new added-value bio-products to fit the current market demands [63]. Indeed, the extraction and concentration of these compounds can induce the enhancement of functional food ingredients and nutraceutical products that can help improve human health [63]. As waste and byproducts are not only a potential source of energy but also a source of bioactive molecules that can benefit human health, researchers are increasingly focusing on their study. Certain bioactive compounds found in these waste materials have properties, for example, antioxidant, anti-inflammatory, and anticancer effects, and can modulate metabolic processes [64].

Hence, there is a need to give attention to using the tons of pomace produced each year to address environmental problems and generate new sources of income. Pomace is the residual pulp, skins, and leaves that are left over after juice extraction from fruits and vegetables. It is a significant source of bioactive compounds, for example, phytochemicals, carotenoids, and fiber, and has the potential to be used as a source for the production of food ingredients and functional foods. Despite this intense research activity, the current destination of fruit and vegetable waste is mainly represented by landfilling, composting, anaerobic digestion, and carbonization [65].

Lettuce waste has potential for utilization in various ways. It has the potential to serve as a feedstock for the production of energy and fertilizers. Additionally, it can also be processed into value-added products such as animal feed, compost, or food additives. Recently, Chen et al. [66] have introduced a new, affordable, and efficient method for producing biochar from lettuce waste using H_3PO_4 as an acidic activation agent at a hydrothermal carbonization process conducted at a low temperature of around 200°C. The results of a conjoint analysis carried out by Plazzotta et al. [67] indicate that it is possible to boost consumer preference for bread made from lettuce waste flour by making appropriate nutritional and sustainability claims. The study found that the addition of lettuce waste flour to bread significantly increased its polyphenolic content (up to 3.4 g GAE kg^{-1}) and enhanced its antioxidant activity by 200%. Bread with 170 and 575 g kg^{-1} of lettuce waste flour showed sensory properties and consumer acceptability like those of commercial whole meal bread, compared to rye bran content. Bread comprising at least 170 g kg^{-1} of lettuce waste flour could be related to nutritional claims associated with its high fiber content (>30 g kg^{-1}). Therefore, lettuce flour could be applied as an efficient ingredient in bread, enhancing the nutritional quality of the bread and increasing its fiber content. Several studies correlate the use of agri-food waste for the production of lettuce with an increased content of phenolic compounds and flavonoids, as well as an increase in the antioxidant capacity by DPPH [68].

In conclusion, lettuce-based products offer several health benefits, as lettuce is a low-calorie food that is rich in vitamins and minerals. Furthermore, lettuce serves as a valuable source of dietary fiber, which can facilitate digestion and contribute to regulating blood sugar levels. The versatility of lettuce makes it a useful ingredient in a wide range of food products, and its mild flavor makes it a good base for many dishes.

5.8 ANTIOXIDANT ACTIVITY

Lettuce is a significant natural source of secondary metabolites that exhibit potent antioxidant activity. These molecules, which include glycosylated flavonoids, carotenoids, phenolic acids, tocopherols, vitamin B groups, AA, and sesquiterpene lactones, are necessary bioactive nutritional components [69, 70]. The benefit of lettuce intake relies on its composition, particularly antioxidants, which can perform as nutrients [8]. Certain epidemiological studies have indicated that vegetable consumption is related to a decreased risk of chronic diseases, such as cancer and CVD [37, 71, 72].

Antioxidant activity refers to the ability of a substance to prevent or slow down the oxidation of other molecules. Oxidation is a reaction producing ROS, which can cause damage to cells and contribute to the development of various diseases, like cancer, CVD, and Alzheimer's disease. Antioxidants work by neutralizing ROS and preventing oxidative damage.

According to a study by Sularz et al. [73], the application of iodine compounds in lettuce cultivation can increase the concentration of certain vitamins, polyphenolic compounds, and antioxidant enzymes, resulting in enhanced antioxidant capacity of the lettuce. Higher antioxidant activity in transplantation may be associated with plant stress at this critical stage of development, since plants respond to environmental conditions through physiological changes as defense mechanisms against different stress conditions [74].

The antioxidant activity of lettuce varies depending on the method used to evaluate it. For example, the FRAP and ABTS methods showed that the antioxidant capacity of lettuce was approximately 1.79 mg TEAC g^{-1} FM and 1.15 mg TEAC g^{-1} FM, respectively [75, 76]. On the other hand, the ORAC method indicated a lower antioxidant capacity of approximately 0.0250 mg TE g^{-1} [13]. The antioxidant activity also depends on the species, geographical origin, and agricultural practices used, as well as the cultivation system, which gives reason for the differences in the results found in the current study with the literature. With studied sampling factors such as high temperatures, protein content, and higher levels of phenolics, these lead to significant antioxidant activity [46]. In a study conducted by Lanza Volpe et al. [77], it was found that the antioxidant activity of lettuce is influenced by growing conditions, and there is significant variability in the antioxidant activity of different lettuce types. The study found that the Butterhead, Latin, and Iceberg types of lettuce grown in a greenhouse system had higher antioxidant capacity than the other lettuce types. Negrão et al. [46] found that there was a significant difference ($p < 0.05$) in antioxidant activity among different cultivation systems under study. Lettuce cultivated in an organic production system had the highest capacity for reducing the DPPH radical (4075.28 µmolTEACg^{-1}), indicating the highest antioxidant activity. In comparison, lettuce grown in hydroponic and conventional cultivation systems had lower contents of 3074.30 and 2136.88 µmol TEAC g^{-1}, respectively. Zambrano-Moreno et al. [78] conducted a study to measure the antioxidant activity of organic and conventional green vegetables, including lettuce. The study showed that organic vegetables had higher antioxidant activity, including the ability to reduce the DPPH radical, compared to those grown in conventional systems. Similar findings were reported by Zhan et al. [79] in the hydroponic cultivation system, where organic lettuce was obtained to have higher antioxidant activity in comparison to conventionally grown lettuce.

Other studies have demonstrated that the genotype of lettuce can have an impact on its antioxidant properties. According to Liu et al. [12], leaf lettuce had the highest total phenolic content (TPC) and DPPH scavenging ability, followed by romaine, butterhead, and Batavia (crisphead subtype) lettuce. Additionally, within a specific type of lettuce, red-pigmented cultivars were found to have a higher TPC and antioxidant capacity compared to green cultivars produced under the same conditions. In the study by Tiveron [46], which assessed the antioxidant activity of plant foods including legumes and vegetables consumed in Brazil, lettuce registered an antioxidant activity value of 54.9 μmol TEAC g^{-1}.

Regarding hydrophilic and lipophilic activity, romaine lettuce showed the highest levels of both hydrophilic and lipophilic antioxidant activity, with its phenolic content also being higher than that of Iceberg and Baby head lettuces. The leaf lipophilic antioxidant activity increased sharply from the stem to the outermost leaves, signifying a protective role in mature or light-exposed leaves, while hydrophilic antioxidant activity showed a non-specific distribution [80].

According to Cho et al. [81], the hydrophilic ORAC antioxidant activity was significantly correlated with total phenols ($r = 0.893, p < 0.0001$) but not with ascorbic acid ($r = 0.009, p = 0.929$) in a total of 95 samples. This suggests that carotenoids and tocopherols are major contributors to lipophilic antioxidant activity, while total phenols contribute significantly to hydrophilic antioxidant activity. Therefore, it is important to consider both hydrophilic and lipophilic components of fruits and vegetables in determining the actual total antioxidant activity.

5.9 HEALTH BENEFITS

Lactuca sativa L. is one of the most effective medicinal herbs [82, 58]. It is an abundant natural source of antioxidants, which contributes to numerous pharmacological properties [83, 84]. These effects were attributed to the high concentrations of quercetin glycosides in the lettuce extract and a significant amount of other phenolics [25].

Lettuce has been shown to have potential in cancer prevention due to its content of some phytochemicals. Lettuce biofortified with iodine induces cancer cells to generate ROS, which leads to anticancer effects through the activation of programmed cell death in cancer cells [73]. Awan et al. [85] found that regular consumption of lettuce in the daily diet may be linked to a reduced incidence of lung cancer in non-smoking patients. This potential benefit of lettuce is believed to be due to the presence of carotenoids, beta-carotene, and retinol. Potential nutrients in lettuce showing also a significant relation with decreased colorectal cancer (CRC) were β-carotene and ascorbic acid (AA), whereas Ca, vitamin E, and folate did not demonstrate a significant relationship with CRC risk [85].

Health benefits of lettuce have also been reported, including its cardio-protective effect. CVDs encompass a group of chronic non-infectious diseases that pose a significant threat to public health and are associated with many complex risk factors, including hypertension, diabetes, hyper-lipidemia, and metabolic syndrome [86]. Certain previous epidemiological studies provide evidence that consuming some green leafy vegetables, such as lettuce, is positively linked to treating CVDs. Nicolle et al. [87] found that incorporating lettuce into the diet of rats at a proportion of 20% may have a protective effect against CVDs.

Other investigations showed that the cholesterol-lowering effects of lettuce are attributed to the synergistic effects of its bioactive compounds, such as alpha-tocopherol, beta-carotene, anthocyanins, and phenolic compounds. A preclinical study found that male rats fed a diet enriched with 20% red oak leaf lettuce had a decreased plasma LDL to HDL cholesterol ratio and liver cholesterol content [87].

It has also been suggested that lettuce has anti-diabetic activity. Cheng et al. [88] conducted an experiment on obese hyperglycemic mice, where they were fed with a diet supplemented with red lettuce varieties at a dose of 100 or 300 mg kg^{-1}. The results showed that the addition of red lettuce with high polyphenol content helped alleviate diabetes symptoms. Dietary anthocyanins derived from bilberry and lettuce have been shown to have potential benefits in promoting hyperglycemia and insulin sensitivity in mice with type 2 diabetes [8]. Gopal et al. [89] discovered that Lactuca xanthin extracted from lettuce with 96% purity, as confirmed by HPLC and LCMS, has the ability to inhibit the activities of alpha-amylase and alpha-glucosidase. This inhibition provides a potential mechanism to decrease post-prandial hyperglycemia by regulating the breakdown of starch during the digestion process. Additionally, as mentioned earlier, plants rich in polyphenols also exhibit anti-diabetic effects through the inhibition of intestinal glycosidases and glucose transporters.

Nie et al. [90] indicated that the polysaccharides obtained from stem lettuce have the potential to be used as immunomodulatory agents in the pharmaceutical and functional food sectors. This means that they could be investigated further for their ability to positively impact the immune system. According to the study conducted by Awan et al. [85], it was found that the methanolic extract of *Lactuca serriola* comprises polyphenolic compounds and showed antioxidant activity, contributing to its potential as a hepatoprotective agent.

Other research suggested that lettuce had a positive impact on human health and protected against neurodegenerative diseases. Im et al. [91] found that isochlorogenic acid, one of the phenolic compounds found in romaine lettuce, showed greater neuroprotective effects compared to three other caffeic acid derivatives found in romaine lettuce.

It has also been reported that romaine lettuce, despite containing minor levels of phenolics and antioxidant activity in comparison to other commonly consumed vegetables, can make a greater contribution to overall antioxidant activity and anti-neurodegenerative effect in human diets. This is because romaine lettuce is consumed at a higher daily per capita intake compared to other frequent vegetables, as observed in the study conducted by Zhan et al. [79]. According to Kim et al. [92], lettuce is considered a notable and valuable source of sleep-potentiating materials, containing antioxidant phenolics that provide protection against oxidative stress caused by sleep disorders.

Additionally, epidemiological, clinical, and *in vitro* studies have demonstrated various biological effects associated with the phenolic components of the diet, like antimicrobial, anti-inflammatory, and antistroke [93].

5.10 CONCLUSION AND PROSPECTS

Lettuce (*Lactuca sativa*) is an annual leaf vegetable of the daisy family, Asteraceae. There are many types or cultivars of lettuce. The most common are leaf lettuce, romaine, iceberg, and butterhead. Lettuce and its derived products are an excellent source of vitamins A, K, and C, some B vitamins, and other secondary metabolites such as phenolic compounds, flavonols, and phenolic acids, which can enhance health promotion. Several studies have shown the benefits of the consumption of lettuces.

KEYWORDS

- **antioxidants**
- **bioactive compounds**
- **lettuce**
- **nutritional value**
- **phytochemicals**
- **waste utilization**

REFERENCES

1. Funk, V. A., Bayer, R. J., Keeley, S., Chan, R., Watson, L., Gemeinholzer, B., Schilling, E., Panero, J. L., Baldwin, B. G., & Garcia-Jacas, N. (2005). Everywhere but Antarctica: Using a super tree to understand the diversity and distribution of the *Compositae*. *Biol Skr*, *55*, 343–374.
2. FAOSTAT. (2023). *Database Collections*. Food and Agriculture Organization of the United Nations, Rome. https://www.fao.org/faostat/en/#home (accessed on 1st January 2026).
3. Mou, B. (2008). Lettuce. In J. Prohens & F. Nuez (Eds.), *Handbook of Plant Breeding*, Vol. I *Vegetables I: Asteraceae, Brassicaceae, Chenopodicaceae, and Cucurbitaceae* (pp. 75–115).
4. Noumedem, J. A. K., Djeussi, D. E., Hritcu, L., Mihasan, M., & Kuete, V. (2017). *Lactuca sativa*. In *Medicinal Spices and Vegetables from Africa* (pp. 437–449). Academic Press.
5. Henz, G. P., & Suinaga, F. (2009). *Tipos de Alface Cultivados no Brasil*. Comunicado Técnico No. 75. Embrapa Hortaliças, Brasília, 7p.
6. Mampholo, B. M., Maboko, M. M., Soundy, P., & Sivakumar, D. (2016). Phytochemicals and overall quality of leafy lettuce (*Lactuca sativa* L.) varieties grown in closed hydroponic system. *J Food Qual*, *39*, 805–815.
7. Yang, X., Gil, M. I., Yang, Q., & Tomás-Barberán, F. A. (2022). Bioactive compounds in lettuce: Highlighting the benefits to human health and impacts of preharvest and postharvest practices. *Compr Rev Food Sci Food Saf*, *21*, 4–45.
8. Shi, M., Gu, J., Wu, H., Rauf, A., Emran, T. B., Khan, Z., et al. (2022). Phytochemicals, nutrition, metabolism, bioavailability, and health benefits in lettuce–A comprehensive review. *Antioxidants*, *11*(6), 1158.
9. Medina-Lozano, I., Bertolín, J. R., & Díaz, A. (2021). Nutritional value of commercial and traditional lettuce (*Lactuca sativa* L.) and wild relatives: Vitamin C and anthocyanin content. *Food Chem*, *359*, 129–864.
10. Wilson, P., Morrison, S., Hedges, L., Kerkhofs, N., & Lister, C. (2004). Phenolics contribute significantly to higher antioxidant activity of red lettuce compared to green lettuce. In *Proceedings of the XXII International Conference on Polyphenols*, Helsinki, Finland, 25–28 August (pp. 273, 274).

11. Kim, D. E., Shang, X., Assefa, A. D., Keum, Y. S., & Saini, R. K. (2018). Metabolite profiling of green, green/red, and red lettuce cultivars: Variation in health beneficial compounds and antioxidant potential. *Food Research International*, *105*, 361–370.
12. Liu, X., Ardo, S., Bunning, M., Parry, J., Zhou, K., Stushnoff, C., Frank, S., Liangli, Y., & Kendall, P. (2007). Total phenolic content and DPPH radical scavenging activity of lettuce (*Lactuca sativa* L.) grown in Colorado. *LWT-Food Science and Technology*, *40*(3), 552–557.
13. Llorach, R., Martínez-Sánchez, A., Tomás-Barberán, F. A., Gil, M. I., & Ferreres, F. (2008). Characterization of polyphenols and antioxidant properties of five lettuce varieties and escarole. *Food Chem*, *108*, 1028–1038.
14. Zapata-Vahos, I. C., Rojas-Rodas, F., David, D., Gutierrez-Monsalve, J. A., & Castro-Restrepo, D. (2020). Comparison of antioxidant contents of green and red leaf lettuce cultivated in hydroponic systems in greenhouses and conventional soil cultivation. *Revista Facultad Nacional de Agronomía Medellín*, *73*(1), 9077–9088.
15. Patella, A., Palumbo, F., Galla, G., & Barcaccia, G. (2019). The molecular determination of hybridity and homozygosity estimates in breeding populations of lettuce (*Lactuca sativa* L.). *Genes*, *10*(11), 1–12. http://dx.doi.org/10.3390/genes10110916 (accessed on 1st January 2026). PMid:31717592.
16. Sivakumar, D., Jifon, J., & Soundy, P. (2017). Spectral quality of photo-selective shade nettings improves antioxidants and overall quality in selected fresh produce after postharvest storage. *Food Reviews International*, *34*(3), 290–307.
17. Wei, T., van Treuren, R., Liu, X., Zhang, Z., Chen, J., Liu, Y., Dong, S., Sun, P., Yang, T., Lan, T., Wang, X., Xiong, Z., Liu, Y., Wei, J., Lu, H., Han, S., Chen, J. C., Ni, X., Wang, J., & Liu, H. (2021). Whole-genome resequencing of 445 *Lactuca* accessions reveals the domestication history of cultivated lettuce. *Nature Genetics*, *53*(5), 752–760.
18. Lindqvist, K. (1960). On the origin of cultivated lettuce. *Hereditas*, *46*, 319–350.
19. De Vries, I. (1997). Origin and domestication of *Lactuca sativa* L. *Genet. Resour. Crop Evol*, *44*, 165–174.
20. Hart, G. (2005). *The Routledge Dictionary of Egyptian Gods and Goddesses* (2nd ed.). Routledge, London, UK.
21. Subbarao, K. V., & Koike, S. T. (2007). Lettuce diseases: Ecology and control. In D. Pimentel (Ed.), *Encyclopedia of Pest Management* (Vol. 2). CRC Press, Boca Raton, FL.
22. Zohary, D. (1991). The wild genetic resources of cultivated lettuce (*Lactuca sativa* L.). *Euphytica*, *53*, 31–35.
23. Fernandez, E., La Vecchia, C., Davanzo, B., Negri, E., & Franceschi, S. (1997). Risk factors for colorectal cancer in subjects with family history of the disease. *Br J Cancer*, *75*(9), 1381–1384.
24. Judd, W. S., Campbell, C. S., Kellogg, E. A., & Stevens, P. F. (1999). *Plant Systematics: A Phylogenetic Approach*. Sinauer Associates, Sunderland, MA.
25. Adesso, S., Pepe, G., Sommella, E., Manfra, M., Scopa, A., Sofo, A., Gian Carlo, T., Mariateresa, R., Francesca Di, G., Giuseppina, A., Pietro, C., & Marzocco, S. (2016). Anti-inflammatory and antioxidant activity of polyphenolic extracts from *Lactuca sativa* (var. Maravilla de verano) under different farming methods. *Journal of the Science of Food and Agriculture*, *96*(12), 4194–4206.

26. Bentham, G. (1873). Notes on the classification, history and geographical distribution of *Compositae*. *J Linn Soc Bot, 13*, 335–377.
27. Gallie, D. R. (2013). Increasing vitamin C content in plant foods to improve their nutritional value—successes and challenges. *Nutrients, 5*(9), 3424–3446.
28. Cronquist, A. (1955). Phylogeny and taxonomy of the *Compositae*. *Am Mid Naturalist, 53*, 478–511.
29. Bremer, K. (1994). *Asteraceae: Cladistics and Classification*. Timber, Portland, OR.
30. Lebeda, A., Dolezalová, I., Feráková, V., & Astley, D. (2004). Geographical distribution of wild *Lactuca* species (*Asteraceae, Lactuceae*). *Bot Rev, 70*, 328.
31. Křístková, E., Doležalová, I., Lebeda, A., Vinter, V., & Novotna, A. (2008). Description of morphological characters of lettuce (*Lactuca sativa* L.) genetic resources. *Horticultural Science (Prague), 35*(3), 113–129.
32. Doležalová, I., Křístková, E., Lebeda, A., & Vinter, V. (2002). Description of morphological characters of wild *Lactuca* L. spp. genetic resources (English-Czech version). *Hort Sci (Prague), 29*(2), 56–83.
33. Grulich, V. (2004). *Lactuca* L. In B. Slavík & J. Štěpánková (Eds.), *Květena České republiky 7* (pp. 487–497). Academia, Praha.
34. Rubatzky, V. E., & Yamaguchi, M. (1997). *World Vegetables*. Chapman and Hall, New York.
35. George, R. A. T. (1999). *Compositae*. In R. A. T. George (Ed.), *Vegetable Seed Production* (pp. 122–1353). CAB International, Wallingford.
36. Hassan, M. N., Mekkawy, S. A., Mahdy, M., Salem, K. F., & Tawfik, E. (2021). Recent molecular and breeding strategies in lettuce (*Lactuca* spp.). *Genetic Resources and Crop Evolution, 68*, 3055–3079.
37. Coria-Cayupán, Y. S., Sánchez de Pinto, M. I., & Nazareno, M. A. (2009). Variations in bioactive substance contents and crop yields of lettuce (*Lactuca sativa* L.) cultivated in soils with different fertilization treatments. *J Agric Food Chem, 57*, 10122–10129.
38. Kim, M. J., Moon, Y., Tou, J. C., Mou, B., & Waterland, N. L. (2016). Nutritional value, bioactive compounds and health benefits of lettuce (*Lactuca sativa* L.). *Journal of Food Composition and Analysis, 49*, 19–34.
39. Xiao, Z., Lester, G. E., Luo, Y., & Wang, Q. (2012). Assessment of vitamin and carotenoid concentrations of emerging food products: Edible microgreens. *J Agric Food Chem, 60*(31), 7644–7651.
40. Dai, J., & Mumper, R. J. (2010). Plant phenolics: Extraction, analysis and their antioxidant and anticancer properties. *Molecules, 15*, 7313–7352.
41. Karakaya, S. (2004). Bioavailability of phenolic compounds. *Crit Rev Food Sci Nutr, 44*(6), 453–464.
42. Rice-Evans, C. A., Miller, N. J., & Paganga, G. (1996). Structure-antioxidant activity relationships of flavonoids and phenolic acids. *Free Rad Biol Med, 20*, 933–956.
43. López, A., Javier, G. A., Fenoll, J., Hellín, P., & Flores, P. (2014). Chemical composition and antioxidant capacity of lettuce: Comparative study of regular-sized (Romaine) and baby-sized (Little Gem and Mini Romaine) types. *J Food Compos Anal, 33*, 39–48.

44. Zhao, X., Carey, E. E., Young, J. E., Wang, W., & Iwamoto, T. (2007). Influences of organic fertilization, high tunnel environment, and postharvest storage on phenolic compounds in lettuce. *Hort Science*, *42*, 71–76.
45. Tiveron, A. P. (2010). *Atividade Antioxidante E Composição Fenólica De Legumes E Verduras Consumidos no Brasil* (Dissertação de mestrado). Universidade de São Paulo, Piracicaba.
46. Negrão, L. D., Sousa, P. V. D. L., Barradas, A. M., Brandão, A. D. C. A. S., Araújo, M. A. D. M., & Moreira-Araújo, R. S. D. R. (2020). Bioactive compounds and antioxidant activity of crisphead lettuce (*Lactuca sativa* L.) of three different cultivation systems. *Food Science and Technology*, *41*, 365–370.
47. Gazula, A., Kleinhenz, M. D., Scheerens, J. C., & Ling, P. P. (2007). Anthocyanin levels in nine lettuce (*Lactuca sativa*) cultivars: Influence of planting date and relations among analytic, instrumented, and visual assessments of color. *Hort Science*, *42*, 232–238.
48. Heimler, D., Isolani, L., Vignolini, P., Tombelli, S., & Romani, A. (2007). Polyphenol content and antioxidative activity in some species of freshly consumed salads. *J Agric Food Chem*, *55*, 1724–1729.
49. Maiani, G., Periago Castón, M. J., Catasta, G., Toti, E., Cambrodón, I. G., Bysted, A., Granado-Lorencio, F., Olmedilla-Alonso, B., Knuthsen, P., & Valoti, M. (2009). Carotenoids: Actual knowledge.
50. Catherine, N., Carnat, A., Fraisse, D., Lamaison, J. L., Rock, E., Michel, H., Amouroux, P., & Remesy, C. (2004). Characterization and variation of antioxidant micronutrients in lettuce (*Lactuca sativa folium*). *J Sci Food Agric*, *84*, 2061–2069.
51. Mou, B. (2005). Genetic variation of beta-carotene and lutein contents in lettuce. *J Am Soc Hortic Sci*, *130*, 870–876.
52. Lee, S. K., & Kader, A. A. (2000). Preharvest and postharvest factors influencing vitamin C content of horticultural crops. *Postharvest Biology and Technology*, *20*(3), 207–220.
53. Carr, A. C., & Vissers, M. C. (2013). Synthetic or food-derived vitamin C—Are they equally bioavailable? *Nutrients*, *5*(11), 4284–4304.
54. Martínez-Sánchez, A., Luna, M. C., Selma, M. V., Tudela, J. A., Abad, J., & Gil, M. I. (2012). Baby-leaf and multi-leaf of green and red lettuces are suitable raw materials for the fresh-cut industry. *Postharvest Biol Technol*, *63*, 1–10.
55. Ghorbani, A., Rakhshandeh, H., & Sadeghnia, H. R. (2013). Potentiating effects of *Lactuca sativa* on pentobarbital-induced sleep. *Iranian Journal of Pharmaceutical Research*, *12*(2), 401.
56. Xu, F., Zou, G. A., Liu, Y. Q., & Aisa, H. A. (2012). Chemical constituents from seeds of *Lactuca sativa*. *Chemistry of Natural Compounds*, *48*, 574–576.
57. Sayyah, M., Hadidi, N., & Kamalinejad, M. (2004). Analgesic and anti-inflammatory activity of *Lactuca sativa* seed extract in rats. *Journal of Ethnopharmacology*, *92*(2, 3), 325–329.
58. Pour, Z. S., Hosseinkhani, A., Asadi, N., Shahraki, H. R., Vafaei, H., Kasraeian, M., Bazrafshan, K., & Faraji, A. (2018). Double-blind randomized placebo-controlled trial on efficacy and safety of *Lactuca sativa* L. seeds on pregnancy-related insomnia. *J Ethnopharmacol*, *227*, 176–180.

59. Ranjbar, M., Afsharypuor, S., Shakibaei, F., & Mazaheri, M. (2020). Effect of topical lettuce (*Lactuca sativa* L.) seed oil on childhood sleep disorders: A randomized double-blind controlled trial. *Res J Pharmacogn*, *7*(3), 47–54.
60. Rahma, S. S. A., Hossain, M. A., Weli, A. M., Al-Riyami, Q., & Al-Sabahi, J. N. (2013). Chemical composition of essential oils and *in vitro* antioxidant activity of fresh and dry leaves crude extracts of medicinal plant of *Lactuca sativa* L. native to Sultanate of Oman. *Asian Pacific Journal of Tropical Biomedicine*, *2013*, 353–357.
61. Ranjbar, M., Mazaheri, M., Shakibaei, F., & Aslani, A. (2018). Essential oil constituents of seeds and fresh leaves of garden lettuce (*Lactuca sativa* L.) grown in Isfahan, Iran. *Research Journal of Pharmacognosy*, *5*(3), 1–5.
62. Dueñas, M., & García-Estévez, I. (2020). Agricultural and food waste: Analysis, characterization and extraction of bioactive compounds and their possible utilization. *Foods*, *9*(6), 817.
63. Esparza, I., Jiménez-Moreno, N., Bimbela, F., Ancín-Azpilicueta, C., & Gandía, L. M. (2020). Fruit and vegetable waste management: Conventional and emerging approaches. *Journal of Environmental Management*, *265*, 110510.
64. Rodriguez Garcia, S. L., & Raghavan, V. (2022). Green extraction techniques from fruit and vegetable waste to obtain bioactive compounds—A review. *Critical Reviews in Food Science and Nutrition*, *62*(23), 6446–6466.
65. Cristobal, J., Caldeira, C., Corrado, S., & Sala, S. (2018). Techno-economic and profitability analysis of food waste biorefineries at European level. *Bioresour Technol*, *259*, 244–252.
66. Chen, Q., Zhang, T. C., Ouyang, L., & Yuan, S. (2022). Single-step hydrothermal synthesis of biochar from H_3PO_4-activated lettuce waste for efficient adsorption of Cd (II) in aqueous solution. *Molecules*, *27*(1), 269.
67. Plazzotta, S., Sillani, S., & Manzocco, L. (2018). Exploitation of lettuce waste flour to increase bread functionality: Effect on physical, nutritional, sensory properties and on consumer response. *International Journal of Food Science & Technology*, *53*(10), 2290–2297.
68. Santos, F. T., Goufo, P., Santos, C., Botelho, D., Fonseca, J., Queirós, A., Monica, S. S., Cosata, M., Henrique, T., & Trindade, H. (2016). Comparison of five agro-industrial waste-based composts as growing media for lettuce: Effect on yield, phenolic compounds and vitamin C. *Food Chemistry*, *209*, 293–301.
69. Uddin, M. Z., Rana, M. S., Hossain, S., Ferdous, S., Dutta, E., Dutta, M., & Emran, T. B. (2020). *In vivo* neuroprotective, antinociceptive, anti-inflammatory potential in Swiss albino mice and *in vitro* antioxidant and clot lysis activities of fractionated *Holigarna longifolia* Roxb. bark extract. *J Complement Integr Med*, *17*, 1–12.
70. Mitra, S., Rauf, A., Tareq, A. M., Jahan, S., Emran, T. B., Shahriar, T. G., Dhama, K., Alhumaydhi, F. A., Aljohani, A. S., & Rebezov, M. (2021). Potential health benefits of carotenoid lutein: An updated review. *Food Chem Toxicol*, *154*, 112328.
71. Lafarga, T., Villaró, S., Rivera, A., Bobo, G., & Aguiló-Aguayo, I. (2020). Bio accessibility of polyphenols and antioxidant capacity of fresh or minimally processed modern or traditional lettuce (*Lactuca sativa* L.) varieties. *J Food Sci Technol*, *57*, 754–763.
72. Chu, Y. F., Sun, J., Wu, X., & Liu, R. H. (2002). Antioxidant and antiproliferative activities of common vegetables. *Journal of Agricultural and Food Chemistry*, *50*, 6910–6916.

73. Sularz, O., Koronowicz, A., Smoleń, S., Kowalska, I., Skoczylas, Ł., Liszka-Skoczylas, M., Tabaszewska, M., & Pitala, J. (2021). Anti- and pro-oxidant potential of lettuce (*Lactuca sativa* L.) biofortified with iodine by KIO_3, 5-iodo- and 3,5-diiodosalicylic acid in human gastrointestinal cancer cell lines. *RSC Adv*, *11*, 27547–27560.
74. Galieni, A., Di Mattia, C., De Gregorio, M., Speca, S., Mastrocola, D., Pisante, M., & Stagnari, F. (2015). Effects of nutrient deficiency and abiotic environmental stresses on yield, phenolic compounds and antiradical activity in lettuce (*Lactuca sativa* L.). *Sci Hortic*, *187*, 93–101.
75. Chisari, M., Todaro, A., Barbagallo, R. N., & Spagna, G. (2010). Salinity effects on enzymatic browning and antioxidant capacity of fresh-cut baby Romaine lettuce (*Lactuca sativa* L. cv. Duende). *Food Chem*, *119*, 1502–1506.
76. Kapoulas, N., Koukounaras, A., & Ilić, Z. S. (2017). Nutritional quality of lettuce and onion as companion plants from organic and conventional production in north Greece. *Sci Hortic*, *219*, 310–318.
77. Lanza Volpe, M., Soto Vargas, V. C., Morón, A., & González, R. E. (2021). Bioactive compounds, antioxidant activity and growth behavior in lettuce cultivars grown under field and greenhouse conditions. *Proceedings*, *70*, 52.
78. Zambrano-Moreno, E. L., Chávez-Jáuregui, R. N., Plaza, M. D. L., & Wessel-Beaver, L. (2015). Phenolic content and antioxidant capacity in organically and conventionally grown eggplant (*Solanum melongena*) fruits following thermal processing. *Food Science and Technology*, *35*(3), 414–420.
79. Zhan, H., Li, S., Sun, J., Liu, R., Yan, F., Niu, B., Zhang, H., & Wang, X. (2014). Lettuce glycoside B ameliorates cerebral ischemia reperfusion injury by increasing nerve growth factor and neurotrophin-3 expression of cerebral cortex in rats. *Indian J Pharmacol*, *46*, 63.
80. Cano, A., & Arnao, M. B. (2005). Hydrophilic and lipophilic antioxidant activity in different leaves of three lettuce varieties. *International Journal of Food Properties*, *8*(3), 521–528.
81. Cho, Y. S., Yeum, K. J., Chen, C. Y., Beretta, G., Tang, G., Krinsky, N. I., Yoon, S., Lee Kim, Y. C., Blumberg, J. B., & Russell, R. M. (2007). Phytonutrients affecting hydrophilic and lipophilic antioxidant activities in fruits, vegetables and legumes. *Journal of the Science of Food and Agriculture*, *87*(6), 1096–1107.
82. Feyzabadi, Z., Aliasl, F., Qaraaty, M., & Aliasl, J. (2016). Effective materia medica in treatment of insomnia in Iranian traditional medicine. *Med Hist J*, *7*(23), 51–68.
83. Pepe, G., Sommella, E., Manfra, M., De Nisco, M., Tenore, G. C., Scopa, A., Sofo, S., Marzocco, S., Adesso, T., Novellino, A., & Campiglia, P. (2015). Evaluation of anti-inflammatory activity and fast UHPLC-DAD-IT-TOF profiling of polyphenolic compounds extracted from green lettuce (*Lactuca sativa* L. var. Maravilla de verano). *Food Chemistry*, *167*, 153–161.
84. Ismail, H., & Mirza, B. (2015). Evaluation of analgesic, anti-inflammatory, anti-depressant and anti-coagulant properties of *Lactuca sativa* (cv. Grand Rapids) plant tissues and cell suspension in rats. *BMC Complementary and Alternative Medicine*, *15*(1), 1–7.
85. Awan, A. F., Akhtar, M. S., Anjum, I., Mushtaq, M. N., Fatima, A., Mannan, A., & Ali, I. (2020). Antioxidant and hepatoprotective effects of *Lactuca serriola* and

its phytochemical screening by HPLC and FTIR analysis. *Pak J Pharm Sci, 33*, 2823–2830.
86. Praveen, P. A., Roy, A., & Prabhakaran, D. (2013). Cardiovascular disease risk factors: A childhood perspective. *Indian J. Pediatrics, 80*, 3–12.
87. Nicolle, C., Cardinault, N., Gueux, E., Jaffrelo, L., Rock, E., Mazur, A., Amouroux, P., & Rémésy, C. (2004). Health effect of vegetable-based diet: Lettuce consumption improves cholesterol metabolism and antioxidant status in the rat. *Clin Nutr, 23*, 605–614.
88. Cheng, P., Pogrebnyak, N., Kuhn, P., Poulev, A., Waterman, C., Rojas-Silva, P., Johnson, W. D., & Raskin, I. (2014). Polyphenol-rich Rutgers Scarlet Lettuce improves glucose metabolism and liver lipid accumulation in diet-induced obese C57BL/6 mice. *Nutrition, 30*, S52–S58.
89. Gopal, S. S., Lakshmi, M. J., Sharavana, G., Sathaiah, G., Sreerama, Y. N., & Baskaran, V. (2017). Lactucaxanthin—A potential anti-diabetic carotenoid from lettuce (*Lactuca sativa*) inhibits α-amylase and α-glucosidase activity *in vitro* and in diabetic rats. *Food Funct, 8*, 1124–1131.
90. Nie, C., Zhu, P., Ma, S., & Wang, M., & Hu, Y. (2018). Purification, characterization and immunomodulatory activity of polysaccharides from stem lettuce. *Carbohydr Polym, 188*, 236–242.
91. Im, S. E., Yoon, H., Nam, T. G., Heo, H. J., Lee, C. Y., & Kim, D. O. (2010). Antineurodegenerative effect of phenolic extracts and caffeic acid derivatives in romaine lettuce on neuron-like PC-12 cells. *J Med Food, 13*, 779–784.
92. Kim, H. D., Hong, K. B., Noh, D. O., & Suh, H. J. (2017). Sleep-inducing effect of lettuce (*Lactuca sativa*) varieties on pentobarbital-induced sleep. *Food Science and Biotechnology, 26*, 807–814.
93. Tomb, A. S. (1977). Lactuceae–Systematic review. In V. H. Heywood, J. B. Harborne, & B. L. Turner (Eds.), *The Biology and Chemistry of the Compositae, II* (pp. 1067–1079). Academic Press, London, New York.

CHAPTER 6

Spinach (*Spinacia oleracea* L.)

TANVEER AHMAD,[1,*] HAMMAD HUSSAIN,[1]
MUHAMMAD ANWAR-UL-HAQ,[1] QUMER IQBAL,[2] and
MUHAMMAD AMJAD[3]

[1]*Department of Horticulture, MNS University of Agriculture, Multan, Punjab, Pakistan*

[2]*Fiblast, LLC, Tuskegee, Alabama, USA*

[3]*Institute of Horticultural Sciences, University of Agriculture, Faisalabad, Punjab, Pakistan*

*Corresponding author

ABSTRACT

Spinach (*Spinacia oleracea* L.) is a cool-season annual leafy vegetable of significant nutritional and medicinal importance. This chapter outlines its botanical classification, global production, and agronomic practices, including soil requirements, propagation, and postharvest handling. Spinach is a rich source of essential vitamins (A, C, K, folate), minerals (iron, magnesium, potassium), dietary fiber, and bioactive phytochemicals such as carotenoids, flavonoids, and phenolic compounds. These constituents contribute to its diverse health-promoting properties, including antioxidant, anti-inflammatory, cardioprotective, cholesterol-lowering, hypoglycemic, anticancer, and anti-obesity activities. The consumption of spinach supports the prevention of chronic diseases, including cardiovascular disorders, diabetes, and certain cancers. This review highlights the vegetable's role as a functional food and emphasizes the importance of optimal cultivation, processing, and culinary methods to preserve its nutritional and therapeutic value.

Postharvest Quality Attributes of Fresh Vegetables and Grains.
Riadh Ilahy, Imen Tlili, Thouraya R'him, and Muhammad Azam (Eds.)
© 2027 Apple Academic Press, Inc. Co-published with CRC Press (Taylor & Francis)

DOI:10.1201/9781779642851-8

6.1 INTRODUCTION

Spinach, *Spinacia oleracea* L. ($2n = 12$), is an annual (seldom biennial for seed production purposes) herbaceous plant and a valuable vegetable, of which leaves and tender shoots are generally consumed after boiling and cooking either in fresh or frozen form or eaten raw as a fresh salad, but rarely consumed in processed form. It belongs to the family Amaranthaceae and the latest botanical classification is presented in Table 6.1. In general, spinach is a cool-season vegetable, so it can be cultivated in both autumn and spring around the world, yielding high production in a short period [1–3].

TABLE 6.1 Botanical Classification of Spinach

Kingdom	Plantae
Sub-Kingdom	Tracheobionta
Super-Division	Spermatophyta
Division	Magnoliophyta
Class	Magnoliopsida
Sub-Class	Caryophyllidae
Order	Caryophyllales
Family	Amaranthaceae
Sub-family	Chenopodiaceae
Genus	*Spinacia* L.
Species	*Spinacia oleracea* L.

Source: NRCS (Natural Resources Conservation Service, United States Department of Agriculture [USDA]).

The contemporary advancement of baby leaf spinach, joined with a rise in diet interests, has been a substantial factor in its extended utilization. It is extensively declared that dark-green leafy vegetables are a generally required part of human nutrition. Moreover, it is found to be the most advantageous dark-green leafy vegetable, as it is rich in essential vitamins, folate, and beta carotene, as well as a sound source of iron, phosphorous, sodium, calcium, and potassium, and ascorbic acid (AA) [4, 5].

6.1.1 *GLOBAL PRODUCTION SCENARIO*

The countries of the European Union produce a total of about 555,416 tons of spinach yearly on an area under cultivation of 31,770 ha. In Spain, the total annual production is approximately 86,990 tons in an area under cultivation of 5100 ha. Roughly 89% of it is ingested locally, and the rest of the 11% is shipped to other countries. In 2020, the total area under spinach cultivation was 920,804 hectares, and the total production was 30,995,069 tons around the world [6].

6.1.2 HISTORY AND DISTRIBUTION

The origin of spinach is considered to be Central Asia, especially Persia (Iran). Later, it was adapted into Spain by Arabian rulers in the 11th century. Afterward, it was introduced into the remaining parts of Europe [7, 8]. Later, it was distributed in other regions around the world via the trading process [9]. In 1930, a cartoonist named M. Fleischer created a fictitious character, that is, Popeye, who consumed spinach to enhance his power. In addition to the typical artistic license, the perception was raised and widespread that spinach is a plentiful source of iron. At that time, anemia was an extremely prevalent disease because of the iron-deficient food consumption habit, and the artistic cartoon benefited health organizations by encouraging the ingestion of spinach. Afterward, it was discovered that this concept was initiated by a transcription error that happened when the analysis of spinach by E. Von Wolf (the German scientist) in 1870 was translated, and the comma used in the European decimal system was mistakenly placed. Consequently, the actual level of iron had increased by 10 times. Conversely, this discovery was too late to bring down the widespread misconception regarding spinach and the superhuman power of Popeye, a fictitious character [10].

6.2 BOTANICAL DESCRIPTION

Spinacia oleracea L. grows a rosette of leaves soon after germination, and the stem starts to get longer as the plant raises. The branches on the side of the main stem appear from the axis of the leaf. Besides, the flowers are borne axillary and usually begin on the middle portion of the main stem and side branches [11]. The stem part is normally vertical with 30 cm tall, soft, and rounded [12]. However, flowering continues in the direction of the base as well as the top of a plant [13]. There are four kinds of spinach identified with orientation to sex expression: vegetative males, extreme males, females, and monoecious [14]. Spinach is normally dioecious, with occasional monoecious plants [15, 16].

In addition, *Spinacia oleracea* L. can be further classified into two subspecies, *S. oleracea* L. ssp. *glabra* and *S. oleracea* L. ssp. *spinosa*. These are largely present in the marketplace and usually offered not only in raw or frozen form but also canned or processed form. The use of frozen and canned spinach is gradually increasing, making the dietary value as well as post-harvest life of this food commodity significant, maybe as a consequence of the encouraged ingestion of processed food in the previous few years [17]. The edible leaves are organized in a rosette, from which a seed stalk arises. The simple leaves are fairly ovate or triangular and may be sometimes wrinkled but usually flat. The flowers are unobtrusive and form small dry fruits [18]. The sex of spinach is fixed by the XY system. Female and male floral organs vary morphologically; however, plants cannot be identified in the vegetative stage prior to flower emergence [19].

6.3 PRODUCTION TECHNOLOGY

6.3.1 SOIL AND CLIMATIC CONDITIONS

Usually, spinach is cultivated in sandy loam to clay loam soils, but it can also be raised in a wide range of soils with a pH between 6.5 and 9.0 [14]. If soil is acidic, lime should be utilized to keep the soil pH at 6.5–6.8, as spinach is susceptible to extremely acidic soils. Moreover, the tolerance of cold temperature in the plant is due to the existence of cold-regulated proteins in it [20].

6.3.2 PROPAGATION

Normally, spinach is raised by seed, which are sown in the hilly areas during August to September and in the plains during October to December. However, the crop cultivated in spring-summer generally has a greater tendency to bolt. The seeds are either spread in the field usually, but it can be seeded by dibbling method in rows with 20 cm of row-to-row distance to boost the growth of plants. Generally, a seed depth of approximately 3.5 to 4.0 cm is recommended. The seed rate differs with the time, method, and region of sowing, as well as cultivars of species. Normally, around 11.5 kg of seed rate of spinach is recommended for one-acre area [14]. Because of its dioecious character, the male plants raise weakly; spinach needs more seed rate as compared to beet. In addition, appropriate land preparation along with sufficient soil moisture is desirable to confirm adequate emergence.

6.3.3 IRRIGATION AND PLANT NUTRITION

Generally, the first irrigation is provided instantly after sowing; it is a light one. However, following irrigations should be provided consistent with the crop need. Normally, one irrigation in every 4 to 6 days and 10 to 12 days in summer and winter, respectively, can offer moisture levels sufficient for optimal plant growth.

Generally, on the basis of soil analysis, the accurate quantity of fertilizer and manure should be utilized. If spinach has a nutrient deficiency, the plants become susceptible to bolting, and ultimately, the vegetative growth is reduced. Usually, 18–20 tons of well-rotten FYM (farmyard manure) are applied as a basal dose, after that 150:100:100 as inorganic NPK fertilizers per hectare is recommended [4]. The fertility from manure is most obtainable when soil temperatures are warm. The FYM is employed before sowing and mixed into the soil appropriately, and 1/3 of the nitrogen and the full dose of phosphorus and potassium is applied before sowing, with the rest of 2/3 of the nitrogen side-dressed or broadcast in two equal split doses after the first and second harvest. Moreover, urea (1.5%) may be used as a foliar spray to fulfill the nitrogen requirements, especially after every cutting.

Additionally, foliar spray of magnesium (Mg) can boost the production of dark green leaves. Normally, we can use Epsom salts 12 to 15 kg per hectare. Similarly, we can correct micronutrient deficiencies, such as manganese and boron deficiency, by foliar spray of these nutrients.

6.3.4 INSECT-PEST AND DISEASE MANAGEMENT

Cutworms, beet armyworm, and aphids are the most common pests of spinach around the world. They can be managed by the application of organic chemicals or inorganic pesticides, which are less harmful to the crop as well as the consumer. Bacterial soft rot is usually due to *Erwinia carotovora*, and downy mildew is caused by *Peronospora effusa* (farinosa), which are the main biotic factors that trigger severe production losses of spinach [21]. In bacterial soft rot, the infected leaves look water-soaked and gray-green in color. The bacteria normally enter either via fungal lesions or wounds. The deterioration progresses at a quicker rate at high humidity and at temperatures above 7°C. The infection of downy mildew arises mainly in the field, but it grows into conspicuousness after crop harvest, particularly above 85% relative humidity and at temperatures ranging from 14 to 24°C.

6.3.5 *HARVESTING AND HANDLING OF SPINACH*

Spinach is normally cut from the time plants have 5–6 leaves until prior to the emergence of seed stalks. A higher yield is obtained when the plants are allowed to grow to full size [11]. For marketing of fresh-cut spinach, it is usually cut by trimming the taproot near the lowest leaves; however, the plants are trimmed around an inch above the soil surface for processing purposes. Well-grown, tender, and succulent leaves will be cut at the time of harvesting. In addition, the crop is multi-cut, and normally it can yield 4–6 cuts in a season [22].

It is not recommended to harvest spinach precipitously after rain, as the leaves are crisp and break easily when wet. After harvesting, disease-affected and discolored leaves should be isolated, and the product managed prudently to prevent bruising or injury of leaves. Ryall et al. [23] suggested that it should be cut only during early morning or afternoon to reduce wilting. Even at high humidity, leaves of spinach lose water more quickly when warm than in cool weather [24].

6.4 POSTHARVEST PRACTICES

6.4.1 *STORAGE OF SPINACH*

Rapid wilting of spinach leaves is due to low humidity and fast air movement to accelerate the loss of AA that ultimately reduces the quality of the food commodity [25]. The following are two methods of storage commonly used around the world.

6.4.1.1 *Low-Temperature Storage*

Normally, it is recommended to preserve spinach near 0°C from packaging to final selling. However, it can be package-iced and vacuum-cooled, as well as hydrocooled. If treated with cold water, it should be centrifuged for around three minutes to eliminate excessive water to reduce degeneration under higher than ideal temperatures. It should be moistened before it is vacuum-cooled if it is kept at 24°C. The application of 100 mg/L chlorinated water for washing spinach prevents the accumulation of microbes, although it has less effect on killing microbes [23].

Vitamin C and chlorophyll contents in spinach kept at 20°C for 6 days reduced significantly, while yellowing of leaves increased. Approximately 83% of vitamin C was lost within 3 days at 20°C [3]. However, thiamine content increased abruptly during storage, and riboflavin content remained unchanged. Peroxidase lowered chlorophyll in the presence of hydrogen peroxide, but degradation was inhibited by the addition of L-ascorbic acid. The lowering of chlorophyll was gentler at 0°C than 20°C [26]. After 2 days of storage, vitamin A content lessened by 21.6% on storage in a closed vessel at 29 to 32°C and 23.7% on storage at 5°C in a refrigerator [27].

6.4.1.2 CONTROLLED ATMOSPHERE STORAGE (CAS)

A carbon dioxide (10%) supplemented atmosphere reduced the yellowing of leaves and preserved the product in good quality at 5°C for three weeks. The CA, thus, extends the post-harvest life of spinach at 5°C for one week. These workers observed no injuries in spinach held in 4% of CO_2 for 3 weeks, but above 10% of CO_2 has no advantageous effect [28].

6.4.2 PROCESSING OF SPINACH

The volume of water used to wash the spinach did not have an effect on the nutritional value, as well as the quality attributes, of the preserved spinach [28]. Dipping spinach leaves in a detergent solution before washing lessened the greenness, optical density of liquor, and vitamin C. The color was rated a little higher for spinach that was blanched in steam. Water-blanching leached out more riboflavin, nitrate, and vitamin C than steam-blanching; however, the carotene content was not changed. Thawing frozen spinach in a freezer after steaming or cooking in a microwave oven results in higher residual vitamin C than other approaches of defrosting [29]. Steam blanching is recognized to change 20% of the calcium (Ca) into non-absorbable and insoluble Ca oxalate [30]. Magnesium (Mg), in comparison, is less sensitive, and iron (Fe) is relatively resistant to cooking [31, 32]. Canning and freezing at –18°C also affect ions, both cations and anions. These methods can lead to cell rupture and thus ion loss. 0.03 mg/L of cadmium and 0.13 mg/L of lead contents in canned spinach have been testified [17].

6.5 NUTRITIONAL AND MEDICINAL IMPORTANCE

Spinacia oleracea is conventionally used for antioxidant, antianemia, antidiabetic, anthelmintic, antihyperlipidemic, anti-inflammatory, antibacterial, antiulcer, antiepileptic, and antiviral activities. It is also consumed in the inhibition of liver and neurodegenerative as well as lung diseases [33].

6.5.1 HEALTH BENEFITS AND BIOLOGICAL ACTIVITIES

In the recent past, consumers have progressively become conscious regarding a nutritious and healthy diet. On the other hand, many people are conscious of the recommendations associated with the consumption of particular diets, but some people follow the instructions seriously. Routines that cover healthy intake have been linked with decreasing the incidence of pathogens and have encouraged the five servings of fruits and vegetables per day program, which endorses that these five servings must be consumed every day on a regular basis.

Other plans in this awareness are the European SFVS (School Fruit and Vegetable Scheme) and APRIFEL (a French Agency for Research and Information on Fruit and Vegetables) with its EGEA (European Geography Association) conferences linked to vegetables and fruit, which aim to train school students to embrace nutritious and healthy intake routines with the intention that they will be potential consumers of this type of healthy food commodities [34]. While, in European countries and North America, the average consumption of these diets is still low regarding recommended levels, even though it varies among and within the countries concerning their socioeconomic levels. This is commonly linked to healthy meals' poor hedonic value, and as a result, attempts are being undertaken to present them as more appealing [35].

People consuming a significant quantity of fruits and vegetables have less possibility of becoming addicted to smoking. If they smoke, they are less reliant on roll-ups and have a reduced rate of depression. Moreover, the cut-off for the controlling consequence on smoking is contrariwise linked to nicotine reliance and is correlated to anxiety and sadness. Interestingly, the consumption of sugary foods like vegetables and fruits stimulates dopamine release that induces feelings of pleasure [36, 37].

The intake of 100 g of spinach contains 150 kcal of energy and offers a substantial quantity of vitamins as well as minerals, covering 49% of the recommended daily amount (RDA) of folates, 47% of vitamin C, 108% of vitamin A, and 604% of vitamin K [38, 39]. In 100 g of spinach, potassium provides 16% of the RDA, and iron and magnesium provide 20% of the RDA equally [38]. Vitamin B-9, accompanied by ascorbic acid (vitamin C), aids in making proteins and generating deoxyribonucleic acid (DNA). In addition, it is essential during pregnancy to stop irregularities in the central nervous system (CNS) and spine medulla, for instance, spina bifida. In recent times, a decrease in the quantity of folate ingested in food has been observed; therefore, it is often obtainable in synthetic form [6]. Nevertheless, the consumption of foliate-rich vegetables like spinach might prevent the possible harmful impacts of medicines. Besides, a lack of vitamin E has been related to several disorders like muscular dystrophy, endocrine disturbances, atherosclerosis, and reduced immunity [17].

Green leafy vegetables, especially spinach, are known to be good sources of vitamin K, particularly K-1, whose deficiency is linked to changes in blood coagulation and undesirable calcifications that lead to arterial sclerosis due to changes in the gamma-carboxy-L-glutamic acid residues that maintain homeostasis [40]. The great amount of Mg in extracts of spinach leaves optimizes low levels of these related to chronic disorders; for example, Alzheimer's, cardiac diseases, high blood pressure, and diabetes [41, 42]. Carotenoids present in spinach, by stopping protein and lipid oxidation in the eye lens, reduce the risk of cataracts [43]. On the other hand, the occurrence of nitrates in its leaves would harm infants, and babies could be produced meta-hemoglobinemia as hemoglobin cannot transport oxygen efficiently [44].

6.5.2 CARDIOPROTECTIVE AND CHOLESTEROL-LOWERING PROPERTIES

Saponins (a triterpene glycoside) in spinach are favorable for maintaining optimal blood pressure and cholesterol levels [45]. Antioxidant-rich food is required by users to decrease the peril of ailments associated with oxidative stress, such as coronary heart diseases, which form a component of a pathology correlated to oxidative stress [40]. The intake of spinach can decrease lipid peroxidation by 5% [46].

As a result of AA as well as fiber content in spinach, it has been inversely associated with metabolic syndrome and circulatory ailment by enhancing arterial pressure [47, 48]. The existence of lutein, violaxanthin, and beta-carotene in spinach upsurges carotenoid levels in lipoproteins, ultimately carrying out a significant character in arterial sclerosis [49, 50].

6.5.3 ANTI-OBESITY PROPERTIES

The consumption of spinach reduces the incidence of diabetes and obesity, as thylakoids and enzymes located in the chloroplasts and rubisco help to reduce weight [51, 52]. In addition, foods copious in spinach, with its low-fat level, high fiber content, and complex carbohydrates, decrease insulin needs, regulate blood glucose concentrations, maintain optimal weight, and reduce the number of people affected by diabetes [20].

6.5.4 HYPOGLYCEMIC ACTIVITY

Spinach is consumed for diseases related to insulin resistance, like diabetes mellitus, as a result of the advantageous role of vitamin B-7, or inositol, and sugary alcohol. Inositol facilitates the liver to dissolve fat; besides, choline along with inositol produces lecithin, which impedes the fat from accumulating [53]. Moreover, it is advantageous for reducing depression as well as acting as a stimulant for serotonin production. Additionally, these complexes have comparable physical properties to carbohydrates while having a less glycemic index, reduced-calorie content, as well as a low insulin response. It comprises 1.2 mg of myo-inositol, 112 mg of glucaric acid, 0.4 mg of mannitol, and 0.22 mg of glycosyl-glycerol per 100 g each [54, 55].

Ulcers are one of the ailments associated with diabetes. Additionally, it has been confirmed that spinach can accelerate the remedial as well as regeneration process [54, 56]. Besides spinach consumption, can enhance antihyperglycemic activity, preventing intestinal α-glucosidase, pancreatic α-amylase, and tyrosine in the liver, then secretagogue insulin activities [57, 58].

6.5.5 ANTI-INFLAMMATORY EFFECTS

The anti-inflammatory effect of a natural antioxidant in aqueous extract of spinach leaves is because of it lessening the reactive oxygen species (ROS) and products of lipid peroxidation, reduction of cyclooxygenase expression and improved formation of antioxidants, and reduced expression of the cyclooxygenase which stimulates the production of arachidonic acid leading to inflammation [59, 60]. Moreover, the galactolipids, flavonoids, and glycoglycerolipids occur in extracts of spinach, consequently preventing inflammations [61–63]. Another constituent from the tissue of spinach, such as budesonide, has been extracted, which has anti-inflammatory activity as well as increases the effect of other anti-inflammatory complexes [64].

6.5.6 ANTICANCER PROPERTIES

The NAO in spinach extract prevents prostate cancer and hyperplasia in the four lobules in investigations conducted in rats [65]. In addition, Mono-galactosyl di-acylglycerol was observed to have a cytotoxic impact in cancer cells of the pancreas [66].

The consumption of folates is linked to a lesser risk of cancer in healthy people [67]. Despite the fact, high concentrations may boost the prevalence as well as the progression of cancer in the body [68]. The lutein in spinach can favor reducing the hazard of cancer in the colon, particularly in young people [69]. Moreover, the World Cancer Research Fund (WCRF) indicates that it can reduce the risk of cancer of the human elementary canal, especially in the larynx, pharynx, esophagus, mouth, and stomach [70, 71].

6.5.7 ANTIOXIDANTS PROPERTIES

6.5.7.1 ANTIOXIDANT COMPOUNDS

In spinach, antioxidant compounds are recognized to be vital for oxidative stress in the body; however, the processes by which these chemicals (ingested as food or supplements) decrease the free radicals are insufficiently understood [68]. Endogenous defense antioxidant compounds can be activated or synthesized by fruits and vegetables. Spinach, for example, causes the creation of enzymes like G-glutamyl-cysteine synthetase, which is responsible for detoxifying chemicals [1].

6.5.7.2 CAROTENOIDS, VITAMINS, FLAVONOIDS, AND PHENOLIC CONSTITUENTS

In spinach, carotenoids, AA, and vitamin E are present in high quantity. Spinach comprises around 8.3% β-carotene as well as 7.3% lutein, which is an effective neuroprotector in transitory cerebral ischemia, and it decreases the levels of malonaldehyde related to cell apoptosis and lipid peroxidation [72, 73].

In spinach extracts, flavonoid and phenolic constituents found are spinacetin, quercetin, jaceidin, and patuletin, which are mostly observed in spinach leaves while they vanish during processing (especially during chopping), ferulic acid (as an anti-inflammatory compound), and caffeic acid (an antioxidant, immunomodulatory, and anti-inflammatory) [74, 75]. Further constituents are flavone and derivatives of spinacetin, spinal side, jaceidin, patuletin, and lignans [38]. Most of these are secondary metabolites, produced for plant defense against insects and pathogens, as well as environmental stress [45]. Vegetables like spinach receive much solar radiation at high altitude areas, which is linked to changes in the phenolic contents and antioxidant properties [12].

6.6 CONCLUSION

Spinach is usually grown as an annual plant and considered a valuable leafy vegetable, generally consumed after boiling and cooking, either fresh or frozen, or eaten raw as a fresh salad. It is interestingly grown in a wide range of soils with a pH ranging between 6.5 and 9.0 in the cool season. Generally, it is recommended to preserve spinach near 0°C from packaging to final selling. However, it can be ice-packed and vacuum-cooled, as well as hydrocooled. The intake of 100 g of spinach contains few calories and offers a prolific range of vitamins and minerals, containing high percentages of the RDA of magnesium, iron, and potassium, as well as folates, vitamin K, vitamin A, and AA.

Spinacia oleracea L. is conventionally used for antioxidant, antianemia, antidiabetic, anthelmintic, antihyperlipidemic, anti-inflammatory, antibacterial, antiulcer, antiepileptic, and antiviral activities. It is also consumed in the inhibition of liver and neurodegenerative as well as lung diseases. The ingestion of spinach decreases the risk of CVDs because of its AA and dietary fibers. Furthermore, it prevents cancer, apparently because of having ample nutritional antioxidants as well as an immense quantity of non-nutritional compounds like antioxidant enzymes, flavonoids, phenolic and carotenoid contents. When spinach is included in the diet, it has antioxidant and health benefits as well as advantageous effects. Finally, unusual efforts must be engrossed by consumers to the processing as well as culinary approaches that may modify or benefit to keep its nutritional and wholesome properties.

KEYWORDS

- **antioxidant activity**
- **functional food**
- **leafy vegetables**
- **nutritional value**
- **phytochemicals**
- ***Spinacia oleracea***

REFERENCES

1. Blomhoff, R. (2005). Dietary antioxidants and cardiovascular disease. *Curr. Opin. Lipidol.*, *16*, 47–54.
2. Alessa, O., Najla, S., & Murshed, R. (2017). Improvement of yield and quality of two *Spinacia oleracea* L. varieties by using different fertilizing approaches. *Physiology and Molecular Biology of Plants*, *23*(3), 693–702.
3. Yoon, Y. E., Kuppusamy, S., Cho, K. M., Kim, P. J., Kwack, Y. B., & Lee, Y. B. (2017). Influence of cold stress on contents of soluble sugars, vitamin C and free amino acids including gamma-aminobutyric acid (GABA) in spinach (*Spinacia oleracea*). *Food Chemistry*, *215*, 185–192.
4. Nemadodzi, L. E. (2015). *Growth and Development of Baby Spinach (Spinacia oleracea L.) in Response to Mineral Nutrition* (Master's dissertation). University of South Africa, Johannesburg.
5. Dicoteau, D. R. (2000). *Vegetable Crops* (pp. 221–237). Prentice Hall.
6. Fajardo, V., Alonso-Aperte, E., & Valera-Moreiras, V. (2015). Folate content in fresh-cut vegetable packed products by 96-well microtiter plate microbiological assay. *Food Chem.*, *169*, 283–288.
7. Maroto, J. (2002). *Horticulture Herbácea Especial* (5th ed.). Editorial: S.A. Mundi-Prensa, Madrid.
8. Kaur, P., Bains, K., & Kaur, H. (2012). Effect of hydrothermal treatments on free radical scavenging potential of selected green vegetables. *Indian J. Nat. Prod. Resour.*, *3*(4), 563–569.
9. Hu, J., Mou, B., & Vick, B. (2007). Genetic diversity of 38 spinach (*Spinacia oleracea* L.) germplasm accessions and 10 commercial hybrids assessed by TRAP markers. *Genetic Resources and Crop Evolution*, *54*, 1667–1674.
10. Lovett, L. (2005). The Popeye principle: Selling child health in the first nutrition crisis. *J. Health Polit. Policy Law*, *30*(5), 803–838.
11. Tiwari, A. K. (2014). Revisiting "vegetables" to combat the modern epidemic of imbalanced glucose homeostasis. *Pharmacogn. Mag.*, *10*, 207–213.
12. Zhou, K., & Yu, L. (2006). Phenolic content and antioxidant properties of commonly consumed vegetables grown in Colorado. *LWT*, *39*, 1155–1162.

13. Yawalkar, K. S. (1980). *Vegetable Crops of India* (2nd ed., p. 370). Agri Horticultural Pub. House, Nagpur.
14. Kale, P. N., & Kale, S. (1984). *Bajipala Utpadan* (Marathi). Continental Press, Pune.
15. Yadav, A., Kumari, R., Yadav, A., Mishra, J. P., Srivatva, S., & Prabha, S. (2016). Antioxidants and its functions in human body: A review. *Res. Environ. Life Sci.*, *9*(11), 1328–1331.
16. Saraswat, M., Muthenna, P., Suryanarayana, P., Petrash, J. M., & Reddy, G. B. (2008). Dietary sources of aldose reductase inhibitors: Prospects for alleviating diabetic complications. *Asia Pac. J. Clin. Nutr.*, *17*(4), 558–565.
17. Murcia, M. A., Vera, A., & García-Carmona, F. (1992). Determination by HPLC of changes in tocopherol levels in spinach after industrial processing. *J. Sci. Agric.*, *60*, 81–84.
18. Hassandokht, M. R. (2012). *Technology of Vegetable Production.* Selseleh Publication, Tehran, Iran. (In Persian).
19. Fujita, N., Ayukawa, Y., Fuke, M., Teraoka, T., Watanabe, K., Arie, T., & Komatsu, K. (2017). Rapid sex identification method of spinach (*Spinacia oleracea* L.) in the vegetative stage using loop-mediated isothermal amplification. *Planta*, *245*(1), 221–226.
20. Kaneko, K., Mizushige, T., Miyazaki, Y., Lazarus, M., Urade, Y., Yoshikawa, M., et al. (2014). δ-opioid receptor activation stimulates normal diet intake but conversely suppresses high-fat diet intake in mice. *Am. J. Physiol. Regul. Integr. Comp. Physiol.*, *306*(4), R265–R272.
21. Hardenburg, R. E., Watada, A. E., & Wang, C. Y. (1986). *The Commercial Storage of Fruits, Vegetables and Florist and Nursery Stocks* (Handbook 66, p. 69). U.S. Department of Agriculture.
22. Singh, G., Kawatra, A., & Sehgal, S. (2001). Nutritional composition of selected green leafy vegetables, herbs and carrots. *Plant Food Hum. Nutr.*, *56*, 359–364.
23. Ryall, A. L., & Lipton, W. J. (1972). *Handling, Transportation and Storage of Fruits and Vegetables* (Vol. 1: Vegetables and melons, p. 473). AVI Publishing Co. Inc., Westport, CT.
24. Ramaiyan, B., Kour, J., Nayik, G. A., Anand, N., & Alam, M. S. (2020). Spinach (*Spinacia oleracea* L.). In G. A. Nayik & A. Gull (Eds.), *Antioxidants in Vegetables and Nuts–Properties and Health Benefits*. Springer, Singapore. https://doi.org/10.1007/978-981-15-7470-2_8 (accessed on 1st January 2026).
25. Ezell, B. D., & Wilcox, M. S. (1959). Loss of vitamin C in fresh vegetables as related to wilting and temperature. *J. Agric. Food Chem.*, *7*(7), 507–517.
26. Watts, R. L., & Watts, G. S. (1940). *The Vegetable Growing Business* (p. 520). Orange Judd Pub. Co., Inc., New York.
27. Rao, C. N. (1967). True vitamin A value of some vegetables. *J. Nutr. Dietet.*, *4*(1), 10–17.
28. Thompson, H. C., & Kelly, W. C. (1957). *Vegetable Crops* (5th ed., pp. 611–617). McGraw-Hill Book Co. Inc., New York.
29. Hosotani, K., & Kuratomi, N. (1992). Effect of blanching and thawing process on the content of β-carotene and L-ascorbic acid in raw and frozen spinach. *Wakayama Daigaku Kyoikugakubu Kiyo, Shizen Kagaku*, *41*, 95–102.

30. Richter, E., & Hondke, S. (1973). Influence of blanching and preservation by air drying at different temperatures, deep freezing and freeze drying on oxalic acid of spinach. *Z. Lebensm. Unters. Forsch.*, *153*(1), 31–38.
31. Astier-Dumas, M. (1975). Changes in the content of nitrates, vitamin C, Mg and Fe during the cooking of spinach. *Ann. Nutr. Aliment.*, *29*(3), 239–246.
32. Hanson, L. P. (1975). *Commercial Processing of Vegetables* (p. 305). Noyes Data Corporation, London.
33. Jiraungkoorskul, W. (2016). Review of neuro-nutrition used as anti-Alzheimer plant, spinach, *Spinacia oleracea. Pharmacognosy Reviews*, *10*(20), 105–109.
34. Dijkstra, S. C., Neter, J. E., van Stralen, M. M., Knol, D. L., Brouwer, I. A., Huisman, M., et al. (2015). The role of perceived barriers in explaining socio-economic status differences in adherence to the fruit, vegetable and fish guidelines in older adults: A mediation study. *Public Health Nutr.*, *18*(5), 797–808.
35. McMorrow, A. M., Connaughton, R. M., Lithander, F. E., & Roche, H. M. (2015). Adipose tissue dysregulation and metabolic consequences in childhood and adolescent obesity: Potential impact of dietary fat quality. *Proc. Nutr. Soc.*, *74*(1), 67–82.
36. Chajra, H., Amstutz, B., Schweikrt, K., Auriol, D., Redziniak, G., & Lefèvre, F. (2015). Opioid receptor delta as a global modulator of skin differentiation and barrier function repair. *Int. J. Cosmet. Sci.*, *37*(4), 386–394.
37. Hitsman, B., Papandonatos, G. D., McChargue, D. E., DeMott, A., Herrera, M. J., Spring, B., et al. (2013). Past major depression and smoking cessation outcome: A systematic review and meta-analysis update. *Addiction*, *108*(2), 294–306.
38. Roberts, J. L., & Moreau, R. (2016). Functional properties of spinach (*Spinacia oleracea* L.) phytochemicals and bioactive. *Food Funct.*, *7*, 3337–3353.
39. Singh, J. P., Kaur, A., Shevkani, K., & Singh, N. (2016). Composition, bioactive compounds and antioxidant activity of common Indian fruits and vegetables. *J. Food Sci. Technol.*, *53*(11), 405–64066.
40. Yamauchi, K., Shimizu, K., & Kamiya, T. (1980). Emulsifying properties of whey protein. *J. Food Sci.*, *45*(5). https://doi.org/10.1111/j.1365-2621.1980.tb06529.x (accessed on 1st January 2026).
41. Boga, M., Haçibekiroğlu, İ., & Kolak, U. (2011). Antioxidant and anticholinesterase activities of eleven edible plants. *Pharmac. Biol.*, *49*(3), 290–295.
42. Vutharadi, S., Jolapuram, U., & Kodidhela, L. D. (2017). Nutraceutical inherent of *Spinacia oleracea* Linn methanolic leaf extract ameliorates isoproterenol-induced myocardial necrosis in male albino Wistar rats via mitigating inflammation. *Biomed. Pharmacother.*, *85*, 239–247.
43. Manayi, A., Abdollahi, M., Raman, T., Nabavi, S. F., Habtemariam, S., Daglia, M., et al. (2016). Lutein and cataract: From bench to bedside. *Crit. Rev. Biotechnology.*, *36*(5), 829–839.
44. Chan, T. Y. (2011). Vegetable-borne nitrate and nitrite and the risk of methaemoglobinaemia. *Toxicol. Lett.*, *200*(1, 2), 107, 108.
45. Simko, I., Hayes, R. J., Mou, B., & McCreight, J. D. (2014). Lettuce and spinach. *Yield Gains Major US Field Crops*, *33*(4), 53–85.

46. McDonagh, S. T. J., Wylie, L. J., Thompson, C., Vanhatalo, A., & Jones, A. M. (2019). Potential benefits of dietary nitrate ingestion in healthy and clinical populations: A brief review. *Eur. J. Sport. Sci.*, *19*(1), 1529–1534.
47. Bondonno, C. P., Yang, X., Croft, K. D., Considine, M. J., Ward, N. C., Rich, L., et al. (2012). Flavonoid-rich apples and nitrate-rich spinach augment nitric oxide status and improve endothelial function in healthy men and women: A randomized controlled trial. *Free Radic. Biol. Med.*, *52*, 95–102.
48. Hermsdorff, H. H. M., Barbosa, K. B. F., Volp, A. C. P., Puchau, B., Bressan, J., Zulet, M. A., et al. (2012). Vitamin C and fibre consumption from fruits and vegetables improves oxidative stress markers in healthy young adults. *Br. J. Nutr.*, *107*, 1119–1127.
49. Bub, A., Watlz, B., Abrahamse, L., Delincée, H., Adam, S., Wever, J., et al. (2000). Moderate intervention with carotenoid-rich vegetable products reduces lipid peroxidation in men. *J. Nutr.*, *130*(9), 2200–2206.
50. Affourtit, C., Bailey, S. J., Jones, A. M., Smallwood, M. J., & Winyard, P. G. (2015). On the mechanism by which dietary nitrate improves human skeletal muscle function. *Front. Physiol.*, *6*, 18–25.
51. Hassan, A., Elobeid, T., Kerkadi, A., Medhat, M., & Suheil, G. (2010). Glycemic index of selected carbohydrate-based foods consumed in Qatar. *Int. J. Food Sci. Nutr.*, *61*(5), 512–518.
52. Pessione, E., & Cirrincione, S. (2016). Bioactive molecules released in food by lactic acid bacteria: Encrypted peptides and biogenic amines. *Front. Microbial.*, *7*, 876–883.
53. Hernandez-Hernandez, O., Ruiz-Aceituno, L., Sanz, M. L., & Martínez-Castro, I. (2011). Determination of free inositols and other low molecular weight carbohydrates in vegetables. *J. Agric. Food Chem.*, *59*(6), 2451–2455.
54. Rahati, S., Eshraghian, M., Ebrahimi, A., & Pishva, H. (2016). Effects of spinach aqueous extract on wound healing in experimental model diabetic rats with streptozotocin. *J. Sci. Food Agric.*, *96*(7), 2337–2343.
55. Da Silva Dias, J. C., & Imai, S. (2017). Vegetable consumption and its benefits on diabetes. *J. Nutr. Ther.*, *6*(1), 110–117.
56. Shiota, A., Hada, T., Baba, T., Sato, M., Yamanaka-Okumura, H., Yamamoto, H., et al. (2010). Protective effects of glycoglycerolipids extracted from spinach on 5-fluorouracil-induced intestinal mucosal injury. *J. Med. Invest.*, *57*(3, 4), 314–320.
57. Volpe, S. L. (2013). Magnesium in disease prevention and overall health. *Adv. Nutr.*, *4*(3), 378S–383S.
58. Li, T., Lu, X., Sun, Y., & Yang, X. (2016). Effects of spinach nitrate on insulin resistance, endothelial dysfunction markers and inflammation in mice with high-fat and high-fructose consumption. *Food Nutr. Res.*, *60*, 32010.
59. Jaime, L., Vázquez, E., Fornari, T., López-Hazas, M. del C., García-Risco, M. R., et al. (2015). Extraction of functional ingredients from spinach (*Spinacia oleracea* L.) using liquid solvent and supercritical CO_2 extraction. *J. Sci. Food Agric.*, *95*, 722–729.
60. Watada, A. E., Kim, S. D., Kim, K. S., & Harris, T. C. (1987). Quality of green beans, bell peppers and spinach stored in polyethylene bags. *J. Food Sci.*, *52*(6), 16–37.
61. Bissonnette, E. Y., Proulx, L. I., Turmel, V., Drouin, R., & Purcell, M. (2004). PCT-233, a novel modulator of pro- and anti-inflammatory cytokine production. *Clin. Exp. Immunol.*, *135*, 440–447.

62. Christensen, L. P. (2009). Galactolipids as potential health-promoting compounds in vegetable foods. *Recent Pat. Food Nutr. Agric.*, *1*(1), 50–58.
63. Shivaranjani, V. L., Poornima, H., Umamaheswari, J. D., & Devi, K. L. (2014). Preliminary phytochemical screening and quantification of bioactive compounds in the leaves of spinach (*Spinacia oleracea* L.). *J. Pharm. Res.*, *8*(8), 1113–1119.
64. Sistrunk, W. A., Mohan, M. K., & Freeman, D. W. (1977). Relationship of processing methodology to quality attributes and nutritional value of spinach. *Hort Science*, *12*, 59.
65. Nyska, A., Suttie, A., Bakshi, S., Lomnitski, L., Grossman, S., Bergman, M., et al. (2003). Slowing tumorigenic progression in TRAMP mice and prostatic carcinoma cell lines using natural antioxidant from spinach, NAO—A comparative study of three antioxidants. *Toxicol. Pathol.*, *31*(1), 39–51.
66. Akasaka, H., Mizushina, Y., Yoshida, K., Ejima, Y., Mukumoto, N., Wang, T., et al. (2016). MGDG extracted from spinach enhances the cytotoxicity of radiation in pancreatic cancer cells. *Radiat. Oncol.*, *11*(1), 153–160.
67. Gedi, M. A., Briars, R., Yuseli, F., Zainol, N., Darwish, R., Salter, A. M., et al. (2017). Component analysis of nutritionally rich chloroplasts: Recovery from conventional and unconventional green plant species. *J. Food Sci. Technol.*, *54*(9), 2746–2757.
68. Weibenborn, A., Ehlers, A., Hirsch-Ernst, K. I., Lampen, A., & Niemann, B. (2017). A two-face vitamin: Folic acid—Prevention or promotion of colon cancer? *Bundesgesundheitsblatt-Gesund.*, *60*(3), 332–340.
69. Salehi, B., Tumer, T. B., Ozleyen, A., Peron, G., Dall'Acqua, S., Rajkovic, J., et al. (2019). Plants of the genus *Spinacia*: From bioactive molecules to food and phytopharmacological applications. *Trends Food Sci. Technol.*, *88*, 260–273.
70. Ozawa, Y., Nagai, N., Suzuki, M., Kurihara, T., Shinoda, H., Watanabe, M., et al. (2016). Effects of constant intake of lutein-rich spinach on macular pigment optical density: A pilot study. *Nippon Ganka Gakkai Zasshi*, *120*(1), 41–48.
71. Mills, C. E., Khatri, J., Maskell, P., Odongerel, C., & Webb, A. J. (2017). It is rocket science—Why dietary nitrate is hard to 'beet'! Part II: Further mechanisms and therapeutic potential of the nitrate–nitrite–NO pathway. *Br. J. Pharmacol.*, *83*(1), 140–151.
72. Reif, C., Arrigoni, E., Schärer, H., Nyström, L., & Hurrell, R. F. (2013). Carotenoid database of commonly eaten Swiss vegetables and their estimated contribution to carotenoid intake. *J. Food Comp. Anal.*, *29*, 64–72.
73. Ya-Xuan, S., Ting, L., Xue-Ling, D., Qiu-Sheng, Z., Bo-Di, H., & Zhao-Feng, J. (2014). Treatment with lutein provides neuroprotection in mice subjected to transient cerebral ischemia. *J. Asian Nat. Prod. Res.*, *16*(11), 1084–1093.
74. Naczk, M., & Shahidi, F. (2003). Phenolic compounds in plant foods: Chemistry and health benefits. *Nutrac Food*, *8*, 200–218.
75. Bajpai, M., Mishra, A., & Prakash, D. (2005). Antioxidant and free radical scavenging activities of some leafy vegetables. *Int. J. Food Sci. Nutr.*, *56*(7), 473–481.

CHAPTER 7

Basil (*Ocimum basilicum*)

RAOUDHA KHANFIR BEN JENANA*

Department of Horticultural Science, Higher Institute of Agronomy of Chott Mariem, University of Sousse, Chott Mariem, Sousse, Tunisia

*Corresponding author

ABSTRACT

Basil (*Ocimum basilicum* L.) is a globally significant aromatic herb, valued for its culinary uses, medicinal properties, and essential oils. Originating from South Asia, it is cultivated worldwide and contains a diverse array of bioactive compounds, including phenolic acids (e.g., rosmarinic acid), flavonoids, terpenoids, and essential oils rich in linalool, methyl chavicol, and eugenol. These constituents contribute to its potent antioxidant, anti-inflammatory, antimicrobial, antiviral, insecticidal, analgesic, anticonvulsant, hypoglycemic, and cardioprotective activities. Basil extracts and essential oils exhibit broad-spectrum efficacy against various pathogens, including bacteria, fungi, and viruses, and show potential as natural food preservatives and alternatives to synthetic agrochemicals. This chapter reviews the phytochemical profile, traditional applications, and scientifically validated health benefits of basil, highlighting its role in functional foods, pharmaceuticals, and sustainable agriculture. Further research is encouraged to optimize its cultivation, extraction, and therapeutic applications.

7.1 INTRODUCTION

Basil (*Ocimum basilicum* L.) is one of the most popular and healthy culinary herbs in the world. Originating from India and South Asia, the

Postharvest Quality Attributes of Fresh Vegetables and Grains.
Riadh Ilahy, Imen Tlili, Thouraya R'him, and Muhammad Azam (Eds.)
© 2027 Apple Academic Press, Inc. Co-published with CRC Press (Taylor & Francis)

DOI:10.1201/9781779642851-9

plant provides several positive health benefits when consumed [1, 2]. This chapter focuses on the functional properties, traditional use, antioxidant phytochemicals of fruit, seeds, peels, wastes, and byproducts, as well as on antioxidant properties, health benefits, and biological activities of basil oil in food science, especially in food flavoring, preservation, and production. It is an attempt to collect and convey information on different aspects of the plant and to highlight the need for research and development.

7.1.1 BOTANICAL NAME, COMMON NAME

Botanical name: *Ocimum basilicum* L. [3]
Common name: Basil

Kingdom	Plantae
Subkingdom	Tracheobionta
Super-Division	Spermatophyta
Division	Magnoliophyta
Class	Magnoliopsida
Subclass	Asteridae
Order	Lamiales
Family	Lamiaceae
Genus	*Ocimum*
Species	*Ocimum basilicum*

7.1.2 HISTORY

According to one of the earliest Age of Discord books on herbal medicines, De Materia Medica, basil is used as a powerful antidote to scorpion stings. Several interesting beliefs are attributed to the historical use of basil. The plant is associated with bad luck and even death in Europe [4]. According to Hindu beliefs, basil leaves allow passage to heaven if they are buried with dead bodies. At the beginning of the 1600s, it entered the culinary recipes of the English, who also used it to ward off flies but also evil spirits by hanging the leaves on their doors [5]. It wasn't until the late 18th century that basil was grown and sold in New York State cities [6].

7.1.3 PRODUCTION

Ocimum basilicum L. is a spontaneous annual plant of tropical, subtropical, and temperate zones. It is grown mainly in Asia, especially in Western

India and Pakistan, Africa, Malaysia, and the Pacific Islands. It is also grown in tropical and warm temperate regions of India and Pakistan. Above 100 herbs and shrubs of genus *Ocimum* originated from tropical and sub-tropical regions and are cultivated around the world [7]. Basil is also cultivated in Punjab and in the lower part of Pakistan [8, 9]. The plant is also cultivated in tropical Africa, Arabia, and the Himalayas as far as Nepal, Sri Lanka, and Malaysia [10, 11]. Basil is highly cultivated in Iran, Japan, China, and Turkey [11]. Basil products are commercially used all over the world for their nutritive properties and medicinal values. Sweet basil (*Ocimum basilicum*) is also an economically important herb in several Mediterranean countries. Approximately 80 ha are grown annually in Italy for fresh and processed consumption [12]. The most important companies manufacturing basil products in the world are located in India, producing leaves for flavoring purposes, seeds, and essential oils. Basil oils are also produced in China and Belgium, while the USA is producing leaves for flavors [13]. The whole world is expecting an increase in the consumption of basil and basil products. The global basil leaves market is expected to grow from 57 million USD to 62 million USD between 2021 and 2026 [14]. The demand for locally grown produce has increased due to new culinary trends that favor fresh herbs over traditional dried leaves. Recently, the production of basil in closed controlled environments such as greenhouses and vertical farms has contributed to the expansion of the market for fresh culinary basil.

7.1.4 BOTANICAL DESCRIPTION

Ocimum basilicum L. is an aromatic, branched herbaceous plant, 60 to 90 cm high. The leaves are simple opposite with a typical aroma and numerous glands which secrete strongly scented volatile oil [15]. The stems are 2.5–5 cm or more in length, oval, pointed, entire or more or less toothed lanceolate, ovate or oblong, slightly toothed. The petiole is 1.3–2.5 cm long. The blade is pointed or angular at the extremity. The whorled inflorescence consists of white or pale purple flowers. The latter contains five sepals fused into a 2-lipped calyx. The fruit is an ellipsoid, black nutlet. Ovary is superior and there is a 2-carpellary, 4-locular and a 4-partite fruit of four achenes. The bracts are stalked, shorter than the calyx, ovate and acute [16]. Calyx is five mm long, enlarging in fruit and very shortly pedicelled. The upper lip is shorter than the lower lip containing two central teeth [5]. Corolla is white or purplish pink and 8–13 mm long, glabrous or variously pubescent. Nutlets are about two mm long, ellipsoid, black, and pitted [17].

7.1.5 TRADITIONAL USE

More than 150 herbaceous and shrubby species belong to the genus *Ocimum* [18, 19]. *O. basilicum* is used all over the world in the fields of catering and in Mediterranean cuisine, but also in the pharmaceutical industry and in the field of aromatherapy and perfumery [20–22]. *O. basilicum* infusions are widely used in Mediterranean regions such as Eastern Morocco to help reduce plasma lipids [23]. *O. basilicum* is used as a traditional medicine to prevent and treat diabetics in Turkey, and it is also used for treating pimples on the face. In India [24], it is used as a folk medicine for the treatment of aches and pains, in Bulgaria [25], and as a sedative in Spain [26]. It has also been recommended for the treatment of headaches, coughs, infections of the upper respiratory tract, kidney malfunction, and to eliminate toxins [27, 28].

7.2 ANTIOXIDANT PHYTOCHEMICALS OF FRUIT, SEEDS, PEELS, WASTES, AND ITS BYPRODUCTS

According to Luca et al. [29], the byproducts of the hydrodistillation of basil can be regarded as rich sources of bioactive metabolites, mainly phenolic compounds. Kačániová [30] identified a total of 28 compounds in *O. basilicum* essential oils, which accounted for 99.8% of the total volatiles [30]. The obtained data revealed that the main component of the *O. basilicum* essential oils was methyl chavicol (88.6%), followed by 1,8-cineole (4.2%) and α-trans-bergamotene (1.7%). In another study on the essential oil composition and biological activities of two varieties of *O. basilicum* (Fino Verde variety and Genovese varieties), Guedri Mkaddem et al. (2022) found that both varieties of *O. basilicum* are rich in essential oils and phenolic acids, which is related to the strong antioxidant and antifungal properties [31].

Previous research performed on *Ocimum basilicum* L. phytochemical plant has identified several active compounds. Vieira et al. [32] showed that *O. basilicum* flowers are richer in EO compounds than leaves. Khare [33] established that the green leaves contain a high concentration of vitamins, minerals, and oils. Bihari et al. [34] found that the phytochemical screening of *O. basilicum* revealed the presence of glycoside, gums, mucilage, proteins, amino acids, tannins, phenolic compounds, triterpenoids, steroids, sterols, saponins, flavones, and flavonoids. A total of 29 compounds were identified in basil essential oil by Hussain et al. [35], and among which, linalool is the main constituent. On the other hand, Bilal et al. [5] cited rosmarinic acid as the predominant phenolic compound. Another study carried out by Javanmardi et al. [36] on the phenolic acids of 23 accessions of basil harvested in Iran, revealed that rosmarinic acid, lithospermic acid B, vanillic acid, p-coumaric acid, hydroxybenzoic acid, syringic acid, ferulic acid, protocatechuic acid, caffeic acid, and gentisic acid were present in the flower and leaf tissues of several basil accessions. Oliveira et al. [37] isolated a number of phytochemical compounds from the *O. basilicum* essential oil and pointed out that 1.8-Cineole, linalool, and geraniol were the main constituents. According to Rubab et al. [15], flavones apogenin, such as luteolin and chrysoeriol, are also present in *O. basilicum*. In addition to these compounds, 6-hydroxy and hydroxyl-flavones in glycosidic combination and lipophilic flavones, such as -trihydroxy- -dimethoxy-flavone, have also been detected.

Rubab et al. [15] also reported that the *Ocimum* genus produced a high quantity of essential oils. Earlier, Shafiq et al. [38] isolated and identified caryophyllene, carvacrol, methyl eugenol, and eugenol, alcohols such as linalool, a minor amount of cineole, sesquiterpenes, and d-terpene from the essential oil of *O. santum*. Keita et al. [39] found that leaves provided basil camphor. Earlier, Derek et al. [40] and Shafiq et al. [38] established that basil leaves contain volatile oil and terpenes.

7.3 ANTIOXIDANT PROPERTIES

It has been established by several researchers that *Ocimum basilicum*, as well as its byproducts, contain several active antioxidant compounds [35, 40–44]. According to Phippen and Simon [45], the antioxidant property has been mainly ascribed to the polyphenoid rosmarinic acid, which is a derivative of cinnamic acid. Solvent extracts of *O. basilicum* have been demonstrated to exhibit a significant effect at the cellular level. One of the typical features of antioxidant actions is the activation of catalase in red cells [46]. In another study, Niture et al. [47] reported that *O. basilicum* extracts increased the level of

O-6-methylguanine-DNA-methyltransferase (MGMT), which consequently increased DNA repair protein activity. As established by Bravo et al. [23], the ethanolic extract of basil exerts a protective effect against oxidative DNA damage, with a decrease in cholesterol synthesis and lipid accumulation in human macrophages. The investigation of Ramesh and Satakopan [44] highlighted the higher antioxidant activity of basil compared to other species of the genus *Ocimum*, such as *Ocimum sanctum*. This result is in accordance with a previous study carried out by Meera et al. [48], who noticed that the ethanolic extract of basil showed powerful antioxidant effects. A similar result was obtained by Hussain et al. [35] while investigating the essential oil from the aerial parts of basil, which exhibited good antioxidant activity.

7.4 HEALTH BENEFITS AND BIOLOGICAL ACTIVITIES

Basil is a widespread herb in many households with a broad spectrum of healing properties. Representing no risk to human health, the components derived from basil oil are generally recognized as safe (GRAS) type by the United States Food and Drug Administration and the European Commission. However, due to the wide variation in the chemical composition of basil essential oil and basil extracts, the accepted daily intake of three major components (linalool, estragole, and eugenol) in basil oil is still subject to regulatory restrictions. Many studies investigated antifungal, antibacterial, and antioxidant activities of essential oils from *Ocimum* spp [49]. Different parts of basil are often used in traditional medicine, especially leaves, seeds, and roots. Its different parts are recommended for their euphoric, expectorant, antiperiodic, and emmenagogue effects. The leaves, being fragrant, are used for their aromatic action, as antiseptics, and can be applied as a paste to inflammations. They are also used as an expectorant. A leaf decoction is recommended against gastric and hepatic disorders, but also against bronchitis and cough. It also acts as a diuretic. The seeds are also recommended against uro-genital affections. The seed oil is used for the treatment of syphilis, otitis, and otorrhea, while the fragrant oil of basil leaves and seeds are used in perfumes and toiletries. A root decoction is recommended against malaria. Leaf extract can be applied as drops to relieve earache [15]. Zhan et al. [50] demonstrated that basil is used for the treatment of various illnesses, from the treatment of cough, headache, worms, and diarrheal symptoms. In addition, basil polysaccharides have been used to treat cancer [51]. According to Brandão et al. [52], basil might have interesting biological properties such as antimicrobial, antifungal, larvicide, insecticide, antiparasitic, antioxidant, and antineoplastic properties.

7.4.1 ANALGESIC ACTIVITY

Choudhury et al. [53] found that the methanolic extract of basil exhibited analgesic activity at 200 mg/kg concentration, which was well comparable with a standard drug such as aspirin.

7.4.2 ANTI-INFLAMMATORY ACTIVITY

According to Benedec et al. [54], basil tincture has important anti-inflammatory effects on bone marrow acute phase response and a reduced one on NO synthesis. The anti-inflammatory effect of basil crude methanolic extract was also described by Bilal et al. [5].

7.4.3 ANTI-CONVULSANT ACTIVITY

As reported by the World Health Organization, about 450 million people in the world have endured mental, neurological, or behavioral problems at some time in their lives [55]. Large-scale research on plants, their derivatives, and their active photochemical compounds has taken place in recent years that could furnish some new alternative treatments and therapeutic uses for various diseases of the central nervous system (CNS) in human beings. Many species of aromatic plants are medicinally used, due to the occurrence of essential oils and phytochemical compounds; some of them possessed certain CNS properties, including anti-epileptic action with a history of usage in folk medicines. Almeida et al. [56] demonstrated that the *O. basilicum* essential oil produced a substantial increase in a dose-dependent reaction time of convulsion of mice. They found that it also reduced lethal effects in response to different stimulants of convulsions used in another study [57]. It was further indicated that the essential oil increased the latent period for the growing of picrotoxin PIC tests. Moreover, it was also observed by these researchers that the essential oil did not intervene with the convulsions induced by strychnine [57, 58].

7.4.4 HYPOGLYCEMIC AND HYPOLIPIDEMIC EFFECTS

Mohan et al. [59] and Hussain et al. [60] indicated the hypoglycemic and hypolipidemic effects of aqueous extracts of various parts of *O. basilicum* on rats. They found that serum lipid-profile and lipid peroxidation products were significantly reduced in blood glucose. They also found that aqueous extract reduced lipid peroxidation (LPO) formation, thiobarbituric acid reactive substances, and increased antioxidant enzymes such as superoxide dismutase, catalase, glutathione peroxidase, and glutathione transferases.

The hypoglycemic effect of the aqueous extract of *Ocimum basilicum* was investigated by Zeggwagh et al. [61]. This team concluded that the aqueous extract exhibits potent anti-hyperglycemic activity in diabetic rats without affecting basal plasma insulin concentrations.

7.4.5 DERMATOLOGIC EFFECTS

According to a study conducted by Balambal et al. [62] on the effects of extracts of different plant parts of basil in humans against acne vulgaris, basil extracts showed fairly high efficacy against acne vulgaris, but the mechanism of their action was not clear.

7.4.6 CARDIO-STIMULANT ACTIVITY

Muralidharan and Dhananjayan [63] tested the alcoholic and aqueous extracts of basil on frog heart and found that the extracts exhibited a heart tonic effect.

7.4.7 ANTIMICROBIAL ACTIVITY

Investigations on the antimicrobial effects of basil have involved different parts of the plant as well as its essential oil, and have been the subject of several experimental laboratory studies and concluded with significant results [35, 64–67]. In India, basil is used for dental ailments due to its antimicrobial effects, but the mechanism of its action is unclear. Furthermore, other studies on the essential oil and its individual phytochemical constituents have shown strong *in vitro* and *in vivo* inhibiting properties of microbes responsible for various diseases and infections in animals and in human beings [35, 68, 69]. Many serious systemic diseases, as well as localized infections, could be efficiently treated with organic solvents and aqueous extracts from various plant parts of basil and its oil [35].

7.4.8 ANTIBACTERIAL ACTIVITY

Many authors have been interested in the antibacterial potential of various parts of *O. basilicum* [69–71]. *O. basilicum* showed significant antibacterial activity against *Salmonella* spp., *Escherichia coli, Campylobacter jejuni*, and *Clostridium perfringens* [71]. On the other hand, Opalchenova and Obreshkova [25] reported that the volatile oil of the aerial parts of *O. basilicum* exhibited significant activity against different species of the genera *Staphylococcus, Enterococcus,* and *Pseudomonas* [68]. The antibacterial activities of various parts and the essential oil of *O. basilicum* have also been reported by many other researchers, including Tomar et al. [68], Gupta et al. [72], Durga et al. [73], and Adigüzel et al. [69]. According to a study by Srivastava et al. [74], sweet basil oil inhibited the Gram-negative bacteria *Salmonella Typhi* and *E. coli* and the Gram-positive bacteria *S. aureus* and *Bacillus subtilis*. Basil oil also showed mild activity against *Streptococcus pneumoniae, Haemophilus influenzae, C. albicans,* and *A. niger*. Among the findings reported by Ouibrahim et al. [75], basil oil exhibited bacteriostatic effects against 20 Gram-positive and Gram-negative bacterial strains.

7.4.9 ANTIFUNGAL ACTIVITY

According to Amer et al. [76], Valeria et al. [77], and Chermette et al. [78], some fungi called dermatophytes often constitute the cause of several infections of the skin, hair, nails, and subcutaneous tissues in animals and humans. Few antifungal agents, derived from plants and their natural products, are licensed for use in veterinary practice or for the treatment of human beings in various countries around the world [79]. The study of the antifungal effect of extracts, phytochemicals, and oil of basil revealed diverging results depending on the type of extracts. Indeed, higher concentrations of the solvent extracts of the aerial parts, root extracts, basil oil, and their isolated phytochemicals are necessary to inhibit the growth of wide ranges of test organisms compared to that of standard antiseptics. Furthermore, the ethanol extract of the aerial parts of the plant and its isolated phytochemicals exhibited a higher degree of antifungal activity than the root extracts. Many phytopathogenic fungi have been inhibited under the effect of essential oils [43, 80–85]. On the other hand, Jakowienko et al. [86] studied the sensitivity of seven fungi isolated from plants to basil oil. They concluded that *Eurotium chevalieri* was the most sensitive, whereas *Aspergillus flavus* was the most resistant. Moreover, Kocić-Tanackov et al. [87] demonstrated that sweet basil oil had an inhibiting effect against the following fungi: *Penicillium aurantiogriseum*, *Penicillium glabrum*, *Penicillium chrysogenum*, and *Penicillium brevicompactum*. Another investigation conducted by Soliman and Badeaa [88] showed different reactions of the studied fungi against the basil extract at different concentrations. Indeed, *Penicillium chrysogenum* was completely inhibited by the application of basil extract at 1.5%. Whereas basil oil at 0.6% v/v inhibited 100% of *Fusarium moniliforme* and *Pyricularia oryzae* but inhibited only 50% of *Fusarium proliferratum*, and was not effective against *Rhizoctonia solani*. According to Soković et al. [89], *Absidia glauca*, *Aspergillus nidulans*, *Aspergillus niger*, *Colletotrichum capsici*, *F. monoliforme*, *Pestalotia psidii*, and *Rhizopus nadssus* were completely inhibited by Eugenol. By testing the Rosmarinic acid in sweet basil, Harsh et al. [90] found that it shows antimicrobial activity against a range of soilborne microorganisms, and it is particularly inhibitory of *Pseudomonas aeruginosa*.

7.4.10 ANTIVIRAL ACTIVITY

Many studies have been done on the behavior of basil (*O. basilicum*) for its antiviral activities. The results concerning antiviral effects of these compounds are divergent. Indeed, the plant parts of *O. basilicum* showed specific inhibitory actions against HIV-1-induced cytopathogenicity in MT-4 cells [85]. Furthermore, according to other studies conducted by Chiang et al. [43], the aqueous and ethanolic extracts of the aerial parts of *O. basilicum*, as well as its phytochemicals, such as apigenin, linalool, and ursolic acid, revealed a wide range of antiviral activity.

7.4.11 INSECTICIDAL EFFECTS

According to Brandão et al. [52], conventional insecticides can be substituted by the essential oils of basil, thanks to their powerful insecticidal activities and their respect for the environment. In a study on the mosquito repellent efficacy of essential oils carried out in Turkey, Erler et al. [91] found that basil essential oil was highly effective against *Culex pipiens*, which was a widespread pest mosquito in most urban and suburban areas of the country. The insecticidal activity of essential oils has been tested at various stages of insect development by several researchers who have deduced that the larvicidal, pupicidal, adulticidal, and repellent efficacy of phytochemicals derived from plant sources could make them alternative agents for the control of mosquitoes [92–96]. Flies and mosquitoes being vectors of diseases and pests, the essential oils of basil were studied to evaluate their repellent and larvicidal activity against these insects. According to Mahmoud et al. [97], the compounds d-limonene, myrcene, and thymol are responsible for the repellent properties, while eugenol and methyl-chavicol have proven larvicidal activity. The essential oil extracted from the *O. basilicum* leaves can be used as a safe, effective, and naturally available larvicidal agent against mosquitoes [98]. On the other hand, the insecticidal activity of the essential oils of *O. basilicum* was tested against Acanthoscelides, a pest of common bean (*Phaseolus vulgaris*), and local application of the essential oil hindered the development of *A. obtectus* [99].

7.5 CONCLUSION AND PROSPECTS

This chapter contributed to summarize most of the research work on the application and use of basil extracts for therapeutic and phytosanitary purposes in general. This chapter has mainly focused on the valuation and uses of the different parts of *O. basilicum* thanks to their spectacular virtues. Moreover, it also provides those chemical constituents which are very useful for sustainable development of medicines and human consumption. In recent years, the appearance of new diseases and the increase in chronic, degenerative, and environmental pollution-related diseases, lifestyle, and stress are partly the result of the use of the traditional system of drugs. Basil has been used in traditional medicine to treat a number of ailments. The effectiveness of basil may be the result of a synergistic effect of its phytochemical constituents. Several studies have also underscored that basil EOs could be used as the candidate for the development of natural antibiotics and disinfectants to control pathogenic microbes in crop fruit and vegetables. Thus, the basil EOs could be used as a naturally sourced food preservative to reduce and substitute or avoid the use of chemical preservatives. All these results and others encourage and prompt larger-scale, in-field research on basil EOs application, in particular for organic farming.

KEYWORDS

- **antioxidant properties**
- **essential oils**
- **natural food preservatives**
- ***Ocimum basilicum***
- **phytochemicals**
- **traditional medicine**

REFERENCES

1. Kopsell, D. A., Kopsell, D. E., & Curran-Celentano, J. (2005). Carotenoid and chlorophyll pigments in sweet basil grown in the field and greenhouse. *Hort Science, 40*(4), 1230–1233.
2. Soran, M. L., Cobzac, S. C., Varodi, C., Lung, I., Surducan, E., & Surducan, V. (2009). The extraction and chromatographic determination of the essential oils from *Ocimum basilicum* L. by different techniques. *Journal of Physics: Conference Series, 182*, 1–5.
3. Linnaeus, C. (1753). Ocimum. In *Species Plantarum* (ed. 1, pp. 597–598). Holmiae [Stockholm]: Laurentii Salvii.
4. Dymock, W., Warden, C. J. H., & Hooper, D. (2005). *Pharmacographica Indica. A History of the Principal Drugs of Vegetable Origin* (Vol. III, pp. 82–85). New Delhi: Shrishti Book Distributors.
5. Bilal, A., Jahan, N., Ahmed, A., Bilal, S. N., Habib, S., & Hajra, S. (2012). Phytochemical and pharmacological studies on *Ocimum basilicum* Linn – A review. *International Journal of Current Research and Review, 4*(23).
6. Sperry, J. (2009). *Garden Seeds*. New York: Daily Gazette. NewsBank InfoWeb America's Historical Newspapers.
7. Libera, Z., Carovic-Stanko, K., Politeo, C., Strikic, F., Kolak, I., Milos, M., & Satovic, Z. (2011). Chemical characterization and genetic relationships among *Ocimum basilicum* L. cultivars. *Chemistry & Biodiversity, 8*(11), 1978–1989.
8. Sabnis, T. S. (1986). *The Flora of Punjab Plains and the Associated Hill Region* (p. 352). Jodhpur: Scientific Publishers.
9. Kirtikar, K. P., & Basu, B. D. (1985). *Indian Medicinal Plants* (2nd ed., Vol. 1, p. 1961). India: International Book Distributor.
10. Tanker, N., Islisulu, F., Koyuncu, M., & Cosscum, M. (1986). Phyto-chemical screening of plants from the Ermenek-Mut Gulnar (Turkey) area III Labiatae. *International Journal of Crude Drug Research, 24*, 177–182.
11. Alves-Silva, J. M., Dias dos Santos, S. M., Pintado, M. E., Perez-Alvarez, J. A., Fernandez-Lopez, J., & Viuda Martos, M. (2013). Chemical composition and *in vitro*

antimicrobial, antifungal and antioxidant properties of essential oils obtained from some herbs widely used in Portugal. *Food Control, 32*, 371–378.

12. Kwon, D. Y., Kim, Y. B., Kim, J. K., & Park, S. U. (2021). Production of rosmarinic acid and correlated gene expression in hairy root cultures of green and purple basil (*Ocimum basilicum* L.). *Preparative Biochemistry & Biotechnology, 51*(1), 35–43.
13. Purushothaman, B., Prasanna Srinivasan, R., Suganthi, P., Ranganathan, B., Gimbun, J., & Shanmugam, K. (2018). A comprehensive review on *Ocimum basilicum*. *Journal of Natural Remedies, 18*(3), 71–85.
14. Sipos, L., Balázs, L., Székely, G., Jung, A., Sárosi, S., Radácsi, P., & Csambalik, L. (2021). Optimization of basil (*Ocimum basilicum* L.) production in LED light environments–a review. *Scientia Horticulture, 289*, 110486.
15. Rubab, S., Hussain, I., Khan, B. A., Unar, A. A., Abbas, K. A., Khichi, Z. H., & Khan, H. (2017). Biomedical description of *Ocimum basilicum* L. *Journal of Islamic International Medical College (JIIMC), 12*(1), 59–67.
16. Jayaweera, D. M. A. (1981). *Medicinal Plants, (Indigenous and Exotic) Used in Ceylon* (Part III, pp. 101–103). Colombo: The National Science Foundation of Sri Lanka.
17. Kirtikar, K. R., & Basu, B. D. (2003). *Indian Medicinal Plants with Illustrations* (2nd ed., Vol. VIII, pp. 2701–2705). Uttaranchal: Oriental Enterprises.
18. Lachowicz, K. J., Jones, J. P., Briggs, D. R., Bienvenu, F. E., Palmer, M. V., Mishra, V., & Hunter, M. M. (1997). Characteristics of plants and plant extracts from five varieties of basil (*Ocimum basilicum* L.) grown in Australia. *Journal of Agricultural and Food Chemistry, 45*(7), 2660–2665.
19. Khazaei, N., Esmaeili, M., Djomeh, Z. E., Ghasemlou, M., & Jouki, M. (1997). Characterization of new biodegradable edible film made from basil seed (*Ocimum basilicum* L.) gum. *Carbohydrate Polymers, 102*, 199–206.
20. Marotti, M., Piccaglia, R., & Giovanelli, E. (1996). Differences in essential oil composition of basil (*Ocimum basilicum* L.) Italian cultivars related to morphological characteristics. *Journal of Agricultural and Food Chemistry, 44*, 3926–3929.
21. Zheljazkov, V. D., Cantrell, C. L., Tekwani, B., & Khan, S. I. (2008). Content, composition, and bioactivity of the essential oils of three basil genotypes as a function of harvesting. *Journal of Agricultural and Food Chemistry, 56*, 380–385.
22. Tohti, I., Tursun, M., Umar, A., Turddi, S., Imin, H., & Nicholas, M. (2006). Aqueous extracts of *Ocimum basilicum* L. (sweet basil) decrease platelet aggregation induced by ADP and thrombin *in vivo* arteriovenous shunt thrombosis *in vivo*. *Thrombosis Research, 118*, 733–739.
23. Bravo, E., Amrani, S., Ariz, M., Harnafi, H., & Napolitano, M. (2008). *Ocimum basilicum* ethanolic extract decreases cholesterol synthesis and lipid accumulation in human macrophages. *Fitoterapia, 79*, 515–523.
24. Tsai, K. D., Lin, B. R., Perng, D. S., Wei, J. C., Yu, Y. W., & Cherng, J. M. (2011). Immunomodulatory effects of aqueous extract of *Ocimum basilicum* (Linn.) and some of its constituents on human immune cells. *Journal of Medicinal Plants Research, 5*, 1873–1883.
25. Opalchenova, G., & Obreshkova, D. (2003). Comparative studies on the activity of basil—an essential oil from *Ocimum basilicum* L. against multidrug-resistant clinical

isolates of the genera *Staphylococcus* and *Pseudomonas* by using different test methods. *Journal of Microbiological Methods, 54*, 105–110.
26. Freire, M. M., Marques, O. M., & Costa, M. (2006). Effects of seasonal variation on the central nervous system activity of *Ocimum gratissimum* L. essential oil. *Journal of Ethnopharmacology, 105*, 161–166.
27. Gupta, S. K., Prakash, J., & Srivastav, S. (2002). Validation of claim of Tulsi, *Ocimum sanctum* Linn as a medicinal plant. *Journal of Experimental Biology, 40*, 765–773.
28. Harborne, J. B. (1973). *Phytochemical Methods*. London: Chapman and Hall Ltd.
29. Luca, S. V., Zengin, G., Sinan, K. I., Skalicka-Woźniak, K., & Trifan, A. (2023). Post-distillation byproducts of aromatic plants from *Lamiaceae* family as rich sources of antioxidants and enzyme inhibitors. *Antioxidants, 12*, 210.
30. Kačániová, M., Galovičová, L., Borotová, P., Vukovic, N. L., Vukic, M., Kunová, S., & Kowalczewski, P. Ł. (2022). Assessment of *Ocimum basilicum* essential oil anti-insect activity and antimicrobial protection in fruit and vegetable quality. *Plants, 11*(8), 1030.
31. Guedri Mkaddem, M., Zrig, A., Ben Abdallah, M., Romdhane, M., Okla, M. K., Al-Hashimi, A., et al. (2022). Essential oil composition and antioxidant and antifungal activities of two varieties of *Ocimum basilicum* L. (*Lamiaceae*) at two phenological stages. *Agronomy, 12*, 825.
32. Vieira, R. F., Grayer, R. J., Paton, A., & Simon, J. E. (2001). Genetic diversity of *Ocimum gratissimum* L. based on volatile oil constituents, flavonoids and RAPD markers. *Biochemical Systematics and Ecology, 29*, 287–304.
33. Khare, C. P. (2008). *Indian Medicinal Plants: An Illustrated Dictionary*. Springer Science & Business Media.
34. Bihari, G. C., Manaswini, B., Prabhat, J., & Kumar, T. S. (2011). Pharmacognostical and phytochemical investigation of various tulsi plants available in south-eastern Odisha. *International Journal of Research in Pharmaceutical and Biomedical Sciences, 2*(2), 605–610.
35. Hussain, A. I., Anwar, F., Sherazi, S. T. H., & Przybylski, R. (2008). Chemical composition, antioxidant and antimicrobial activities of basil (*Ocimum basilicum*) essential oils depend on seasonal variations. *Food Chemistry, 108*(3), 986–995.
36. Javanmardi, J., Khalighi, A., Kashi, A., Bais, H. P., & Vivanco, J. M. (2002). Chemical characterization of basil (*Ocimum basilicum* L.) found in local accessions and used in traditional medicines in Iran. *Journal of Agricultural and Food Chemistry, 50*, 5878–5883.
37. Oliveira, J. S., Porto, L. A., Estevam, C. S., Siqueira, R. S., Alves, P. B., & Niculau, E. S. (2009). Phytochemical screening and anticonvulsant property of *Ocimum basilicum* L. leaf essential oil. *Boletín Latinoamericano y del Caribe de Plantas Medicinales y Aromáticas, 8*, 195–202.
38. Shafiq, M. M., Rafoq, A., Shahid, M., Khan, S. A., & Bhatty, M. K. (1986). Essential oils of species of *Labiatae*. Part II. Studies on the essential oil of *Ocimum sanctum*. *Pakistan Journal of Industrial Research, 29*, 111–112.
39. Keita, S. M., Vincent, C., Schmitt, J. P., & Bélanger, A. (2000). Essential oil composition of *Ocimum basilicum* L., *O. gratissimum* L. and *O. suave* L. in the Republic of Guinea. *Flavor and Fragrance Journal, 15*, 339–341.

40. Derek, V. B., Bilyard, H. J., & Brown, G. D. (1989). Stars of caffeic acid in several genera of *Labiatae*. *Phytochemistry, 28*, 2109–2114.
41. Marinova, E. M., & Yanishlieva, N. V. (1997). Antioxidative activity of extracts from selected species of the family of *Lamiaceae* in sunflower oil. *Food Chemistry, 58*, 245.
42. Politeo, O., Jukić, M., & Miloš, M. (2007). Chemical composition and antioxidant capacity of free volatile aglycones from basil (*Ocimum basilicum* L.) compared with its essential oil. *Food Chemistry, 101*, 379–385.
43. Chiang, L. C., Ng, L. T., Cheng, P. W., Chiang, W., & Lin, C. C. (2005). Antiviral activities of extracts and selected pure constituents of *Ocimum basilicum*. *Clinical and Experimental Pharmacology and Physiology, 32*(10), 811–816.
44. Ramesh, B., & Satakopan, V. N. (2010). *In vitro* antioxidant activities of *Ocimum* species; *Ocimum basilicum* L. and *Ocimum sanctum*. *Journal of Cell and Tissue Research, 10*, 2145–2150.
45. Phippen, W. B., & Simon, J. E. (1998). Anthocyanins in basil (*Ocimum basilicum* L.). *Journal of Agricultural and Food Chemistry, 46*, 1734–1738.
46. Siurin, S. A. (1997). Effects of essential oil on lipid peroxidation and lipid metabolism in patients with chronic bronchitis. *Klinicheskaia Meditsina (Moskva), 75*, 43–45.
47. Niture, S. K., Rao, U. S., & Srivenugopal, K. S. (2006). Chemo preventative strategies targeting the MGMT repair protein; augmented expression in human lymphocytes and tumor cells by ethanolic and aqueous extracts of several Indian medicinal plants. *International Journal of Oncology, 29*, 1269–1278.
48. Meera, R., Devi, P., Kameswari, B., Madhumitha, B., & Merlin, N. J. (2009). Antioxidant and hepatoprotective activities of *Ocimum basilicum* Linn. and *Trigonella foenum-graecum* Linn. against H_2O_2 and CCl_4 induced hepatotoxicity in goat liver. *Indian Journal of Experimental Biology, 47*(7), 584–590.
49. Li, Q., Chiou, L., & Chang. (2016). Essential oils in food preservation, flavor and safety. In V. R. Preedy (Ed.), *Basil (Ocimum basilicum L.) oils* (Chapter 25, pp. 231–238). Academic Press.
50. Zhan, Y., An, X., Wang, S., Sun, M., & Zhou, H. (2020). Basil polysaccharides: A review on extraction, bioactivities and pharmacological applications. *Bioorganic & Medicinal Chemistry, 28*(1), 115–179.
51. Twilley, D., Rademan, S., & Lall, N. (2018). Are medicinal plants effective for skin cancer? In *Medicinal Plants for Holistic Health and Well-Being* (pp. 13–75). Academic Press.
52. Brandão, L. B., Santos, L. L., Martins, R. L., Rodrigues, A. B. L., da Costa, A. L. P., Faustino, C. G., & de Almeida, S. (2022). The potential effects of species *Ocimum basilicum* L. on health: A review of the chemical and biological studies. *Pharmacognosy Reviews, 16*(31), 23.
53. Choudhury, G. B., Prabhat, K. J., Nayak, B. S., Panda, S. K., & Tripathy, S. K. (2010). Phytochemical investigation and evaluation of analgesic activity of leafy extracts of various (*Ocimum tulsi*) species. *The Indian Pharmacist, 8*(12), 67–70.
54. Benedec, D., Pârvu, A. E., Oniga, I., Toiu, A., & Tiperciuc, B. (2007). Effects of *Ocimum basilicum* L. extract on experimental acute inflammation. *Revista Medico-Chirurgicală a Societății de Medici și Naturaliști din Iași, 111*(4), 1065–1069.

55. Sayers, J. (2001). The world health report—Mental health: New understanding, new hope. *Bulletin of the World Health Organization, 79*(11), 1085.
56. Almeida, R. N., Motta, S. C., & Leite, J. R. (2003). Óleos essenciais com propriedades anticonvulsivantes. *Boletín Latinoamericano y del Caribe de Plantas Medicinales y Aromáticas, 2*, 3–6.
57. Almeida, R. N., Agra, M. F., Maior, F. N. S., & Pergentino de Sousa, D. (2011). Essential oils and their constituents: Anticonvulsant activity. *Molecules, 16*, 2726–2742.
58. Ismail, M. (2006). Central properties and chemical composition of *Ocimum basilicum* L. essential oil. *Pharmaceutical Biology, 44*, 619–626.
59. Mohan, L., Amberkar, M. V., & Kumari, M. (2011). *Ocimum sanctum* L. (Tulsi)–An overview: A review article. *International Journal of Pharmaceutical Sciences Review and Research, 7*, 51–53.
60. Hussain, E. H., Jamil, K., & Rao, M. (2001). Hypoglycemic, hypolipidemic and antioxidant properties of Tulsi (*Ocimum sanctum* Linn) on streptozotocin-induced diabetes in rats. *Indian Journal of Clinical Biochemistry, 16*(2), 190–194.
61. Zeggwagh, N. A., Sulpice, T., & Eddouks, M. (2007). Antihyperglycemic and hypolipidemic effect of *Ocimum basilicum* aqueous extract in diabetic rats. *American Journal of Pharmacology and Toxicology, 2*(3), 123–129.
62. Balambal, R., Thiruvengadam, K. V., Kameswaran, L., Janaki, V. R., & Thambiah, A. S. (1985). *Ocimum basilicum* L. in acne vulgaris–A controlled comparison with a standard regime. *Journal of the Association of Physicians of India, 33*, 507–508.
63. Muralidharan, A., & Dhananjayan, R. (2004). Cardiac stimulant activity of *Ocimum basilicum* Linn. extracts. *Indian Journal of Pharmacology, 36*, 163–166.
64. Patel, V. K., & Bhatt, V. H. (1988). Folklore therapeutic indigenous plants in periodontal disorders in India (review, experimental and clinical approach). *International Journal of Clinical Pharmacology, Therapy and Toxicology, 26*, 176–184.
65. Chand, S., Lusunzi, I., Veal, D. A., Williams, L. R., & Karuso, P. (1994). Rapid screening of the antimicrobial activity of extracts and natural products. *The Journal of Antibiotics, 47*(11), 1295–1304.
66. Navarro, V., Villarreal, M. L., Rojas, G., & Lozoya, X. (1996). Antimicrobial evaluation of some plants used in Mexican traditional medicine for the treatment of infectious diseases. *Journal of Ethnopharmacology, 53*(3), 143–147.
67. Yano, Y., Satomi, M., & Oikawa, H. (2006). Antimicrobial effect of spices and herbs on *Vibrio parahaemolyticus*. *International Journal of Food Microbiology, 111*, 6–11.
68. Tomar, U. S., Daniel, V., Shrivastava, K., & Panwar-Mangal, S. (2010). Comparative evaluation and antimicrobial activity of *Ocimum basilicum* L. (*Labiatae*). *Journal of Global Pharma Technology, 2*, 49–53.
69. Adigüzel, A., Güllüce, M., Şengül, M., Öğütcü, H., Şahin, F., & Karaman, İ. (2005). Antimicrobial effects of *Ocimum basilicum* (*Labiatae*) extract. *Turkish Journal of Biology, 29*(3), 155–160.
70. Bozin, B., Mimica-Dukić, N., Simin, N., & Anackov, G. (2006). Characterization of the volatile composition of essential oil of some *Lamiaceae* species and the antimicrobial and antioxidant activities of the entire oils. *Journal of Agricultural and Food Chemistry, 54*, 1822–1828.

71. Wannissorn, B., Jarikasem, S., Siriwangchai, T., & Thubthimthed, S. (2005). Antibacterial properties of essential oils from Thai medicinal plants. *Fitoterapia, 76*, 233–236.
72. Gupta, P. C., Batra, R., Chauhan, A., Goyal, P., & Kaushik, P. (2009). Antibacterial activity and TLC bioautography of *Ocimum basilicum* L. against pathogenic bacteria. *Journal of Pharmaceutical Research, 2*, 407–409.
73. Durga, R. K., Karthikumar, S., & Jegatheesan, K. (2010). Isolation of potential antibacterial and antioxidant compounds from *Acalypha indica* and *Ocimum basilicum*. *African Journal of Plant Science, 4*(5), 163–166.
74. Srivastava, S., Cahill, D. M., Conlan, X. A., & Adholeya, A. (2014). A novel *in vitro* whole plant system for analysis of polyphenolics and their antioxidant potential in cultivars of *Ocimum basilicum*. *Journal of Agricultural and Food Chemistry, 62*(41), 10064–10075.
75. Ouibrahim, A., Tlili-Ait-Kaki, Y., Bennadja, S., Amrouni, S., Djahoudi, A. G., & Djebar, M. R. (2013). Evaluation of antibacterial activity of *Laurus nobilis* L., *Rosmarinus officinalis* L., and *Ocimum basilicum* L. from northeast of Algeria. *African Journal of Microbiology Research, 7*(42), 4968–4973.
76. Amer, S., Aly, M. M., & Sabbagh, S. (2006). Bio-control of dermatophytes using some plant extracts and actinomycetes filtrates. *Egyptian Journal of Biotechnology*, 315–330.
77. Valeria, F. M., Preve, L., & Tullio, V. (1996). Fungi responsible for skin mycoses in Turin (Italy). *Mycoses, 39*, 141–150.
78. Chermette, R., Ferreiro, L., & Guillot, J. (2008). Dermatophytoses in animals. *Micropathological, 166*, 385–405.
79. Araujo, C. R., Miranda, K. C., Fernandes, O. F. L., Soares, A. J., & Silva, M. R. R. (2009). *In vitro* susceptibility testing of dermatophytes isolated in Goiania, Brazil, against five antifungal agents by broth microdilution method. *Revista do Instituto de Medicina Tropical de São Paulo, 51*, 9–12.
80. Ji-Wen, Z., Sheng-Kun, L., & Wen-Jun, W. (2009). The main chemical composition and *in vitro* antifungal activity of the essential oils of *Ocimum basilicum* L. var. *pilosum* (Willd.) Benth. *Molecules, 14*, 273–278.
81. Bokhari, F. M. (2009). Antifungal activity of some medicinal plants used in Jeddah, Saudi Arabia. *Mycopath, 7*, 51–57.
82. Abdulmoniem, M. A., & Saadab, A. M. A. (2016). Antifungal activity of some Saudi plants used in traditional medicine. *Asian Journal of Plant Sciences, 5*, 907–909.
83. Agwa, A., Aly, M. M., & Bonaly, R. (2000). Isolation and characterization of two *Streptomyces* species producing non-polyenic antifungal agents. *Journal of Union Arab Biologists, 7*, 62–84.
84. Martin, K. W., & Ernst, E. (2004). Herbal medicines for treatment of fungal infections: A systematic review of controlled clinical trials. *Mycoses, 47*(3–4), 87–92.
85. Yamasaki, K., Nakano, M., Kawahata, T., Mori, H., Otake, T., & Ueba, N. (1998). Anti-HIV-1 activity of herbs in *Labiatae*. *Biological and Pharmaceutical Bulletin, 21*, 829–833.
86. Jakowienko, P., Wojcik-Stopczynska, B., & Jadczak, D. (2011). Antifungal activity of essential oils from two varieties of sweet basil (*Ocimum basilicum* L.). *Vegetable Crops Research Bulletin, 74*, 97–106.

87. Kocić-Tanackov, S. D., Dimić, G. R., Pejin, D. J., Mojović, L. V., Pejin, J. D., & Tanackov, I. J. (2012). Antifungal activity of the basil (*Ocimum basilicum* L.) extract on *Penicillium aurantiogriseum*, *P. glabrum*, *P. chrysogenum*, and *P. brevicompactum*. *Acta Periodica Technologica, 43*, 247–256.
88. Soliman, K. M., & Badeaa, R. I. (2002). Effect of oil extracted from some medicinal plants on different mycotoxigenic fungi. *Food and Chemical Toxicology, 40*, 1669–1675.
89. Soković, M. D., Glamočlija, J. M., & Ćirić, A. D. (2013). Natural products from plants and fungi as fungicides. In Nita, M. (Ed.), *Fungicides—Showcases of Integrated Plant Disease Management from Around the World*. InTech.
90. Harsh, P. B., Travis, S. W., Herbert, P. S., & Jorge, M. V. (2002). Root-specific elicitation and antimicrobial activity of rosmarinic acid in hairy root cultures of *Ocimum basilicum*. *Plant Physiology and Biochemistry, 40*(11), 983–995.
91. Erler, F., Ulug, I., & Yalcinkaya, B. (2007). Repellent activity of five essential oils against *Culex pipiens*. *Fitoterapia, 77*, 491–494.
92. Ansari, M. A., Vasudevan, P., Tandon, M., & Razdan, R. K. (2000). Larvicidal and mosquito repellent action of peppermint (*Mentha piperita*) oil. *Bioresource Technology, 71*(3), 267–271.
93. Ansari, M. A., Razdan, R. K., Tandon, M., & Vasudevan, P. (2000). Larvicidal and repellent actions of *Dalbergia sissoo* Roxb. (*F. Leguminosae*) oil against mosquitoes. *Bioresource Technology, 73*(3), 207–211.
94. Vahitha, R., Venkatachalam, M. R., Murugan, K., & Jebanesan, A. (2002). Larvicidal efficacy of *Pavonia zeylanica* L. and *Acacia ferruginea* DC against *Culex quinquefasciatus* Say. *Bioresource Technology, 82*(2), 203, 204.
95. Jeyabalan, D., Arul, N., & Thangamathi, P. (2003). Studies on effects of *Pelargonium citrosa* leaf extracts on malarial vector, *Anopheles stephensi* Liston. *Bioresource Technology, 89*(2), 185–189.
96. Rajkumar, S., & Jebanesan, A. (2004). Mosquitocidal activities of octacosane from *Moschosma polystachyum* Linn. (*Lamiaceae*). *Journal of Ethnopharmacology, 90*(1), 87–89.
97. Mahmoud, H. E., Bashir, N. H. H., & Assad, Y. O. H. (2017). Effect of basil (*Ocimum basilicum*) leaves powder and ethanolic extract on the 3rd larval instar of *Anopheles arabiensis* (*Culicidae*: *Diptera*). *International Journal of Mosquito Research, 4*(2), 52–56.
98. Govindarajan, M., Sivakumar, R., Rajeswary, M., & Yogalakshmi, K. (2013). Chemical composition and larvicidal activity of essential oil from *Ocimum basilicum* (L.) against *Culex tritaeniorhynchus*, *Aedes albopictus* and *Anopheles subpictus* (*Diptera*: *Culicidae*). *Experimental Parasitology, 134*(1), 7–11.
99. Rodríguez-González, Á., Álvarez-García, S., González-López, Ó., Da Silva, F., & Casquero, P. A. (2019). Insecticidal properties of *Ocimum basilicum* and *Cymbopogon winterianus* against *Acanthoscelides obtectus*, insect pest of the common bean (*Phaseolus vulgaris*, L.). *Insects, 10*(5), 151.

PART III
Roots and Tubers

CHAPTER 8

Carrot (*Daucus carota* L.)

NAJLA MEZGHANI,[1,2,*] JIHEN BEN AMOR,[2] IMEN ZAOUALI,[1] ANISSA RIAHI,[3] RYM FEKIH,[4] and NEJI TARCHOUN[2]

[1]*National Gene Bank of Tunisia, Tunis, Tunis Governorate, Tunisia*

[2]*Research Laboratory "Management of Horticultural Species in Organic and Conventional System," High Agronomic Institute of Chott Mariem, University of Sousse, Sousse, Sousse Governorate, Tunisia*

[3]*Laboratory of Horticulture, National Agricultural Research Institute of Tunisia (INRAT), University of Carthage, Ariana Governorate, Tunisia; Higher School of Agriculture of Kef, University of Jendouba, Kef Governorate, Tunisia*

[4]*Food Resources Education and Research Center, Graduate School of Agricultural Science, Kobe University, Kasai, Hyōgo Prefecture, Japan*

*Corresponding author

ABSTRACT

Carrot (*Daucus carota* L.) is among the most important vegetables in the world. It is recognized as a functional food because of its diverse nutritional composition, which includes vitamins and minerals, as well as the phytochemicals and bioactives that promote health beyond basic nutrition. Among the functional properties of carrot, the antioxidant, anti-inflammatory, antimicrobial, and hypoglycemic activities of carrot roots and carrot-derived extracts are thought to be important in the prevention of chronic diseases. We summarize the nutrient and phytochemical content of carrots and carrot products, as well as how these bioactive components may interact to modulate disease development or to provide specific health benefits to plants, animals, and humans.

Postharvest Quality Attributes of Fresh Vegetables and Grains.
Riadh Ilahy, Imen Tlili, Thouraya R'him, and Muhammad Azam (Eds.)
© 2027 Apple Academic Press, Inc. Co-published with CRC Press (Taylor & Francis)

DOI:10.1201/9781779642851-11

8.1 INTRODUCTION

8.1.1 BOTANICAL NAME, COMMON NAME

Carrot (*Daucus carota* L., $2n = 18$) is a globally distributed, outcrossing species of the Apiaceae family. Because of its wide variation, several subspecies have been identified, based primarily on inflorescence and fruit characteristics [1–4]. Domesticated carrots, taxonomically segregated from wild carrots, are classified as *D. carota* var. *sativus* Hoffm., *D. sativus* Hoffm. Roehl, and *D. carota* subsp. *sativus* (Hoffm.) Arcangeli [5]. They have several common names around the world, including *gazar* in Arabic, *hu luo bo* in Chinese, *carotte* in French, *gajar* in India, and *cenoura* in Portuguese [6].

8.1.2 HISTORY AND ORIGIN

Cultivated carrots can be divided into two principal groups based on root pigmentation: the eastern-type or anthocyanin carrot (e.g., yellow or purple carrot), and the western-type or carotene carrot (e.g., orange or red carrot) [7, 8]. It is widely recognized that wild carrots were domesticated in the area around Afghanistan in southwestern Asia to produce eastern-type cultivated carrots approximately 1100 years ago. However, the western-type cultivated carrot did not regularly appear in historical documents until six centuries following domestication, and multiple suggestions have been proposed to explain their origin. According to Vilmorin [9], orange carrots originated directly from wild carrots in Europe. The earliest molecular study by Iorizzo et al. [10] on carrot domestication showed that the western-type cultivated carrot may have originated directly from eastern-type carrots, in agreement with Banga's [7] hypothesis based on historical painting. This was further supported by the transcriptome analyses of Rong et al. [11], while Heywood [12] suggested that introgression from wild carrots might have had an impact.

8.1.3 PRODUCTION

Carrots are now grown all over the world, with an extensive adaptation to temperate growing regions in Asia, Europe, and the Americas, as well as more recent cultivar development for sub-tropical and even tropical climates [13]. Carrots are classified among the top ten vegetables in terms of global production, following tomatoes, onions, cabbage, cucumber, and eggplant. Global production (carrot and turnip combined) is estimated to be 42,814,538 tons per year, with China being the biggest producer (48%), followed by Uzbekistan (5%) and the Russian Federation (4.8%).

8.1.4 BOTANICAL DESCRIPTION

Carrot is a biennial plant that requires vernalization in order to flower. It forms a basal rosette of leaves in the first year and stores carbohydrates in its hypertrophic root [14]. Floral stem elongation and flowering begin after a vernalization period at temperatures between 0 and 10°C and a long day photoperiod [15, 16]. Fast-growing cultivars reach maturity in three months (90 days) after seed sowing, whereas slow-growing cultivars take an extra month (120 days).

The transition from vegetative to reproductive development begins when the flat vegetative meristem producing leaves switches to a conical aspect, allowing the stem elongation and the development of a flower cluster forming a compound umbel, with multiple umbellets in each umbel. The primary umbel grows at the tip of the main flowering stem, whereas smaller secondary umbels grow from the main branch, and these branches develop into third, fourth, and even later-flowering umbels [17].

Plants are andromonoecious (they have both male and female flowers) and have protandrous hermaphroditic flowers (pollen is shed before the stigmas become receptive). Some plant populations are reported to exhibit gynodioecy as male sterile plants (functionally female plants) co-occur with hermaphroditic individuals [18].

The fruits are oval schizocarps with two mericarps, with each mericarp being a true seed. When fully ripe, the produced seeds are brown, small in size, with an average size of between 2 and 3 mm, and number approximately 10,500 per gram. Harvesting of carrot seeds crops begins when the earliest maturing seed on the primary umbel is mature and starts to fall.

Wild carrots play an important role in the ecosystem because their leaves and roots provide food for a variety of animals, including larvae of *Lepidoptera* species. Furthermore, carrot flowers produce nectar, which attracts bees to pollinate the plants. Domesticated carrots, on the other hand, are a valuable source of nutritious food for humans, and many people enjoy their distinct pleasant flavor, texture, and color.

8.2 CHEMICAL AND PHYTONUTRIENTS COMPOSITION

Carrots are an important source of moisture, vitamins, minerals, trace elements, and fiber. They also contain a variety of phytochemicals with significant health-promoting properties [19].

8.2.1 *CHEMICAL COMPOSITION*

According to Holland et al. [20], the chemical composition of carrots is as follows: moisture (88.8%), carbohydrate (6%), total sugars (5.6%), crude fiber (2.4%), protein (0.7%), fat (0.5%), K (240 mg/100 g), Na (40 mg/100 g), Ca (34 mg/100 g), P (25 mg/100 g), Mg (9 mg/100 g), Fe (0.4 mg/100 g), Zn (0.2 mg/100 g), carotenes (5.33 mg/100 g), thiamine (0.04 mg/100 g), Cu (0.02 mg/100 g), riboflavin (0.02 mg/100 g), niacin (0.2 mg/100 g), vitamin C (4 mg/100 g), and energy value (126 kJ/100 g). The edible portion of carrot cultivars from various geographic regions was found to contain approximately 10% carbohydrates, with protein ranging from 0.8 to 1.1 g/100 g and soluble carbohydrates from 6.6 to 7.7 g/100 g [21], 1.67–3.35% reducing sugars, 1.02–1.18% non-reducing sugars, and 2.71–4.53% total sugars [22]. Carrots have the highest moisture content (84.1%) when compared to pawpaw, coconut, turmeric, and ash. Pawpaw has the highest fiber content (23.8%), crude protein content (12.8%), and ash content (9.6%). Turmeric has the highest recorded carbohydrate content (68.59%), while coconuts have the highest fat content (43.0%) [23]. The phytochemical analysis of the aerial portions of Nigerian carrots revealed that the study material contained carbohydrate (51.81%), crude protein (14.59%), lipid (10.37%), fiber (9.07%), carbohydrate (51.81%), moisture (10.23%), and the ash content (12.99%) as well as macro and trace elements such as P (11.00 mg/L), Na (5.38 mg/L), Fe (3.19 mg/L), K (2.25 mg/L), Ca (2.02 mg/L), Mn (1.15 mg/L), Mg (1.02 mg/L), As (0.83 mg/L), Se (0.40 mg/L), Zn (0.26 mg/L), Cu (0.13 mg/L), Cd (0.02 mg/L), Cr (0.02 mg/L), Ni (0.17 mg/L), and Pb (0.04 mg/L). The crude fiber in carrot roots includes 13.0, 15.2, and 71.7% of hemicellulose, lignin, and cellulose, respectively [24].

8.2.2 *PHYTONUTRIENTS*

Phytonutrients, commonly referred to as phytochemicals, are bioactive substances that are stored in various plant parts for the purpose of disease defense or prevention [25]. Flavonoids, phenolic acids, carotenoids, polyacetylene, and ascorbic acid (AA) are among the phytonutrients found in large amounts in carrots [26–28].

8.2.2.1 CAROTENOIDS

Carotenoids are a group of isoprenoid molecules present in all photosynthetic plants, including carrots. Additionally, some non-photosynthetic bacteria and fungi contain carotenoids. Depending on its carotenoids content, the carrot root can be white, yellow, orange, or purple [29]. Orange carrots contain higher amounts of total carotenoids than yellow carrots [30–32]. Among various colored carrot genotypes, red carrot has the highest total carotenoids content (7.37 mg, 1.36-fold), followed by rainbow (7.14 mg, 1.90-fold), orange (3.94 mg, 1.09-fold), and yellow carrot (0.64 mg, 1.42-fold) [33]. The accumulation of carotenoids begins at the second month of the plant development, and it reaches its highest level in the third month [34]. The most abundant carotenoids in carrot are α- and β-carotene (Figure 8.1), which account for 80–90% of total carotenoid content in most carrot genotypes [35]. Carrots also contain lutein (1–5%), zeaxanthin, cryptoxanthin, lycopene, phytoene, and phytofluene, and the composition of carotenoids is severely influenced by root color [36]. Purple carrots have 2.2 and 2.3 more α- and β-carotene than orange carrots; yellow carrot contained only trace amounts of them [30, 32, 37], and white carrot contained none [31]. Except for yellow carrot, the purple-yellow carrot had the highest lutein content. Lycopene levels in red carrots were significantly higher than in other carrots [31].

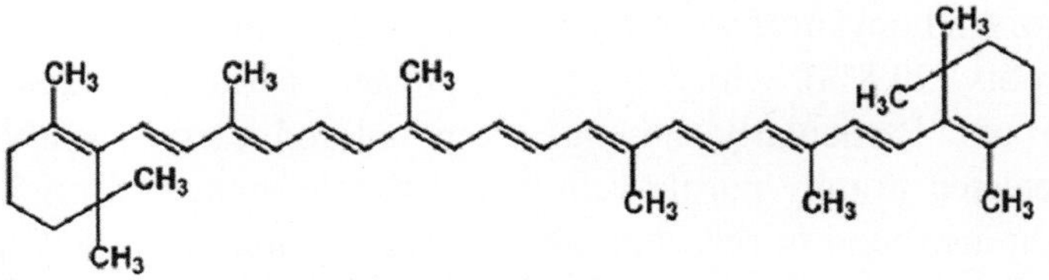

FIGURE 8.1 Chemical structure of β-carotene.

8.2.2.2 PHENOLIC COMPOUNDS

There are several subgroups of phenolic compounds, including flavonoids, phenolic acids, lignans, tannins, curcuminoids, and stilbenoids. Phenolic compounds play an important role in plant resistance to physiological mechanisms, and their accumulation in carrots is caused by cold, injury, or ethylene [17]. Carrots are essentially rich in phenolic acids as well as anthocyanins, a class of flavonoids. Chlorogenic acid, caffeic acid, ferulic acid, p-hydroxybenzoic acid, and other isomers of hydroxycinnamic acid are the major phenolic acids found in carrots [31, 38]. The concentration of these compounds varies with the color of the root. The major phenolic acid is chlorogenic acid (Figure 8.2), which accounts for 51.4, 52.4, 57.1, and 72.5% of total phenolic compounds in white, orange, yellow, and purple carrots, respectively. Caffeic acid content ranked the same as chlorogenic acid content but with much lower amounts in all carrots [31].

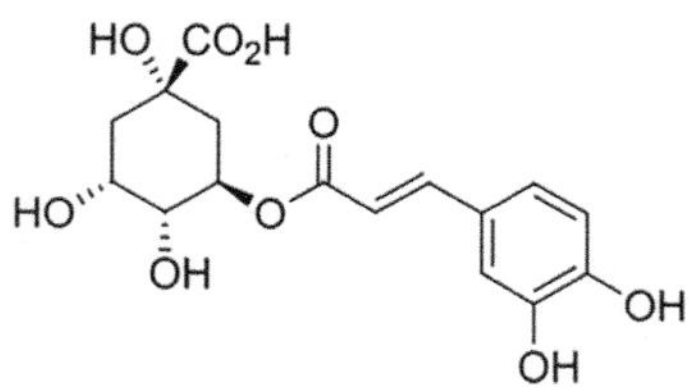

FIGURE 8.2 Chemical composition of chlorogenic acid.

The compositions of the various carrot tissues are similar, but the content of individual phenolic compounds varies and decreases from the outside to the inside [39]. Flavonoids mainly consist of anthocyanins, which are found in purple carrots. Anthocyanins are responsible for the unlimited variety of colors found in fruits, vegetables, and plants, ranging from red and orange to purple and blue. Nearly 600 anthocyanins have been isolated from plants; the most well-known are based on six anthocyanidins: delphinidin, cyanidin, pelargonidin, malvidin, petunidin, and peonidin. Anthocyanins were found in significant amounts only in purple carrot roots (20.8%), while they were present in less than 2% of other carrot roots. Particularly, among seven colored carrots, anthocyanins were detected in only purple-yellow and purple-orange carrots, with no anthocyanins found in red, dark orange, typical orange, yellow, or white carrots. Purple-yellow carrots had the highest amount of chlorogenic acid, followed by purple-orange carrots, and it was significantly higher than in other carrot colors. Chlorogenic acid levels were 2.5, 13.9, 29.8, 16.3, 61.5, and 19.2 times higher in purple-yellow carrots than in purple-orange, red, dark orange, typical orange, yellow, and white carrots, respectively. Caffeic acid content ranked the same as chlorogenic acid content, but it was much lower than chlorogenic acid content in all carrots

[31]. HPLC analysis of over 30 carrot lines, including gene bank accessions, open-pollinated (OP), and hybrid commercial cultivars, revealed a total anthocyanin content range of 0.5–191 mg/100 g fresh weight (fw). Total anthocyanin content is strongly and positively correlated with total phenolic content, with correlation coefficients (r) ranging from 0.85 [40] to 0.99 [41], implying that anthocyanins account for a significant proportion of phenols in purple carrots [42].

8.2.2.3 POLYACETYLENES

Polyacetylenes are a well-known class of non-volatile bioactive phytochemicals with at least two conjugated C-C triple bonds. Apiaceae plants (including carrot) contain aliphatic C17 polyacetylenes of the falcarinol type [43]. 12 polyacetylenes have been isolated from carrot, out of over 1400 identified in plants. Falcarinol (Figure 8.3), falcarindiol, and falcarindiol-3-acetate are three of the 12 essential polyacetylenes found in carrot roots [44].

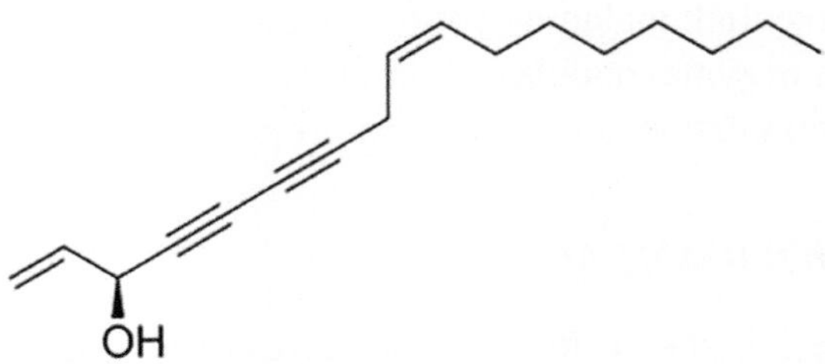

FIGURE 8.3 Chemical composition of falcarinol.

8.2.2.4 VITAMINS

Carrots are a rich source of vitamins A, C, and E. Vitamin A (Figure 8.4) is a fat-soluble vitamin that is essential for immune and organ function, eye health, and visual acuity [45].

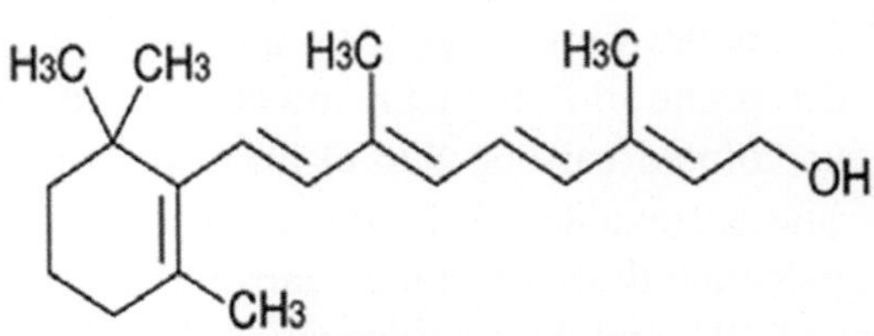

FIGURE 8.4 Chemical composition of vitamin A.

Vitamin C is a water-soluble vitamin also known as L-ascorbic acid. Vitamin C is used as a nutritional supplement and is included in a number of medicines for colds and influenza due to its numerous beneficial effects on the human body. Tincheva et al. [46] discovered that raw carrots had comparable vitamin C levels to eggplants (14.9 and 13.3 mg/100 g sample, respectively) but lower than cauliflower and cabbage (68.5 and 80.3 mg/100 g). Cooking time, temperature, and method all affect vitamin C retention. Long cooking times and elevated temperatures have been found to cause especially severe vitamin C losses [47]. Vitamin C content in carrot decreased from 5.8 mg/100 g row sample to 2.6 mg/100 g cooked sample [48]. Streaming was the most suitable cooking technique in terms of vitamin C retention compared to boiling and microwaving [46].

8.3 HEALTH BENEFITS AND BIOLOGICAL ACTIVITIES

Despite being classified as "non-essential" nutrients, phytochemicals found in carrots and other crops with functional properties have enormous therapeutic potential, including antioxidant activity, antimicrobial effects, detoxification enzyme modulation, immune system stimulation, hormone metabolism modulation, and anticancer properties [49].

8.3.1 ANTIOXIDANT ACTIVITY

Oxidative stress is one of the leading causes of a wide range of modern chronic civilization diseases. Oxidative stress in humans is caused by a decrease in antioxidative potential or an increase in the production of oxygen or nitrogen radicals [50]. The majority of radicals that cause oxidative stress are reactive nitrogen species (RNS) or reactive oxygen species (ROS) [51]. Different methods and assays are used to measure antioxidant activity in plants, such as ABTS radical scavenging, DPPH free radical scavenging, FRAP, ORAC, and CUPRAC [52, 53].

Several studies, Sun et al. [31], Soltoft et al. [35], Algarra et al. [54], Boadi et al. [55], Pereira-Caro et al. [56], and Bartosz et al. [57], were conducted to determine the phytochemical content and antioxidant ability of various colored carrots. Sun et al. [31] found that the anthocyanin content and antioxidant properties of black carrot extracts were significantly higher than those of orange carrot extracts, as measured by the FRAP assay, DPPH, and ABTS scavenging. When compared to other carrot colors, black carrots were found to have an exceptionally high content of total phenolics and flavonoids, resulting in the highest antioxidant ability in terms of FRAP, CUPRAC, cupric ion reducing antioxidant capacity, and ABTS activity, followed by rainbow carrot, and the least in orange, red, and yellow carrot [33]. This was confirmed by Blando et al. [58], who reported that the anthocyanin profiles of polyphenolic extracts of black carrot (BC) and purple carrot (PC) were similar, but quantitative differences were observed. The total anthocyanin content in BC was more than twice that in PC (13.84 ± 0.61 vs. 5.64 ± 0.48 mg K Eq. g^{-1} DW). When comparing anthocyanin-rich carrots to orange carrots, phenolic acids (mostly chlorogenic acid) were found to be abundant. High polyphenol content also accounted for high reducing capacity (as measured by the Folin-Ciocalteu reagent, FCR) and antioxidant capacity (as measured by the TEAC and ORAC assays), which were highest for BC (FCR value: 16.6 ± 1.1 mg GAE. g^{-1} DW; TEAC: 76.6 ± 10.6). Carotenoids did not contribute to total antioxidant capacity, but they correlated with antioxidant capacity of hydrophobic extracts [31]. Shebaby et al. [59] reported that oil extract from Lebanon wild carrot (*D. carota subsp. carota*) had a distinct chemical composition and significant antioxidant activity when tested *in vitro* using the DPPH, FIC, and FRAP assays. This activity was more pronounced in methanolic extracts (leaves and seeds) than in essential oils in Algerian germplasm [60].

8.3.2 ANTICANCER ACTIVITY

In several studies, the cancer-preventive properties of carrots have been mainly attributed to their content of antioxidants, fibers, and minerals. This activity was initially explained by its high carotenoid content [61, 62], with stronger negative correlations of developing cancer associated with the intake of α-carotene rather than β-carotene [63, 64]. However, intervention studies have revealed that the addition of carotenoids does not protect against the development of this disease [65, 66]. α-carotene and β-carotene may serve as biomarkers for the consumption of other bioactive constituents in carrots associated with cancer-preventive properties. Such anticancer compounds are found in carrots, mainly phenylpropanoids [67] and polyacetylenic oxylipins [68–70], with the latter type of bioactive constituent being the most studied *in vitro* and *in vivo*. Particularly, the two major polyacetylenic oxylipins in carrots, falcarinol (FaOH) and falcarindiol (FaDOH), have received substantial attention in recent years due to their anti-inflammatory and cytotoxic activities *in vitro* [71], and recently, the anti-neoplastic effects of these polyacetylenic oxylipins have been proved in cancer-primed rat models for colorectal cancer (CRC) [72].

8.3.3 ANTIFUNGAL AND ANTIBACTERIAL ACTIVITIES

Fungal and bacterial diseases have been linked to significant agricultural losses. Carrot seed oil contains sesquiterpenic allelochemicals including carotol, daucol, and β-caryophyllene, which have antimicrobial and antifungal properties [73, 74]. Jasicka-Misiak et al. [75] investigate the *in vitro* inhibitory effects of carotol and daucol isolated from the surface of carrot seeds cultivars on *Alternaria alternata*. Both compounds exhibited a moderate inhibitory effect on the target fungi's mycelium radial growth (20% inhibition). Carotol-only mixtures exhibited strong inhibitory activity. This activity increased as the carotol content of individual mixtures increased. Sharma et al. [76] demonstrated the antifungal activity of carrot seed essential oil, its constituents, and extracts against three phytopathogenic rice fungi (*Rhizoctonia solani, Bipolaris oryzae,* and *Fusarium moniliforme*). Carrot seed methanol extract and carotol had significantly higher antifungal activity than the other components tested (daucene and daucol). In a study established by John et al. [77], carrot peel extracted with methanol and acetone inhibited the growth of *Shigella flexneri, Escherichia coli, Staphylococcus aureus,* and *Klebsiella pneumoniae*. The antimicrobial activity of the extracts increased linearly as the concentration of the extracts (g/mL) increased. The growth inhibition zone for various bacteria measured in acetone extracts ranged from 11 to 16 mm. The growth inhibition zone measured for methanol extracts ranged between 12 and 17 mm. According to Al-Baarri et al. [78], the addition of carrot extract to Hyphothiocyanite increased its antimicrobial activity against *S. aureus* and *E. coli*.

With regards to wild species, Tavares et al. [79] discovered that *D. carota* subsp. *halophilus* essential oil has antifungal properties against several human pathogenic fungi, including yeasts, dermatophytes, and *Aspergillus* strains. *D. carota* subsp. *maritimus* has been identified as having antibacterial properties. *D. carota* subsp. *carota* essential oil has also been described as an antimicrobial agent against a variety of

Gram-negative and Gram-positive bacteria [80, 81]. These biological activities can be attributed to the high amounts of geranyl acetate (29.0%), α-pinene (27.2%), and 11αH-himachal-4-en-1β-ol (9.2%). In Poland, oils extracted from cultivated carrot (*D. carota* subsp. *carota* var. Koral and Perfekcja) seed were more effective against all tested organisms, including four species of bacteria and two species of fungi. This is because the main compounds in cultivated carrot oil were oxygenated sesquiterpenes (44%–51%), whereas monoterpene hydrocarbons (71.9–83.8%) were predominant in wild material [80].

8.3.4 ANTI-INFLAMMATORY AND IMMUNOMODULATORY EFFECTS

Inflammation is defined as the body's immune system's normal physiological response to harmful stimuli such as infections, damaged cells, and toxic compounds. Acute or chronic inflammatory responses can damage or disrupt the heart, pancreas, liver, kidneys, lungs, brain, gastrointestinal tract, and reproductive system. Carrots and their phytochemical compounds have been shown to have the potential to act as a natural immunomodulator via antioxidant and anti-inflammatory properties, as well as modulation of other immune components. Kamiloglu et al. [82] investigated the ability of polyphenol-rich black carrot and its byproducts, peel and pomace, to modulate the inflammatory response in tumor necrosis factor (TNF-α) treated endothelial cells after gastrointestinal digestion, as well as in a co-culture of intestinal Caco-2 and endothelial EA.hy926 cells. The results showed that digestion increased absorption significantly; the transported polyphenols from black carrots and byproducts significantly reduced the secretion of certain pro-inflammatory markers, such as IL-8, monocyte chemoattractant protein 1, vascular endothelial growth factor, and intercellular adhesion molecule 1, particularly in TNF-treated endothelial cells. According to Olejnik et al. [83], purple carrot anthocyanin extract affects the mRNA expression of proinflammatory interleukins IL-1 negatively (91% and IL-6, respectively), as well as of inflammatory mediators like inducible nitric oxide synthase (iNOS) and cyclooxygenase-2 (COX-2) in lipopolysaccharide-activated RAW264.7 cells. In their investigation on male Wistar rats, Ekam et al. [84] showed that (i) carotenoid extracts of carrots at a dose of 559 mg/kg for 21 days increased levels of lymphocytes, eosinophils, monocytes, and platelets; and (ii) the immunomodulatory effect of the Golden Neo-Life Dynamite (GNLD) carotenoid complex, which contains α-carotene, β-carotene, lutein, zeaxanthin, β-cryptoxanthin, lycopene, and alpha-tocopherol, is greater than that of carrot extracted carotenoid, which contains α-carotene and β-carotene alone. In weanling mice, β-carotene supplementation at a dose of 50 mg/kg feed for 14 weeks increased the synthesis of immunoglobulin A (IgA) in the jejunum or ileum. An improvement in the immune response to gastrointestinal infections in neonates and children was also observed [85].

8.3.5 ANTI-DIABETIC EFFECT

Diabetes is a chronic disease that occurs when the pancreas does not produce enough insulin or when the body does not use the insulin that is produced effectively. Insulin is a hormone that regulates blood glucose levels by stimulating the production of amylase protein [86]. The inhibition of α-glucosidase and α-amylase can significantly reduce post-prandial blood glucose increases and thus can be an important strategy in the management of blood glucose levels in type 2 diabetic and borderline patients [87]. Black carrot has been shown to have a positive effect in the treatment of Type 2 diabetes. This property is due to the phenolic compounds found in black carrot. Among the anthocyanin molecules found in black carrot, cyanidin 3-xylosyl galactoside was discovered to be a potential drug for *in vitro* inhibition of α-amylase and α-glucosidase [88]. Carrot juice exhibited anti-hyperglycemic and anti-cholesterol activity in albino rats injected with alloxan monohydrate to induce type 1 diabetes; however, this activity was less effective in the carrot juice treatment group than in the insulin treatment group [89].

8.4 CONCLUSION AND PROSPECTS

Carrots and their derivatives are a good source of phytochemicals, including carotenoids, polyphenols, and vitamins. These phytochemicals, either alone and/or in combination, have remarkable therapeutic potential in several ailments. Such substances may range from isolated nutrients, dietary supplements, and specific diets to herbal products, genetically engineered designer foods, beverages, and processed foods.

KEYWORDS

- **antioxidant activity**
- **carotenoids**
- ***Daucus carota***
- **chronic disease prevention**
- **functional food**
- **phytochemicals**

REFERENCES

1. Sáenz-Laín, C. (1981). Research on *Daucus* L. (Umbelliferae). *Actas III Congr OPTIMA. A Jard Bot Madr*, *37*, 481–534.
2. Arbizu, C., Reitsma, K. R., Simon, P. W., Spooner, D. M., et al. (2014). Morphometrics of *Daucus* (Apiaceae): A counterpart to a phylogenomic study. *American Journal of Botany*, *101*, 2005–2016.
3. Mezghani, N., Zaouali, I., Bel Amri, W., Rouz, S., Simon, P. W., Hannachi, C., Ghrabi, Z., Neffati, M., Bouzbida, B., & Spooner, D. M. (2014). Fruit morphological descriptors as a tool for discrimination of *Daucus* L. germplasm. *Genetic Resources and Crop Evolution*, *61*, 499–510.
4. Mezghani, N., Ben Amor, J., Spooner, D. M., Boubaker, H., Najmi, A. M., Rouz, S., Hannachi, C. H., Neffati, M., & Tarchoun, N. (2017). Multivariate analysis of morphological diversity among closely related *Daucus* species and subspecies in Tunisia. *Genetic Resources and Crop Evolution*, *64*, 2145–2159.
5. Nguyen, H. H. V., & Nguyen, L. T. (2015). Carrot processing. In *Handbook of Vegetable Preservation and Processing* (pp. 449–478).
6. Al-Snafi, A. E. (2017). Nutritional and therapeutic importance of *Daucus carota*—A review. *IOSR Journal of Pharmacy*, *7*(2), 72–88.
7. Banga, O. (1957). The development of the original European carrot material. *Euphytica*, *6*, 64–76.
8. Banga, O. (1963). *Main Types of the Western Carotene Carrot and Their Origin* (153 p.). W. E. J. Tjeenk Willink, Zwolle.
9. Vilmorin, M. (1859). L'hérédité dans les végétaux. In M. Vilmorin (Ed.), *Notice sur l'amélioration Des Plantes Pars La Semis* (pp. 5–29). Librairie Agricole, Paris, France.
10. Iorizzo, M., Senalik, D. A., Ellisson, S. L., Grzebelus, D., Cavagnaro, P. F., Allender, C., et al. (2013). Genetic structure and domestication of carrot (*Daucus carota* subsp. *sativus*) Apiaceae. *American Journal of Botany*, *100*(5), 930–938.
11. Rong, J., Lammers, Y., Strasburg, J. L., Schidlo, N. S., Ariyurek, Y., de Jong, T. J., Klinkhamer, P. G. L., Smulders, M. J. M., & Vrieling, K. (2014). New insights into domestication of carrot from root transcriptome analyses. *BMC Genomics*, *15*, 895.

12. Heywood, V. (1983). Relationships and evolution in the *Daucus carota* complex. *Israel Journal of Ecology and Evolution, 32*, 51–65.
13. Simon, P. W., Freeman, R. E., Vieira, J. V., Boiteux, L. S., Briard, M., Nothnagel, T., Michalik, B., & Kwon, Y. S. (2008). Carrot. In J. Prohens & F. Nuez (Eds.), *Handbook of Plant Breeding: Vegetables II: Fabaceae, Liliaceae, Solanaceae, and Umbelliferae* (pp. 327–357). Springer, New York.
14. Whitaker, T. W., Sherf, A. F., Lange, W. H., Nicklow, C. W., & Radewald, J. D. (1970). Carrot production in the United States. *Agricultural Handbook No. 375*, 41 pp.
15. Dickson, M. H., & Peterson, C. E. (1958). Hastening greenhouse seed production for carrot breeding. *Proceedings of the American Society for Horticultural Science, 71*, 412–415.
16. Atherton, J. G., & Basher, E. A. (1984). The effects of photoperiod on flowering in carrot. *Journal of Horticultural Science, 59*, 213–215.
17. Rubatzky, V. E., Quiros, C. F., & Simon, P. W. (1999). *Carrots and Related Vegetable Umbelliferae*. CABI Publishing.
18. Ronfort, J., Saumitou-Laprade, P., Cugen, J., & Couvet, D. (1995). Mitochondrial DNA diversity and male sterility in natural populations of *Daucus carota* ssp. *carota. Theoretical and Applied Genetics, 91*, 150–159.
19. Sharma, K. D., Karki, S., Thakur, N. S., & Attri, S. (2012). Chemical composition, functional properties and processing of carrot—A review. *Journal of Food Science and Technology, 49*(1), 22–32.
20. Holland, B., Unwin, J. D., & Buss, D. H. (1991). *Vegetables, Herbs and Spices: Fifth Supplement to McCance and Widdowson's*. 172 p. London.
21. Howard, F. D., MacGillivary, J. H., & Yaaguchi, M. (1962). Nutrient composition of fresh California grown vegetables. *Bulletin No. 788*, California Agricultural Experiment Station, University of California, Berkeley.
22. Kaur, G., Jaiswal, S. P., Barr, K. S., & Kumar, J. C. (1976). Physico-chemical characteristics of some important varieties of carrot. *Indian Food Packer, 30*(2), 5–8.
23. Ajenu, C. O., Imoisi, C., Emomighahwe, O. F., & Irede, L. E. (2021). Comparative evaluation of the proximate and micro-nutritional benefits of pawpaw, carrot, turmeric and coconut. *Journal of Chemical Society of Nigeria, 46*(5), 870–878.
24. Kochar, G. K., & Sharma, K. K. (1992). Fiber content and its composition in commonly consumed Indian vegetables and fruits. *Journal of Food Science and Technology, 29*, 187–190.
25. Gupta, C., & Prakash, D. (2014). Phytonutrients as therapeutic agents. *Journal of Complementary and Integrative Medicine, 11*(3), 151–169.
26. Bozalan, K., & Karadeniz, F. (2011). Carotenoids profile, total phenolic content and antioxidant activity of carrots. *International Journal of Food Properties, 14*, 1060–1068.
27. Karabacak, Ç. E., & Karabacak, H. (2019). Factors affecting carotenoid amounts in carrots. *Ecological Life Sciences, 14*(2), 29–39.
28. Anjani, G., Ayustaningwarno, F., & Eviana, R. (2022). Critical review on the immunomodulatory activities of carrot's β-carotene and other bioactive compounds. *Journal of Functional Foods, 99*, 1–11.

29. Baranski, R., Allender, C., & Klimek-Chodacka, M. (2012). Towards better tasting and more nutritious carrots: Carotenoid and sugar content variation in carrot genetic resources. *Food Research International*, *47*(2), 182–187.
30. Alasalvar, C., Grigor, J. M., Zhang, D., Quantick, P. C., & Shahidi, F. (2001). Comparison of volatiles, phenolics, sugars, antioxidant vitamins, and sensory quality of different colored carrot varieties. *Journal of Agricultural and Food Chemistry*, *49*(3), 1410–1416.
31. Sun, T., Simon, P. W., & Tanumihardjo, S. A. (2009). Antioxidant phytochemicals and antioxidant capacity of biofortified carrots (*Daucus carota* L.) of various colors. *Journal of Agricultural and Food Chemistry*, *57*(10), 4142–4147.
32. Ben Amor, J., Mezghani, N., Periago, M. J., Navarro-Gonzalez, I., Elvira-Torales, L. I., Mezghani, N., Oukrimi, Y., & Tarchoun, N. (2020). Variations in the sugars and antioxidant compounds related to root color in Tunisian carrot (*Daucus carota* subsp. *sativus*) landraces. *Italian Journal of Food Science*, *32*, 654–673.
33. Singh, B. K., Koley, T. K., Maurya, A., Singh, P. M., & Singh, B. (2018). Phytochemical and antioxidative potential of orange, red, yellow, rainbow and black colored tropical carrots (*Daucus carota* subsp. *sativus* Schubl. & Martens). *Physiology and Molecular Biology of Plants*, *24*(5), 899–907.
34. Simpson, K., Cerda, A., & Stange, C. (2016). Carotenoid biosynthesis in *Daucus carota*. In C. Stange (Ed.), *Carotenoids in Nature. Subcellular Biochemistry*, *79*. Springer, Cham.
35. Soltoft, M., Bysted, A., Madsen, K. H., Mark, A. B., Bugel, S. G., Nielsen, J., & Knuthsen, P. (2011). Effects of organic and conventional growth systems on the content of carotenoids in carrot roots, and on intake and plasma status of carotenoids in humans. *Journal of the Science of Food and Agriculture*, *91*(4), 767–775.
36. Bhandari, S. R., Rhee, J., Choi, C. S., Jo, J. S., Shin, Y. K., Song, J. W., Kim, S. H., & Lee, J. G. (2022). Morphological and biochemical variation in carrot genetic resources grown under open field conditions: The selection of functional genotypes for a breeding program. *Agronomy*, *12*, 553.
37. Jourdan, M., Gagne, S., Dubois-Laurent, C., Maghraoui, M., Huet, S., Suel, A., Hamama, L., Briard, M., Peltier, D., & Geoffriau, E. (2015). Carotenoid content and root color of cultivated carrot: A candidate-gene association study using an original broad unstructured population. *PLOS ONE*, *10*(1), 1–19.
38. Simoes, A. D. N., Allende, A., Tudela, J. A., Puschmann, R., & Gil, M. I. (2011). Optimum controlled atmospheres minimize respiration rate and quality losses while increase phenolic compounds of baby carrots. *LWT–Food Science and Technology*, *44*, 277–283.
39. Gonçalves, E. M., Pinheiro, J., Abreu, M., Brandão, T. R. S., & Silva, C. L. M. (2010). Carrot (*Daucus carota* L.) peroxidase inactivation, phenolic content and physical changes kinetics due to blanching. *Journal of Food Engineering*, *97*(4), 574–581.
40. Alegria, C., Gonçalves, E. M., Moldão-Martins, M., Cisneros-Zevallos, L., & Abreu, M. (2016). Peel removal improves quality without antioxidant loss, through wound-induced phenolic biosynthesis in shredded carrot. *Postharvest Biology and Technology*, *120*, 232–239.
41. Cuéllar-Villarreal, M. R., Ortega-Hernández, E., Becerra-Moreno, A., Welti-Chanes, J., Cisneros-Zevallos, L., & Jacobo-Velázquez, D. A. (2016). Effects of ultrasound

treatment and storage time on the extractability and biosynthesis of nutraceuticals in carrot (*Daucus carota*). *Postharvest Biology and Technology*, *119*, 18–26.
42. Iorizzo, M., Curaba, J., Pottorff, M., Ferruzzi, M. G., Simon, P., & Cavagnaro, P. F. (2020). Carrot anthocyanins genetics and genomics: Status and perspectives to improve its application for the food colorant industry. *Genes*, *11*, 906.
43. Schmiech, L., Carole, A., Witulski, B., & Hofmann, T. (2009). Structure determination of bisacetylenic oxylipins in carrots (*Daucus carota* L.) and enantioselective synthesis of falcarindiol. *Journal of Agricultural and Food Chemistry*, *57*, 11030–11040.
44. Dawid, C., Dunemann, F., Schwab, W., Nothnagel, T., & Hofmann, T. (2015). Bioactive C17-polyacetylenes in carrots (*Daucus carota* L.): Current knowledge and future perspectives. *Journal of Agricultural and Food Chemistry*, *63*, 9211–9222.
45. Grune, T., Lietz, G., Palou, A., Ross, A. C., Stahl, W., Tang, G., Thurnham, D., Yin, S. A., & Biesalski, H. K. (2010). β-Carotene is an important vitamin A source for humans. *Journal of Nutrition*, *27*, 1–18.
46. Tincheva, P. A. (2019). The effect of heating on the vitamin C content of selected vegetables. *World Journal of Advanced Research and Reviews*, *3*(3), 27–32.
47. Igwemmar, N. C., Kolawole, S. A., & Imran, I. A. (2013). Effect of heating on vitamin C content of some selected vegetables. *International Journal of Science and Technology Research*, *2*(11), 209–212.
48. Hussain, M. I., Hamza, A., & Rashid, M. A. (2016). Estimation of vitamin C in carrot before cooking and after cooking. *Journal of Food and Nutrition Sciences*, *4*(4), 108–112.
49. Shrinet, K., Singh, R. K., Chaurasia, A. K., Tripathi, A., & Kumar, A. (2021). Bioactive compounds and their future therapeutic applications. *Natural Bioactive Compounds*, 337–362.
50. Fang, Y. Z., Yang, S., & Wu, G. (2002). Free radicals, antioxidants, and nutrition. *Nutrition*, *18*, 872–879.
51. Domej, W., Oettl, K., & Renner, W. (2014). Oxidative stress and free radicals in COPD–Implications and relevance for treatment. *International Journal of Chronic Obstructive Pulmonary Disease*, *9*, 1207–1214.
52. Moharram, H. A., & Youssef, M. M. (2014). Methods for determining the antioxidant activity: A review. *Alexandria Journal of Food Science and Technology*, *11*, 31–42.
53. Munteanu, I. G., & Apetrei, C. (2021). Analytical methods used in determining antioxidant activity: A review. *International Journal of Molecular Sciences*, *22*(7), 3380.
54. Algarra, M., Fernandes, A., Mateus, N., de Freitas, V., Esteves da Silva, J. C. G., & Casado, J. (2014). Anthocyanin profile and antioxidant capacity of black carrots (*Daucus carota* L. ssp. *sativus* var. *atrorubens* Alef.) from Cuevas Bajas, Spain. *Journal of Food Composition and Analysis*, *33*, 71–76.
55. Boadi, N. O., Badu, M., Kortei, N. K., Saah, S. A., Annor, B., Mensah, M. B., Okyere, H., & Fiebor, A. (2021). Nutritional composition and antioxidant properties of three varieties of carrot (*Daucus carota*). *Scientific African*, *12*, 1–8.
56. Pereira-Caro, G., Ordóñez-Díaz, J. L., de Santiago, E., Moreno-Ortega, A., Cáceres-Jiménez, S., Sánchez-Parra, M., et al. (2021). Antioxidant activity and bio-accessibility of polyphenols in black carrot (*Daucus carota* L. ssp. *sativus* var. *atrorubens* Alef.)

and two derived products during simulated gastrointestinal digestion and colonic fermentation. *Foods*, *10*, 457.
57. Bartosz, G., Baran, S., Grzesik-Pietrasiewicz, M., & Sadowska-Bartosz, I. (2022). Antioxidant capacity and hydrogen peroxide formation by black and orange carrots. *Journal of the Science of Food and Agriculture*, *31*, 71–77.
58. Blando, F., Marchello, S., Maiorano, G., Durante, M., Signore, A., Laus, M. N., Soccio, M., & Mita, G. (2021). Bioactive compounds and antioxidant capacity in anthocyanin-rich carrots: A comparison between the black carrot and the Apulian landrace "Polignano" carrot. *Plants*, *10*, 564.
59. Shebaby, W. N., El-Sibai, M., Smith, K. B., Karam, M. C., Mroueh, M., & Daher, C. F. (2013). The antioxidant and anticancer effects of wild carrot oil extract. *Phytotherapy Research*, *27*(5), 737–744.
60. Ksouri, A., Dob, T., Belkebir, A., Krimat, S., & Chelghoum, C. (2015). Chemical composition and antioxidant activity of the essential oil and the methanol extract of Algerian wild carrot *Daucus carota* L. ssp. *carota* (L.) Thell. *Journal of Materials and Environmental Science*, *6*(3), 784–791.
61. Burton, G. W., & Ingold, K. U. (1984). Beta-carotene: An unusual type of lipid antioxidant. *Science*, *224*, 569–573.
62. Biesalski, H. K., & Obermueller-Jevic, U. C. (2001). UV light, beta-carotene and human skin: Beneficial and potentially harmful effects. *Archives of Biochemistry and Biophysics*, *389*, 1–6.
63. Michaud, D. S., Feskanich, D., Rimm, E. B., Colditz, G. A., Speizer, F. E., Willett, W. C., & Giovannucci, E. (2000). Intake of specific carotenoids and risk of lung cancer in 2 prospective US cohorts. *American Journal of Clinical Nutrition*, *72*, 990–997.
64. Knekt, P., Järvinen, R., Teppo, L., Aromaa, A., & Seppänen, R. (1999). Role of various carotenoids in lung cancer prevention. *Journal of the National Cancer Institute*, *91*, 182–184.
65. Omenn, G. S., Goodman, G. E., Thornquist, M. D., Balmes, J., Cullen, M. R., Glass, A., et al. (1996). Effects of a combination of β-carotene and vitamin A on lung cancer and cardiovascular disease. *New England Journal of Medicine*, *334*, 1150–1155.
66. Deding, U., Baatrup, G., Christensen, L. P., & Kobaek-Larsen, M. (2020). Carrot intake and risk of colorectal cancer: A prospective cohort study of 57,053 Danes. *Nutrients*, *12*, 332.
67. Yang, R. L., Yan, Z. H., & Lu, Y. (2008). Cytotoxic phenylpropanoids from carrot. *Journal of Agricultural and Food Chemistry*, *56*, 3024–3027.
68. Purup, S., Larsen, E., & Christensen, L. P. (2009). Differential effects of falcarinol and related aliphatic C17-polyacetylenes on intestinal cell proliferation. *Journal of Agricultural and Food Chemistry*, *57*, 8290–8296.
69. Kobaek-Larsen, M., Baatrup, G., Notabi, M. K., El-Houri, R. B., Pipó-Ollé, E., Christensen, E. A., & Christensen, L. P. (2019). Dietary polyacetylene oxylipins falcarinol and falcarindiol prevent inflammation and colorectal neoplastic transformation: A mechanistic and dose-response study in a rat model. *Nutrients*, *11*, 2223.
70. Zaini, R. G., Brandt, K., Clench, M. R., & Le Maitre, C. L. (2012). Effects of bioactive compounds from carrots (*Daucus carota* L.), polyacetylenes, beta-carotene and lutein

on human lymphoid leukemia cells. *Anti-Cancer Agents in Medicinal Chemistry, 12,* 640–652.

71. Young, J. F., Duthie, S. J., Milne, L., Christensen, L. P., Duthie, G. G., & Bestwick, C. S. (2009). Biphasic effect of falcarinol on Caco-2 cell proliferation, DNA damage, and apoptosis. *Journal of Agricultural and Food Chemistry, 55,* 618–623.
72. Kobaek-Larsen, M., El-Houri, R. B., Christensen, L. P., Al-Najami, I., Fretté, X., & Baatrup, G. (2017). Dietary polyacetylenes, falcarinol and falcarindiol, isolated from carrots prevents the formation of neoplastic lesions in the colon of azoxymethane-induced rats. *Food & Function, 8,* 964–974.
73. Barrero, A. F., Oltra, J. E., Alvarez, M., Raslan, D. S., Saude, D. A., & Aksira, M. (2000). New sources and antifungal activity of sesquiterpene lactone. *Fitoterapia, 71,* 60–64.
74. Bruno, M., Bancheva, S., Roselli, S., & Maggio, A. (2013). Sesquiterpenoids in subtribe Centaureinae (Cass.) Dumort. (tribe Cardueae, Asteraceae): Distribution, 13C NMR spectra data and biological properties. *Phytochemistry, 95,* 19–93.
75. Jasicka-Misiaka, I., Lipoka, J., Nowakowska, E. M., Wieczorek, P. P., Młynarz, P., & Kafarski, P. (2004). Antifungal activity of the carrot seed oil and its major sesquiterpene compounds. *Zeitschrift für Naturforschung C, 59*(11, 12), 791–796.
76. Sharma, M., Chahal, K. K., Kaur, R., Singh, R., & Kataria, D. (2019). Antifungal potential and structure–activity relationship of carrot seed constituents. *Journal of Food Biochemistry, 43*(9), 1–9.
77. John, S., Priyadarshini, S., Monica, S. J., Sivaraj, C., & Arumugam, P. (2017). *In vitro* evaluation of antioxidant and antimicrobial activity of carrot peel. *International Journal of Pharmacognosy and Phytochemical Research, 9*(7), 970–974.
78. Al-Baarri, A. N., Legowo, A. M., Hayakawa, S., & Ogawa, M. (2015). Enhancement of antimicrobial activity of hypothiocyanite using carrot against *Staphylococcus aureus* and *Escherichia coli. Procedia Food Science, 3,* 473–478.
79 Tavares, A. C., Gonçalves, M. J., Cavaleiro, C., et al. (2008). Essential oil of *Daucus carota* subsp. *halophilus*: Composition, antifungal activity and cytotoxicity. *Journal of Ethnopharmacology, 119*(1), 129–134.
80. Staniszewska, M., Kula, J., Wieczorkiewicz, M., & Kusewicz, D. (2005). Essential oils of wild and cultivated carrots–The chemical composition and antimicrobial activity. *Journal of Essential Oil Research, 17,* 579–583.
81. Alves-Silva, J. M., Zuzarte, M., Gonçalves, M. J., Cavaleiro, C., Cruz, M. T., Cardoso, S. M., & Salgueiro, L. (2016). New claims for wild carrot (*Daucus carota* subsp. *carota*) essential oil. *Evidence-Based Complementary and Alternative Medicine,* 1–10.
82. Kamiloglu, S., Grootaert, C., Capanoglu, E., Ozkan, C., Smagghe, G., Raes, K., & Camp, J. V. (2017). Anti-inflammatory potential of black carrot (*Daucus carota* L.) polyphenols in a co-culture model of intestinal Caco-2 and endothelial EA.hy926 cells. *Molecular Nutrition & Food Research, 61*(2), 1–11.
83. Olejnik, A., Kowalska, K., Kidoń, M., Czapski, J., Rychlik, J., Olkowicz, M., & Dembczyński, R. (2016). Purple carrot anthocyanins suppress lipopolysaccharide-induced inflammation in the co-culture of intestinal Caco-2 and macrophage RAW264.7 cells. *Food Function, 7,* 557.

84. Ekam, V. S., Udosen, E. O., & Chigbu, A. E. (2006). Comparative effect of carotenoid complex from Golden Neo-Life Dynamite (GNLD) and carrot-extracted carotenoids on immune parameters in albino Wistar rats. *Nigerian Journal of Physiological Sciences*, *21*(1–2), 1–4.
85. Nishida, K., Sugimoto, M., Ikeda, S., & Kume, S. (2014). Effects of supplemental β-carotene on mucosal IgA induction in the jejunum and ileum of mice after weaning. *British Journal of Nutrition*, *111*(2), 247–253.
86. Kim, S. K., Cuzzort, L. M., & Allen, E. D. (1991). Effects of age on diabetes- and insulin-induced changes in pancreatic levels of alpha-amylase and its mRNA. *Mechanisms of Ageing and Development*, *58*(2, 3), 151–161.
87. Tundis, R., Loizzo, M. R., & Menichini, F. (2010). Natural products as alpha-amylase and alpha-glucosidase inhibitors and their hypoglycaemic potential in the treatment of diabetes: An update. *Mini-Reviews in Medicinal Chemistry*, *10*(4), 315–331.
88. Karkute, S. G., Koley, T. K., Yengkhom, B. K., Tripathi, A., Srivastava, S., Maurya, A., & Singh, B. (2018). Anti-diabetic phenolic compounds of black carrot (*Daucus carota* subsp. *sativus* var. *atrorubens* Alef.) inhibit enzymes of glucose metabolism: An *in silico* and *in vitro* validation. *Medicinal Chemistry*, *14*(6), 641–649.
89. Kumar, S., Kumari, S., Prasad, N., Dubey, N. K., & Dubey, U. (2020). Anti-diabetic, haematinic and anti-cholesterolemic effects of carrot (*Daucus carota* Linn.) juice metabolites to cure alloxan monohydrate induced type-1 diabetes in albino rats. *Journal of Diabetes and Metabolic Disorders Control*, *7*, 1–4.

CHAPTER 9

Potato (*Solanum tuberosum*)

IMEN HENANE,[1,*] RIADH ILAHY,[1] IMEN TLILI,[1] MARWA CHOUIKHI,[1] YOSR LAAYOUNI,[1] HATEM ZGALLAI,[2] and NOURI KHAMASSI[1]

[1]*Laboratory of Horticulture, National Agricultural Research Institute of Tunisia (INRAT), University of Carthage, Ariana Governorate, Tunisia*

[2]*National Agricultural Research Institute of Tunisia (INRAT), University of Carthage, Ariana Governorate, Tunisia*

*Corresponding author

ABSTRACT

Potatoes (*Solanum tuberosum*) are a widely cultivated and consumed staple food crop. Potatoes are not solely recognized for their versatility and affordability as a food choice. They are increasingly being recognized as a "health food" due to their numerous associated health benefits. Potatoes serve as a valuable reservoir of phytochemical compounds and vitamins, which have demonstrated antioxidant properties. These compounds can aid in shielding the body against oxidative stress and lowering the likelihood of chronic illnesses. Regular consumption of potatoes has been linked to better digestion, decreased inflammation, and a reduced likelihood of chronic illnesses like heart disease and cancer. In conclusion, incorporating potatoes into a balanced diet can offer both healthful and therapeutic properties.

9.1 INTRODUCTION

Potato is an important and strategic tuber crop, which ranks third after rice, wheat, and maize. Potato world's production reached a record 381.7 million

Postharvest Quality Attributes of Fresh Vegetables and Grains.
Riadh Ilahy, Imen Tlili, Thouraya R'him, and Muhammad Azam (Eds.)
© 2027 Apple Academic Press, Inc. Co-published with CRC Press (Taylor & Francis)

DOI:10.1201/9781779642851-12

tons in 2021 [1]. In most countries, the potato is increasingly becoming an important staple food, and the per capita consumption per year currently exceeds 30 kg in Tunisia. Depending on the Mediterranean climatic conditions, potatoes are cultivated under various seasons, such as season, late season, extra-early, early, and recently summer-crop in Tunisia [2, 3]. The Spunta cultivar is the most commonly grown variety, accounting for up to 86% of the total production [4], and the Cap Bon region serves as the primary hub for potato production, encompassing both consumption and the cultivation of potato seeds. Potato is mainly used for fresh consumption and processing, but some recent cultivars were bred for double purposes.

Besides starch, potato tubers accumulate various bioactive secondary metabolites with beneficial impacts on human health [4–6]. While white-fleshed potato tubers are mostly preferred by consumers, recently the attention has shifted to colored/pigmented potato cultivars, and many breeding programs for tuber quality improvement are ongoing in various research institutes [4]. Besides, potato is subjected to various processing treatments and home cooking practices with a notable impact on the functional quality, depending on the cultivar and applied treatment. Therefore, this chapter intends to briefly summarize the functional quality in different potato cultivars and the effect of various cooking techniques on the nutritional profile of potato tubers.

9.2 HISTORY

Towards the close of the previous century, the question of where cultivated plants originated has attracted much attention from various scientists. Furthermore, there existed a distinct necessity for the integration of various disciplines, including historical research, language studies, plant biology, and archaeology, to facilitate the tracing of the intricate evolutionary patterns of our heritage crops. Numerous of these crops had already been under cultivation approximately 9 to 10 millennia ago, as was suggested earlier by Candolle, Vavilov, and others.

Many amazing ceramics and ancient handicrafts evidenced the antiquity of potato cultivation and were found on the northern coast of Peru, inherited from the Moche, Chimu, and Inca cultures from the fourth century AD onwards. Almost all the ceramics were found in Peru, and none were from Colombia, Ecuador, Bolivia, Argentina, or Chile, although this culture was popular and largely distributed in those countries as well.

Focused researches using light and scanning electron microscope techniques applied to starch and cell structures pointed potatoes back to the Chilca valley near Lima, radiocarbon dated at an age of c. 7000 years before present [6]. Others have found potatoes dated back to 4000 b.p. at Casma in Peru, reported by Ugent et al. [7], or 13000 b.p. in Chile, as was suggested by Ugent et al. [8].

Besides, the native names of the potato evidenced the distribution and antiquity of potato cultivation [9], based on their total difference in the main Indian languages that were spoken in old areas dedicated to potato cultivation. Potato names were *iomza*, *iomuy* for the Chibcha language in Central Columbia, *ajsu*, *impari*, and then *papa* in Quechua, the language of the Inca empire, *amka* and *choque* were used by the Aymani Indians in Bolivia, whilst in Chile, the Araucanians gave it the name *poni*.

In Europe, the word *papa*, referring to potato, was never adopted, but the actual word was rather extracted from the term batata, which originates from the Caribbean Arawak Indian word for Ipomoea, specifically the sweet potato. This can be attributed to the fact that the Spaniards were familiar with the sweet potato. The potato appeared on the European continent almost at the end of the 16th century. Unfortunately, there is no solid reference about the date of actual introduction, and available data rely only on circumstantial evidence.

Salaman [10] assumed that in 1570, the potato first arrived in southern Spain and was cultivated somewhere in Seville, and then transported to Italy and the whole of Europe after that, suggesting the Spanish source of the potato spread through Europe. Hawkes et al. [11] also proposed a second introduction point of the potato to Europe, which is England, and this was earlier, referring back to 1588–1593. Regardless of the exact introduction date and place of the potato, it seems that from the Spanish introduction, it spread to Europe and parts of Asia, but from England, it spread to Ireland, Scotland, and Wales and parts of northern Europe, and from different British sources to almost all of the British overseas colonies and the USA. Regarding the Slavic nations, it seems that the potato was mainly introduced from Germany since potato names there are derived from German ones, such as Kartoffel and Grundbirne.

9.3 BRIEF OVERVIEW OF POTATO CLASSIFICATION

The genus Solanum, including cultivated potato, is large, comprising up to 1000 species. The cultivation of this crop is widespread across the globe, except for the extreme north and south, with a strong concentration of species diversity in both South and Central America, alongside Australia. Besides *Solanum tuberosum,* 6 other cultivated species and more than 230 wild species of potato are generally recognized. The remaining part of this genus is comprised of non-tuberiferous species such as *Solanum nigrum* and various spiny herbs and shrubs.

The section *Tuberarium*, formerly including most of the tuber-bearing species, has been changed to the older name *Petota*, which seems more appropriate. They are distinguished mostly by several aspects and traits such as actinomorphic flowers, a jointed or articulated pedicel or flower stalk, unbranched glandular or eglandular hairs, no thorns or spines, and generally compound leaves. Potato leaf is interruptedly imparipinnate, and most species in *Petota* possess underground stolons, bearing at their tips potato tubers. However, some morphologically close species are devoid of stolons and tubers such as *Iuglandifolia* and *Etuberosa*. The *Petota* section is comprised of two subsections: ESTOLONIFERA (containing the two non-tuber-bearing series *Etuberosa* and *Juglandifolia*) and POTATOE (containing 19 other series) disregarding herbarium specimen species.

9.4 ANTIOXIDANT PHYTOCHEMICALS OF FLESH AND SKINS OF POTATO TUBERS

9.4.1 POTATO FLESH

Potato is a versatile food, rich in carbohydrates (>95% of the energy in uncooked/unprocessed potato) and water (62–87 g/100 g). The protein content of the potato is similar to that of cereals and notably higher when compared to other root vegetables and tubers, which typically range from 0.85–4.2 g/100 g with an average of 2.12 g/100 g. In addition to vitamins such as B1, B2, B3, and B6, potato tubers contain 3.5–4.3 g/100 g amylose, a high level of phosphorus, and potassium, attaining 126 mg/100 g and 693.8 mg/100 g, respectively [12]. Potato tubers contain an array of phytochemicals such as carotenoids, phenolics, flavonoids, and ascorbic acid (AA), contributing to the sensory attributes as well as the health-promoting properties of potatoes.

9.4.1.1 CAROTENOIDS

Potato is a source of carotenoids, mainly lutein, zeaxanthin, violaxanthin, and neoxanthin. Ezekiel et al. [13] reviewed the main literature related to potato phytochemicals, as well as factors affecting their content. The authors found large variation not only between cultivated and wild potato cultivars, but also among cultivated potatoes in terms of phytochemicals and antioxidant activity. The authors highlighted the need to use this untapped genetic diversity in future breeding programs aiming to improve the functional quality of potato cultivars.

Lutein and zeaxanthin are responsible for the yellow and orange potato flesh, respectively. Research findings indicate that potato tubers with yellow and dark yellow flesh accumulate significantly greater total carotenoid content, ranging from 100 to as high as 1000 μg/100 g fw, in contrast to white-fleshed tubers, which typically contain 50–100 μg/100 g fw [13]. Anthocyanins are present in both purple and red-fleshed potato cultivars, with purple-fleshed potatoes having the greatest concentrations at 368 mg/100 g fw [9–14]. Following evaluating German potato varieties with yellow and white flesh, Breithaupt and Bamedi [15] discovered that the primary carotenoids are violaxanthin, antheraxanthin, lutein, and zeaxanthin, with a cumulative level of 175 μg/100 g fw. Nevertheless, the levels of β-carotene, β-cryptoxanthin, and neoxanthin were quite low, ranging from 41 to 131 μg/100 g fw. The four main carotenoids attained 175 μg/100 g fw compared to only 41–131 μg/100 g fw for the others. The total carotenoid content of immature potato tubers increases and then falls as they grow [16, 17]. Based on their nutritional importance, there is a need to focus on Baby potatoes or immature potatoes in order to promote their consumption.

Lachman et al. [18] investigated the effect of flesh color and growing location on the level of phenolic compounds and antioxidant capacity of different potato genotypes. Among these were the cultivars Ditta, Impala, Karin, and Saturna, all characterized by their yellow flesh, and the purple-fleshed cultivars Valfi and Violette. The authors revealed that pigmented genotypes exhibited the highest total phenolic levels, ranging from 4.55 to 4.81 g/kg dw, and the highest antioxidant activity using the DPPH assay, with Violette ranking top, obtaining 274.4 mg equivalents of AA/kg dw. Yellow-fleshed potato genotypes exhibited lower total phenolic levels and, consequently, lower antioxidant activity values.

Recently, Ilahy et al. [19] focused on the effect of two growing locations on the carotenoid content of different potato genotypes. The authors revealed genotype-dependent accumulation and synthesis of carotenoids. The genotypes UNV and SYN displayed the lowest and greatest total carotenoids content, respectively, at 0 and 13.2 mg β-carotene equivalent/kg fw. These values were 25%–230% greater than those of the cultivar Spunta. Whereas, SYN had the highest level of carotenoids under both growing seasons, suggesting a minor (G × E) interaction with respect to the other genotypes.

9.4.1.2 PHENOLIC AND FLAVONOIDS

Phenolic compounds are a large family of potent secondary metabolites derived from tyrosine and phenylalanine and characterized by an aromatic ring formed by shikimate and acetate pathways. Phenolic compounds are grouped into simple mono- and polyphenols [20] with flavones, flavonols, and their glycoside derivatives as major sub-classes. Phenolic acids, the nonflavonoid polyphenolics containing a carboxylic group attached to phenol with the basic structure C6–C1, are also grouped into two main classes as hydroxyl cinnamic and benzoic acids, whereas polyphenols contain at least two phenolic rings that confer both desirable and undesirable attributes to plant produce [21–24]. Polyphenols were long considered anti-nutritional due to the property of precipitating with macromolecules and some bad tastes associated with some fruits and vegetables. However, consideration has recently changed from anti-nutritional to nutritional, as reported by several investigations supporting their beneficial role in health promotion [24]. The polyphenols are known to possess more radical-inhibition activity than vitamin C and E (*in vitro*) within the body [25]. Additionally, the antioxidant properties of phenolics can serve as possible replacements for harmful synthetic antioxidants such as BHT and BHA, which are widely used in the food industry [26].

In general, consumers' attraction to fruits and vegetables is influenced by both non-visual (taste, flavor, texture, and nutritional content) and visual (size, uniformity, color, and freshness) features, which are important factors in determining their decision to buy a product. According to reports, phenolic chemicals have a significant role in determining the flavor and organoleptic characteristics of a variety of fruits and vegetables [27]. Numerous health benefits, including antimicrobial, anti-inflammatory, antiallergic, antimutagenic, antiviral, antineoplastic, antithrombotic, and vasodilatory activity, have also been linked to the phenolic content [28]. These benefits are primarily attributable to their antioxidant and free radical scavenging qualities [29, 30]. These chemicals are becoming increasingly popular among producers, consumers, and processors.

Potatoes were shown to be the third largest provider of phenolics and antioxidants in the American diet, behind apples and oranges [31, 32]. Chandrasekara and Josheph [33] reported that among six different tuber and starchy crops, the common potato had the highest total phenolics. Potato genotypes with white flesh displayed a phenolic concentration ranging from 369.1 to 527.2 μg/g, whereas those with yellow flesh had a lower amount, ranging from 237.7 to 407.0 μg/g. Purple or red-fleshed cultivars synthesize and store up to 4-fold the phenolic content of white-fleshed genotypes. Various data from different countries suggests that integrating the new colored potato tubers in the daily diet in various forms might contribute to significantly increase the daily intake of various important phytochemicals beneficial for human health.

Flavonoids are also a plant phenolic group mainly associated with sugars (aglycones). Due to their lower redox potential, flavonoids possess the thermodynamic capability to reduce extremely oxidizing free radicals, which have redox potentials within the range of 2.13 to 1.0 V, including radicals like superoxide, peroxyl, alkoxyl, and hydroxyl radicals, through the process of hydrogen atom donation. Van et al. [34] found that flavonoids could lower free radical production by chelating metal ions like iron and copper. Quercetin, specifically, is recognized for its capacity to chelate iron and maintain its stability. Within flavonoid structures, trace metals affix themselves to distinct positions within the multiple rings [35]. The basic structure of flavonoids comprises two aromatic rings, denoted as "A" and "B," joined together by a three-carbon "C" ring (C6–C3–C6), and the distinction changes in the oxygenated heterocyclic C3 ring classifies them into subtypes, with major ones including flavonols, flavanols (3-ol), flavones, flavanones, isoflavanoids, and anthocynadins [22–30]. Galactose, xylose, and rhamnose are among the sugars involved in the production of glycosides, while glucose is the primary sugar [36]. The flavonol ring A is synthesized from the combination of three malonyl-CoA molecules, whereas the aromatic B ring contains a chromanol structure, and the C ring is derived from phenylalanine [37, 38].

The tubers of potatoes assemble flavonoids at a rate of 200–300 g/g fw [39]. Catechin, epicatechin, erodictyol, kaempeferol, and naringenin are the most abundant flavonoids in potato tubers, with rutin accounting for just a small percentage [40]. Due to their high intake, potatoes and potato-derived products are a source of several low-level chemicals, including flavonols [41].

Colored-fleshed potato tubers (red or purple) owe their characteristic coloring to polyphenolic compounds, anthocyanins, occurring both in flesh and/or in the peel [42]. Anthocyanins, which belong to the flavonoid sub-group, are found in significant quantities in potatoes with pigmented flesh. In red and purple potato genotypes, the average phenolic content is 2-3-fold higher with respect to yellow and white-fleshed potato cultivars. Anthocyanins showed high antioxidant activity and numerous biological functions, including neutralizing free radicals, anti-mutagenicity, anti-carcinogen, and anti-hypertensive activity.

In potato tubers, anthocyanin levels ranged from 5.5 to 35 mg/100 g fw [43]. It has been shown that purple and red-fleshed cultivars acquire two-fold higher flavonoid content with respect to white-fleshed cultivars. Flavonoid content differs among potato fractions; the highest levels were detected in potato skin, attaining 900 mg/100 g fw in purple-fleshed and 500 mg/100 g fw in red-fleshed types [39]. Acylated glucosides of pelargonidin are accumulated by both red and purple flesh potatoes; only purple potatoes additionally contain acylated glucosides of malvidin, petunidin, peonidin, and delphinidin [18–40]. Additionally, as stated in Fossen et al. [44], in the purple-fleshed potato cultivar Congo, unique anthocyanins were found to be comprised of feruloyl glucoside and glucoside-pyranosides of malvidin and petunidin, which are novel anthocyanins. Compared to yellow, white, and white/purple fleshed potato genotypes, red purple potato tubers acquire higher levels of neochlorogenic acid, caffeoyl putrescine, chlorogenic acid, cryptochlorogenic acid, caffeic acid, and kaempferol-3-rutinose.

Navarre et al. [45], during their evaluation of more than 50 potato genotypes, including cultivars, breeding lines, primitive germplasm, and wild species, for phenolics and hydrophilic antioxidant activity, concluded that chlorogenic acid was the major phenolic, with the levels ranging from 22 to 473 mg/100 g dw. Rutin and kaempferol-3-rutinose were the most abundant flavonols with total phenolics ranging from 1.8 to 11 mg/g dw and antioxidant capacity from 27 to 219 μmol TE/g DW. Since the highest-phenolic accumulating genotypes are colored-flesh potatoes and American consumers generally prefer white-flesh potatoes, the authors proposed creating high-phytonutrient white-flesh lines appropriate for different processing requirements and prioritized those that function effectively in various growing regions.

Fossen et al. [46] extracted anthocyanins from the sprouts and tubers of an ancient Norwegian potato cultivar and determined their entire structural makeup. The authors documented the existence of two previously unidentified anthocyanins that were acylated with caffeic acid. The caffeic acid-acylated anthocyanins constitute 18% of the total anthocyanin content in the sprouts. The tubers containing a purple zone in the flesh exhibited higher proportions of anthocyanins acylated with p-coumaric acid. Considering colorful potatoes as potential sources of food coloring, Opheim and Andersen [47], Rodriguez-Saona et al. [48], and Reyes et al. [49] assessed how pH and temperature affect the stability and visual color of aqueous anthocyanin extracts from purple and red-fleshed potatoes, contrasting them with commercial grape and purple carrot anthocyanin extracts. The authors highlighted that purple carrot and red-flesh potato extracts revealed better stability with respect to grape and purple-flesh potato extracts. In this context, they suggested using potatoes with purple and red flesh as a natural colorant source when the pH is in the right range.

Recently, Ilahy et al. [19] focused on the effect of two growing locations on the carotenoid content of different potato genotypes. The authors revealed genotype-dependent accumulation and synthesis of phenolics among the tested potato clones within growing seasons. The range of total phenolic content was between 548.6 and 1546.9 mg GAE/kg fw in the late cropping season and between 583 and 133.9 mg GAE/kg fw in the summer cropping season, respectively. The range of flavonoid concentration in the summer cropping season was 693.4 to 2629.2 mg RE/kg fw, while in the late cropping season it varied from 1091 to 3642.2 mg RE/kg fw.

Keutgen et al. [50] also evaluated the antioxidant qualities of potato tubers as affected by genotypic diversity and growing conditions on seven new tetraploid potato lines and three modern cultivars, namely Jelly, Satina, and Tajfun. The authors revealed that the antioxidant content and properties were significantly affected by the cultivation system. In fact, phenolic and flavonoid, vitamin C content, as well as the antioxidant activity (FRAP assay), were higher under organic growing conditions, but the citric acid content was higher under an integrated production system.

Ru et al. [51], assessing 14 potato genotypes differing in flesh color grown in China, reported that total phenolic content was highest in red and purple-fleshed potato tubers, ranging from 7.72 to 40.45 in the bound fraction and 51.36 to 70.67 mg GAE/g of potato powder in the free fraction, but lower values were recorded for white and yellow-fleshed potato genotypes.

Cebulak et al. [52] reported generally higher phenolic acids, chlorogenic acid, gallic acid, p-coumaric acid, and caffeic acid content in colored potato genotypes with respect to white potato genotypes. The results of Rasheed et al. [53] indicated that potato tubers possess hydrophilic antioxidants, which encompass phenols, AA, anthocyanins, and flavanols. Additionally, chlorogenic acid, gallic acid, caffeic acid, and catechin significantly contribute to the antioxidant activity in yellow-fleshed potato tubers, whereas in red- and purple-fleshed tubers, the primary contributors are chlorogenic acids and anthocyanins.

9.4.1.3 VITAMIN C

Based on its biological action, vitamin C possesses reductive characteristics. It provides two main functions: antioxidant activity and a cofactor role in oxygen-catalyzed hydroxylation reactions. Vitamin C is characterized by a very high antioxidant potential. It is mainly produced by fresh fruits and vegetables and their byproducts [54]. Vitamin C is present in significant quantities in several root crops. The content can be lowered while cooking. A medium-sized potato has a large amount of vitamin C: 65–420 mg/kg fw [12, 41, 42, 55]. In the United States, potatoes are thought to have 20% as much vitamin C as fruits (18%), according to McCay et al. [56]. The vitamin C content of potatoes from the high Tiber valley area was evaluated by Tamasi et al. [57]. Compared to Daytona and Volare (1.04–1.10 g/kg fdw), Kennebec and Sifra had higher vitamin C contents (1.27 to 1.38 g/kg f dw). Although the high vitamin C content in Kennebec, cooking reduced the content by 50%. Recently, Ilahy et al. [19] focused on the effect of two growing locations on the carotenoid content of different potato genotypes and showed differences between 144.3 and 231.2 mg/kg fw and between 211.51 and 289.33 mg/kg fw during the summer and late cropping seasons, respectively.

9.4.2 POTATO SKIN

Nowadays, consumers are becoming more interested in healthy foods and fresh fruits and vegetables, providing the required intake of beneficial antioxidants. Natural antioxidants found in potato peels include polyphenols, glycoalkaloids, polysaccharides, proteins, and vitamins [58]. It has been found that half of the tuber's phenolic content is contained in the skin proportion and adjacent tissues, decreasing toward the center. According to a recent study, potato peels accumulate different phenolic compounds such as chlorogenic acid, gallic acid, rutin, quercetin, p-coumaric acid, and anthocyanins. The concentrations varied from 0.17 to 4.10 mg/g dw for chlorogenic acid, from 0.16 to 3.33 mg/g dw for caffeic acid, from 0.03 to 0.85 mg/g dw for ferulic acid derivative, from 0.07 to 5.30 mg/g dw for p-coumaric acid, from 0.002 to 0.068 mg/g dw for total anthocyanin, with a total phenolic content ranging from 4.54 to 13.9 mg GAE/100 g dw for different potato genotypes varying in flesh color [58]. Regarding the antioxidant power, it has been reported that values varied from 560.93 to 1907.26 mg GAE/g dw using the FRAP assay, from 0.28 to 2.14 mM/g dw using the TEAC assay, and from 82 to 90% using the FRSA.

Besides, potato skin has attracted great interest for its large health-giving properties, and here below is an overview of the recent findings about the medical properties of potato peel. It has been reported that potato peel extract is characterized by high antioxidant capacity and has hepato-protective effects, reduces toxicity of cholesterol oxidation products, and protects against atopic dermatitis in a mouse model and erythrocytes against oxidative damage [59–61]. Potato peel powders were also tested for mutagenic capacity against *Salmonella tiphymurium-Esherechia coli* using microsome assay and revealed its safety for use as a natural antioxidant for food processing and preservation. Peel powders exhibited a potent activity in reducing diabetic alterations and hypocholesterolaemic effect [61]. Dietary supplementation with 10% of potato peel powder significantly the reduced oxidative stress, glycemia, lipid oxidative products, and protects eye lens in a rat model of streptozotocin-induced diabetes.

9.5 ANTIOXIDANT ACTIVITY

Potato tubers accumulate and store various secondary metabolites with strong antioxidant activity. Those compounds contribute to various processes such as plant defense against pathogens and free-radical-mediated reactions. The levels of antioxidants in potato tubers vary not only depending on genotypes (flesh colors) but also depending on the considered fraction (skin/flesh). In fact, lower antioxidant activity (FRAP assay) was observed in yellow and white-fleshed potato tubers, ranging from 3.22 μmol Trolox/g dw in cultivar Lord to 3.67 μmol Trolox/g dw in cultivar Bella Rosa, with respect to red and purple tubers of potatoes varying from 3.36 μmol Trolox/g dw in Blue Congo to 6.89 μmol Trolox/g dw in Vittelotte [52]. Additionally, greater antioxidant activity was detected in the skin, ranging from 13.88 to 46.40 in white and yellow potato skin with respect to purple potato skin with 17.10 μmol TE/g dw.

Similarly, for flavonoids, quercetin was reported as 3-fold more potent with respect to kaempferol, eridictyol, and catechin. Ilahy et al. [19], assessing the behavior of different potato genotypes under different growing locations, found that the HAA ranged from 216.3 to 381.2 μM Trolox/100 g fw in Spunta and UNV, respectively, when grown under the new cropping season, and from 237.9 to 405.9 μM Trolox/100 g fw in NMA and Spunta, respectively, when grown under the late cropping season. Regarding the LAA, the data revealed variation also between genotypes, from 107.9 μM Trolox/100 g fw for BDA and SYN to 196.3 μM Trolox/100 g fw grown under the new cropping season, and from 101.75 to 182.41 μM Trolox/100 g fw for cultivars grown under the late cropping season. In comparison to Spunta and BDA during the (NCS), hydrophilic and lipophilic antioxidant activity were 24%–76% and 3%–83% higher, respectively.

Reyes et al. [62] observed a high positive correlation between antioxidant capacity and anthocyanin and phenolic content, and concluded that these compounds are the major contributing factor to the antioxidant activity. Potatoes can help boost antioxidant intake significantly. The reported hydrophilic ORAC values vary from 200 to 1400 μmol Trolox equivalents/100 g FW, while lipophilic ORAC values range from 5 to 30 nmol α-tocopherol equivalents/100 g FW [43].

9.6 THERAPEUTIC VALUE

Consumers are increasingly interested in foods that provide health advantages in addition to the fundamental nutrition they require. Various health-promoting actions were generally linked to the antioxidant and free radical scavenging characteristics of phenolics and were the topic of numerous earlier studies and in-depth reviews [28, 29, 63, 64].

Potatoes serve as a prominent dietary source of phenolic compounds and various antioxidants. These antioxidants obtained from potatoes are characterized by strong free radical scavenging activity and lower the blood's cholesterol buildup in order to reduce the development of coronary heart disease. From a nutritional standpoint, potatoes are the second-highest source of polyphenol intake for humans, following only tomatoes. Potato peel contains phenolic acids, mainly chlorogenic acid (50.31%).

Additionally, other polyphenols like gallic acid (41.67%), protocatechuic (7.81%), and caffeic (0.21%) acids were also present in potato peel. Potato peel extract showed similar antioxidant capacity to widely used commercial antioxidants such as BHA and BHT [43]. In this context, it is recommended to consume potato tubers along with their skin.

Chlorogenic acid is well known for its large health-giving properties, among others: protect against main degenerative diseases, cancer, heart disease [65], hypertension [66], and viral and bacterial diseases; inhibit the matrix metalloproteinase (MMP)-9, an important enzyme responsible for tumor invasion and metastasis [67]; reduce the rate at which glucose is released into the bloodstream [68]; and block nitrosamine formation [5].

Additionally, recent studies highlighted that natural antioxidants in complex matrices consumed with the diet are more efficacious than pure single compounds in preventing degenerative pathologies due to particular interactions and synergism [69]. Anthocyanins are best known to prevent cardiovascular diseases (CVD), cancer, and diabetes [70, 71]. Anthocyanin-enriched fractions and extracts obtained from colored potatoes have demonstrated their ability to alleviate lymph-node carcinoma of the prostate and prostate cancer cell proliferation [71]. Purple potato flakes rich in phenols and anthocyanins were proved to protect against high-cholesterol diet oxidative damage in rats [72], and red potato flakes improved the antioxidant system via enhancing hepatic superoxide dismutase mRNA in rats [72]. Potatoes are also beneficial for eye health due to their high carotenoid content [73, 74].

Numerous studies have demonstrated the efficiency of secondary metabolites in different potato fractions in combating various diseases. Singh et al. [59] studied the potential of potato skin extract to offer protection against carbon tetrachloride-induced damage to the liver in rats. Enzymatic antioxidants and biochemical parameters were monitored to assess hepatic damage. The authors illustrated that pretreatment with potato peel extract substantially mitigates the liver injury induced by carbon tetrachloride in rats, likely attributed to its robust antioxidant activity.

Thompson et al. [75] examined the impact of freeze-dried potato powder on chemically induced breast carcinogenesis in rats, evaluating both dose dependency and variations in anticancer activity among different potato cultivars. The authors noted no adverse effects in rats fed diets containing freeze-dried potato powder ranging from 5% to 50% (w/w). Using a range of dietary concentrations (12.5%, 25%, and 50%, w/w) of a red pigmented potato cv. Mountain Rose, with high content of chlorogenic acid derivatives and anthocyanin content, showed great inhibition of carcinogenesis (23% reduction in cancer incidence and a 49% reduction in cancer multiplicity).

Charepalli et al. [14], taking into account the available evidence on functional compounds like anthocyanins and their potential impact on cancer, emphasized the importance of investigating the anticancer potential of purple-fleshed potato, a widely consumed global food crop, against colon cancer stem cells. Effects of potato anthocyanin/purple fleshed potato extract were compared to a positive control. *In vitro*, potato anthocyanin inhibited the growth and raised colon cancer stem cell programmed apoptosis, independent of p53. *In vivo*, the extract from purple-fleshed potatoes decreased the quantity of crypts housing cells with nuclear β-catenin through the initiation of apoptosis. Additionally, it inhibited tumor occurrence to a degree similar to that observed in the control group. Further studies in the near future will certainly reveal more bioactivities by which potato antioxidants exert their therapeutic effect on human health *in vivo* as well as *in vitro* experiments, leading to a complete shift toward colored potatoes and promote their consumption.

According to a recent survey, Rasheed et al. [53] reported updated data about the beneficial health properties associated with potato polyphenols. In fact, potato tubers exhibit anti-obesity effects by inhibiting lipid metabolism and suppressing adipogenesis; anti-inflammatory activity by reducing wounds, suppressing proliferation of colon cancer cells, and reducing inflammatory diseases; antidiabetic activity by reducing insulinemia and preventing type-2 diabetes; anticancer activity by inhibiting liver and prostate tumor cells; and antioxidant activity by reducing cellular toxicities, reducing CVDs, and inhibiting lipid peroxidation.

It has been shown that potato peel extract protects against liver injury [59] and erythrocyte oxidative damage [60]. The bioactive compounds in freeze-dried potato powder induced an important reduction in induced breast cancer in a rat model. Additionally, various beneficial effects have been attributed to potato phenolics, such as antibacterial, anti-inflammatory, and vasodilatory activity [28–63]. Most of these effects are mainly attributed to the potent radical scavenging properties of phenolic compounds [29, 30].

9.7 CONCLUSIONS AND FUTURE TRENDS

It is not clear which is more beneficial for human health, colored or traditional potato tubers, in terms of functional quality and phytochemical composition. Despite the amount of phenolic compounds, content in potatoes are comparatively low with respect to other fresh produce; high antioxidant activity is generally recorded. Practically consuming either white and/or colored potato cultivars, or both types along with their skin, is rather more beneficial for human health. Besides, various innovative uses of potato or potato derivates are emerging, such as the biofortification of potato for food and nutrition security in developing countries, composite potato-containing flour for various food purposes, and the use of pigmented potatoes as a potential source of natural anthocyanins for use in the food industry, since producing potatoes is less expensive than producing other horticulture crops.

KEYWORDS

- **antioxidant activity**
- **colored potato cultivars**
- **functional quality**
- **nutritional value**
- **phytochemicals**
- ***Solanum tuberosum***

REFERENCES

1. N/A. (2017). http://www.fao.org/faostat/en/#data/QC (accessed on 1st January 2026).
2. Harbaoui, K., Vander Lee, T., Vleeshouwers, G. A. A. V., Khammassy, N., Harrabi, M., & Hamada, W. (2013). Characterization of *Phytophthora infestans* isolates collected from potato and tomato crops in Tunisia during 2006–2008. *Potato Research*, *56*, 11–29.
3. Khamassy, N., Essid, M. F., & Ilahy, R. (2018). A first insight on the effect of climatic change on potato production under Tunisian Highland conditions. In *Proceedings of the 9th World Potato Congress*, Cusco, Peru, May 27–31, 2018.
4. Ilahy, R., Tlili, I., & Khamassy, N. (2013). Total carotenoids and lycopene content in some potato (*Solanum tuberosum* L.) clones under selection in Tunisia. In *Proceedings of the 9th Triennial Conference of the African Potato Association*, June 30–July 4, Great Rift Valley Lodge, Naivasha, Kenya.
5. Friedman, M. (1997). Chemistry, biochemistry, and dietary role of potato polyphenols: A review. *Journal of Agricultural and Food Chemistry*, *45*, 1523–1540.
6. Martins, R. (1976). *New Archaeological Techniques for the Study of Ancient Root Crops in Peru* (Ph.D. thesis). University of Birmingham.

7. Ugent, D., Pozorski, S., & Pozorski, T. (1982). Archaeological potato tuber remains from the Casma Valley of Peru. *Economic Botany, 36*, 182–192.
8. Ugent, D., Dillehay, T., & Ramirez, C. (1987). Potato remains from a late Pleistocene settlement in southcentral Chile. *Economic Botany, 41*, 17–27.
9. Hawkes, J. G. (1947). On the origin and meaning of South American Indian potato names. *Botanical Journal of the Linnean Society, 53*(350), 205–250.
10. Salaman, R. N. (1937). The potato in its early home and its introduction into Europe. *Journal of the Royal Horticultural Society, 62*, 61–77, 112–123, 156–162, 253–266.
11. Hawkes, J. G. (1967). The history of the potato. *Journal of the Royal Horticultural Society, 92*, 207–224, 249–262, 288–302, 364–365.
12. Burlingame, B., Mouille, B., & Charrondiere, R. (2009). Nutrients, bioactive non-nutrients and anti-nutrients in potatoes. *Journal of Food Composition and Analysis, 22*, 494–502.
13. Ezekiel, R., Singh, N., Sharma, S., & Kaur, A. (2013). Beneficial phytochemicals in potato—A review. *Food Research International, 50*, 487–496.
14. Charepalli, V., Reddivari, L., Radhakrishnan, S., Vadde, R., Agarwal, R., & Vanamala, J. K. (2015). Anthocyanin-containing purple-fleshed potatoes suppress colon tumorigenesis via elimination of colon cancer stem cells. *The Journal of Nutritional Biochemistry, 26*(12), 1641–1649.
15. Breithaupt, D. E., & Bamedi, A. (2002). Carotenoids and carotenoid esters in potatoes (*Solanum tuberosum* L.): New insights into an ancient vegetable. *Journal of Agricultural and Food Chemistry, 50*, 7175–7181.
16. Kotikova, Z., Hejtmankova, A., Lachman, J., Hamouz, K., Trnkova, E., & Dvorak, P. (2007). Effect of selected factors on total carotenoid content in potato tubers (*Solanum tuberosum* L.). *Plant, Soil and Environment, 53*, 355–360.
17. Morris, W. L., Ducreux, L., Griffiths, D. W., Stewart, D., Davies, H. V., & Taylor, M. A. (2004). Carotenogenesis during tuber development and storage in potato. *Journal of Experimental Botany, 55*, 975–982.
18. Lachman, J., & Hamouz, K. (2005). Red- and purple-colored potatoes as a significant antioxidant source in human nutrition—A review. *Plant, Soil and Environment, 51*, 477–482.
19. Ilahy, R., Tlili, I., R'Him, T., Homa, F., Lenucci, M. S., & Khamassy, N. (2019). Effect of growing seasons on agronomic and functional traits of Tunisian potato genotypes. *Journal of Postharvest Technology, 7*(2), 1–20.
20. Harborne, J. B. (1989). *Methods in Plant Biochemistry, Plant Phenolics* (Vol. 1). Academic Press Ltd.
21. Babic, I., Amiot, M. J., Nguyen, C., & Aubert, S. (1993). Changes in phenolic content in fresh ready-to-use shredded carrots during storage. *Journal of Food Science, 58*, 351–356.
22. Liu, R. H. (2004). Potential synergy of phytochemicals in cancer prevention: Mechanism of action. *The Journal of Nutrition, 134*(12), 3479S–3485S.
23. Marinova, D., Ribarova, F., & Atanassova, M. (2005). Total phenolics and total flavonoids in Bulgarian fruits and vegetables. *Journal of Chemical Technology and Metallurgy, 40*, 255–260.
24. Bravo, L. (1998). Polyphenols: Chemistry, dietary sources, metabolism and nutritional significance. *Nutrition Reviews, 56*, 317–333.

25. Rice-Evans, C. A., Miller, N. J., & Paganga, G. (1996). Structure–antioxidant activity relationships of flavonoids and phenolic acids. *Free Radical Biology and Medicine, 20*(7), 933–956.
26. Temelli, F., Saldana, M. D. A., & Comin, L. (2012). Application of supercritical fluid extraction in food processing. In J. Pawliszyn (Ed.), *Comprehensive Sampling and Sample Preparation* (Vol. 4, pp. 415–440). Elsevier, Amsterdam.
27. Barba, A. A., Calabretti, A., d'Amore, M., Piccinelli, A. L., & Rastrelli, L. (2008). Phenolic constituents' levels in cv. Agria potato under microwave processing. *Food Science and Technology, 41*, 1919–1926.
28. Alan, L., & Miller, N. D. (1996). Antioxidant flavonoids, structure, function and clinical usage. *Alternative Medicine Review, 1*, 103–111.
29. Amakura, Y., Umino, Y., Tsuji, S., & Tonogai, Y. (2000). Influence of jam processing on the radical scavenging activity and phenolic content in berries. *Journal of Agricultural and Food Chemistry, 48*, 6292–6297.
30. Rice-Evans, C. A., Miller, N. J., & Paganga, G. (1997). Antioxidant properties of phenolic compounds. *Trends in Plant Science, 2*, 152–159.
31. Chun, O. K., Kim, D. O., Smith, N., Schroeder, D., Han, J. T., & Lee, C. Y. (2005). Daily consumption of phenolics and total antioxidant capacity from fruit and vegetables in the American diet. *Journal of the Science of Food and Agriculture, 85*, 1715–1724.
32. Song, W., Derito, C. M., Liu, M. K., He, X., Dong, M., & Liu, R. H. (2010). Cellular antioxidant activity of common vegetables. *Journal of Agricultural and Food Chemistry, 58*(11), 6621–6629.
33. Chandrasekara, A., & Josheph Kumar, T. (2016). Roots and tuber crop as functional foods: A review on phytochemical constituents and their potential health benefits. *International Journal of Food Science, 2016*.
34. Mira, L., Fernandez, M. T., Santos, M., Rocha, R., Florêncio, M. H., & Jennings, K. R. (2002). Interactions of flavonoids with iron and copper ions: A mechanism for their antioxidant activity. *Free Radical Research, 36*(11), 1199–1208.
35. Van Dam, J., Kooman, P. L., & Struik, P. C. (1996). Effects of temperature and photoperiod on early growth and final number of tubers in potato (*Solanum tuberosum* L.). *Potato Research, 39*, 51–62.
36. Markham, K. R. (1982). *Techniques in Flavonoid Identification* (Vol. 31). Academic Press, London.
37. Fatland, B. L., Ke, J., Anderson, M. D., Mentzen, W. I., Cui, L. W., Allred, C. C., & Wurtele, E. S. (2002). Molecular characterization of a heteromeric ATP-citrate lyase that generates cytosolic acetyl-coenzyme A in *Arabidopsis*. *Plant Physiology, 130*(2), 740–756.
38. Tsao, R., & McCallum, J. (2009). Chemistry of flavonoids. In L. A. de la Rosa, E. Alvarez-Parrilla, & G. Gonzalez-Aguilar (Eds.), *Fruit and Vegetable Phytochemicals: Chemistry, Nutritional Value and Stability* (pp. 131–153). Blackwell Publishing, Ames, Iowa.
39. Lewis, C. E., Walker, J. R. L., Lancaster, J. E., & Sutton, K. H. (1998). Determination of anthocyanins, flavonoids and phenolic acids in potatoes. I: Colored cultivars of *Solanum tuberosum* L. *Journal of the Science of Food and Agriculture, 77*, 45–57.
40. Brown, C. R. (2005). Antioxidants in potato. *American Journal of Potato Research, 62*, 163–172.

41. Tudela, J. A., Cantos, E., Espín, J. C., Tomás-Barberán, F. A., & Gil, M. I. (2002). Induction of antioxidant flavonol biosynthesis in fresh-cut potatoes: Effect of domestic cooking. *Journal of Agricultural and Food Chemistry, 50*, 5925–5931.
42. Andre, C. M., Ghislain, M., Bertin, P., Oufir, M., del Rosario Herrera, M., Hoffmann, L., & Evers, D. (2007). Andean potato cultivars (*Solanum tuberosum* L.) as a source of antioxidant and mineral micronutrients. *Journal of Agricultural and Food Chemistry, 55*(2), 366–378.
43. Brown, C. R. (2008). Breeding for phytonutrient enhancement of potato. *American Journal of Potato Research, 85*, 298–307.
44. Fossen, T. R., & Andersen, O. M. (2000). Anthocyanins from tubers and shoots of the purple potato *Solanum tuberosum. Journal of Horticultural Science and Biotechnology, 75*, 360–363.
45. Navarre, D. A., Pillai, S. S., Shakya, R., & Holden, M. J. (2011). HPLC profiling of phenolics in diverse potato genotypes. *Food Chemistry, 127*, 34–41.
46. Fossen, T., Øvstedal, D. O., Slimestad, R., & Andersen, Ø. M. (2003). Anthocyanins from a Norwegian potato cultivar. *Food Chemistry, 81*, 433–437.
47. Opheim, S., & Andersen, O. M. (1992). Anthocyanins in the genus *Solanum. Phytochemistry: Life Sciences Advances, 11*, 239–243.
48. Rodriguez-Saona, L. E., Giusti, M. M., & Wrolstad, R. E. (1998). Anthocyanin pigment composition of red-fleshed potatoes. *Journal of Food Science, 63*(3), 458–465.
49. Reyes, L. F., Villarreal, J. E., & Cisneros-Zevallos, L. (2007). The increase in antioxidant capacity after wounding depends on the type of fruit or vegetable tissue. *Food Chemistry, 101*, 1254–1262.
50. Keutgen, A. J., Wszelaczyńska, E., Pobereżny, J., Przewodowska, A., Przewodowski, W., Milczarek, D., & Keutgen, N. (2019). Antioxidant properties of potato tubers (*Solanum tuberosum* L.) as a consequence of genetic potential and growing conditions. *PLOS ONE, 14*(9), e0222976.
51. Ru, W., Pang, Y., Gan, Y., Liu, Q., & Bao, J. (2019). Phenolic compounds and antioxidant activities of potato cultivars with white, yellow, red and purple flesh. *Antioxidants, 8*(10), 419.
52. Cebulak, T., Krochmal-Marczak, B., Stryjecka, M., Krzysztofik, B., Sawicka, B., Danilčenko, H., & Jarienė, E. (2023). Phenolic acid content and antioxidant properties of edible potato (*Solanum tuberosum* L.) with various tuber flesh colors. *Foods, 12*(1), 100.
53. Rasheed, H., Ahmad, D., & Bao, J. (2022). Genetic diversity and health properties of polyphenols in potato. *Antioxidants, 11*(4), 603.
54. Combris, P., Amiot-Carlin, M., & Cavaillet, F. (2007). *Les Fruits et Légumes Dans L'alimentation. Enjeux et Déterminants de la Consommation.* Rennes, INRR, Expertise scientifique collective Inra.
55. Han, J. S., Kozukue, N., Young, K. S., Lee, K. R., & Friedman, M. (2004). Distribution of ascorbic acid in potato tubers and in home-processed and commercial potato foods. *Journal of Agricultural and Food Chemistry, 52*(21), 6516–6521.
56. McCay, C. M., McCay, J. B., & Smith, O. (1987). The nutritive value of potatoes. In W. F. Talburt & O. Smith (Eds.), *Potato Processing* (pp. 287–332). Van Nostrand Reinhold, New York.

57. Tamasi, G., Cambi, M., Gaggelli, N., Autino, A., Cresti, M., & Cini, R. (2015). The content of selected minerals and vitamin C for potatoes (*Solanum tuberosum* L.) from the high Tiber Valley area, southeast Tuscany. *Journal of Food Composition and Analysis*, *41*, 157–164.
58. Rodríguez-Martínez, B., Gullón, B., & Yáñez, R. (2021). Identification and recovery of valuable bioactive compounds from potato peels: A comprehensive review. *Antioxidants*, *10*(10), 1630.
59. Singh, N., Kamath, V., Narasimhamurthy, K., & Rajini, P. S. (2008). Protective effect of potato peel extracts against carbon tetrachloride-induced liver injury in rats. *Environmental and Toxicological Pharmacology*, *26*, 241–246.
60. Singh, N., & Rajini, P. S. (2008). Antioxidant-mediated protective effect of potato peel extract in erythrocytes against oxidative damage. *Chemical-Biological Interactions*, *173*(2), 97–104.
61. Singh, A., Raigond, P., Lal, M. K., Singh, B., Thakur, N., Changan, S. S., & Dutt, S. (2020). Effect of cooking methods on glycemic index and *in vitro* bio accessibility of potato (*Solanum tuberosum* L.) carbohydrates. *LWT*, *127*, 109363.
62. Reyes, L. F., Miller, J. C., & Cisneros-Zevallos, L. (2005). Antioxidant capacity, anthocyanins and total phenolics in purple- and red-fleshed potato (*Solanum tuberosum* L.) genotypes. *American Journal of Potato Research*, *82*, 271–277.
63. Manach, C., Scalbert, A., Morand, C., Rémésy, C., & Jiménez, L. (2004). Polyphenols: Food sources and bioavailability. *The American Journal of Clinical Nutrition*, *79*, 727–747.
64. Pojer, E., Mattivi, F., Johnson, D., & Stockley, C. S. (2013). The case for anthocyanin consumption to promote human health: A review. *Comprehensive Reviews in Food Science and Food Safety*, *12*(5), 483–508.
65. Nogueira, T., & do Lago, C. L. (2007). Determination of caffeine in coffee products by dynamic complexation with 3,4-dimethoxycinnamate and separation by CZE. *Electrophoresis*, *28*, 3570–3574.
66. Yamaguchi, T., Chikama, A., Mori, K., Watanabe, T., Shioya, Y., Katsuragi, Y., et al. (2007). Hydroxyquinone-free coffee: A double-blind, randomized controlled dose-response study of blood pressure. *Nutritional Metabolism and Cardiovascular Diseases*, *18*, 408–414.
67. Jin, U. H., Lee, J. Y., Kang, S. K., Kim, J. K., Park, W. H., Kim, J. G., Moon, S. K., & Kim, C. H. (2005). A phenolic compound, 5-caffeoylquinic acid (chlorogenic acid), is a new type and strong matrix metalloproteinase-9 inhibitor: Isolation and identification from methanol extract of *Euonymus alatus*. *Life Sciences*, *77*, 2760–2769.
68. Bassoli, B. K., Cassolla, P., Borba-Murad, G. R., Constantin, J., Salgueiro-Pagadigorria, C. L., Bazotte, R. B., de Silva, R. S., & de Souza, H. M. (2008). Chlorogenic acid reduces the plasma glucose peak in the oral glucose tolerance test: Effects on hepatic glucose release and glycemia. *Cell Biochemistry and Function*, *26*, 320–328.
69. Vitaglione, P., Morisco, F., Caporaso, N., & Fogliano, V. (2005). Dietary antioxidant compounds and liver health. *Critical Reviews in Food Science and Nutrition*, *44*(7–8), 575–586.

70. Konczak, I., & Zhang, W. (2004). Anthocyanins—More than nature's colors. *Journal of Biomedicine and Biotechnology*, *5*, 239–240.
71. Reddivari, L., Vanamala, J., Chintharlapalli, S., Safe, S. H., & Miller, J. C. (2007). Anthocyanin fraction from potato extracts is cytotoxic to prostate cancer cells through activation of caspase-dependent and caspase-independent pathways. *Carcinogenesis*, *28*, 2227–2735.
72. Han, K. H., Matsumoto, A., Shimada, K. I., Sekikawa, M., & Fukushima, M. (2007). Effects of anthocyanin-rich purple potato flakes on antioxidant status in F344 rats fed a cholesterol-rich diet. *The British Journal of Nutrition*, *98*, 914–921.
73. Wang, H., Nair, M. G., Strasburg, G. M., Chang, Y. C., Booren, A. M., Gray, J. I., et al. (1999). Antioxidant and anti-inflammatory activities of anthocyanins and their aglycon, cyanidin, from tart cherries. *Journal of Natural Products*, *62*, 294–296.
74. Tan, J. S., Wang, J. J., Flood, V., Rochtchina, E., Smith, W., & Mitchell, P. (2008). Dietary antioxidants and the long-term incidence of age-related macular degeneration: The Blue Mountains Eye Study. *Ophthalmology*, *115*, 334–341.
75. Thompson, M. D., Thompson, H. J., McGinley, J. N., Neil, E. S., Rush, D. K., Holm, D. G., et al. (2009). Functional food characteristics of potato cultivars (*Solanum tuberosum* L.): Phytochemical composition and inhibition of 1-methyl-1-nitrosourea induced breast cancer in rats. *Journal of Food Composition and Analysis*, *22*, 571–576.

PART IV
Cereals

CHAPTER 10

Barley (*Hordeum vulgare* L.)

SONIA MANSOURI,[1,*] SANA MEDIMAGH,[1] KHAOULA NEFZI,[2] MARWA HASSINE,[3] HATEM ZGALLAI,[4] ONS RIAHI,[1] YASSMINE ABIDI,[1] and MOKHTAR BARAKET[2]

[1]*Laboratory of Field Crops, National Agricultural Research Institute of Tunisia (INRAT), University of Carthage, Tunis, Ariana Governorate, Tunisia*

[2]*National Research Institute of Rural Engineering, Water and Forests (INRGREF), Ariana Governorate, Tunisia*

[3]*Laboratory of Genetics and Cereal Breeding, National Agronomic Institute of Tunisia (INAT), Ariana Governorate, Tunisia*

[4]*National Agricultural Research Institute of Tunisia (INRAT), University of Carthage, Ariana Governorate, Tunisia*

*Corresponding author

ABSTRACT

As the fourth-largest grain crop in the world, barley (*Hordeum vulgare* L.) is one of the main cereal grains farmed in temperate regions. It is the major staple crop and a significant source of food and feed for the populations of many countries. The content of the mature grain determines the quality of the grain for these two final uses: feed and food.

Much research on barley focuses on the mature grain and on the major groups of components that determine end-use properties: starch, proteins, vitamins, fiber, and natural bioactive antioxidants such as phenolic compounds and lipids.

Postharvest Quality Attributes of Fresh Vegetables and Grains.
Riadh Ilahy, Imen Tlili, Thouraya R'him, and Muhammad Azam (Eds.)
© 2027 Apple Academic Press, Inc. Co-published with CRC Press (Taylor & Francis)

DOI:10.1201/9781779642851-14

The investigation of barley's nutritional and bioactive qualities, as well as its application as a functional food in the diets of humans and animals, is still ongoing. In addition to its nutritional qualities, barley will support its potential as a functional food for animal and human health, as well as its role in preventing certain chronic diseases.

10.1 INTRODUCTION

10.1.1 BOTANICAL NAME, COMMON NAME

Hordeum vulgare is the term given to barley in botany. It is an annual grass in the Poaceae family, which is cultivated as a cereal crop for its grain. Barley is one of the most widely grown and important cereal crops in the world, with a lengthy cultivation history that dates back to antiquity. It is used for human consumption, animal feed, and also in the production of beer and other alcoholic beverages. Barley is available in a wide range of varieties, each with unique characteristics and uses, but all belonging to the same species, *Hordeum vulgare* [1].

The common name for *Hordeum vulgare*, the species of cereal grain commonly referred to as barley, is simply "barley." The term "barley" is used to refer to the entire plant, as well as its grain, which is employed for many different things, like brewing, animal feed, and human consumption [1].

Barley is a staple crop in many parts of the world and has been farmed for thousands of years. Barley is utilized in many industrial processes, such as the production of malt for making beer and whisky and the creation of food products like breakfast cereals and snack bars, in addition to being a vital source of nutrition for both humans and animals [2].

10.1.2 HISTORY

Barley has been farmed for thousands of years; the Neolithic era is when the earliest records of its cultivation date from. It was among the first crops that people domesticated and has been an important food source for many cultures throughout history. Barley was used in ancient Egypt to make beer and as a form of currency [3].

10.1.3 PRODUCTION

Barley is a widely grown crop around the world, with major producing countries including Russia, Canada, Germany, France, and Australia.

In India, barley is mainly cultivated in the northern states of Punjab, Haryana, and Uttar Pradesh. In 2020, the total global production of barley was around 151 million metric tons, with Russia being the top producer with 21 million metric tons, followed by Germany, Canada, France, and Australia. In India, the total production of barley in 2020–21 was estimated at around 1.6 million metric tons, with Punjab being the largest producer, followed by Haryana and Uttar Pradesh [4–6].

Mostly utilized as animal feed, barley is also a vital component in the making of beer and other alcoholic drinks. It is also used in the food industry to make soups, stews, and baked goods, as well as in the production of some medicines [2].

Belonging to the grass family, barley is a cereal grain. It is one of the oldest cultivated grains, with evidence of its cultivation dating back to ancient civilizations such as the Sumerians and Egyptians. Barley is used as a major food and feed crop around the world and has been an important part of human and animal diets for centuries [7].

Barley is a versatile crop that is grown for both food and feed purposes. Barley is an important source of nutrition for humans and animals due to its high fiber, protein, vitamin, and mineral content. It is a staple food in many cultures and is commonly used to make bread, beer, and other food products [8, 9].

Barley is a nutritious food that provides a good source of energy and lowers the risk of diabetes and heart disease due to its high fiber content. Additionally, it is a good source of potassium, selenium, and magnesium, among other important minerals [10].

As a feed source, barley is an important source of nutrition for livestock. It is commonly used in the feed industry for ruminants, such as cattle and sheep, as it is easily digestible and provides a good source of energy and protein [11].

In addition to its importance as a food and feed source, barley has a significant role in climate change mitigation. Barley is a crop that is well adapted to different climatic conditions and is more resilient to extreme weather events such as droughts and heat waves compared to other cereal crops. This makes it an important crop for adaptation to climate change [12].

Moreover, barley has a unique ability to sequester carbon in the soil through its root system. As one of the main causes of global warming, carbon dioxide emissions from the atmosphere are decreased with the aid of carbon sequestration [12].

In conclusion, barley is an important crop that provides both nutritional and environmental benefits. Its versatility as a food and feed source, as well as its ability to mitigate climate change, make it a valuable crop for agriculture and food systems.

10.2 PHYTOCHEMICAL COMPOSITION

Barley is a basic food in several countries, and the nutritional compounds in barley comprise various types of phytochemicals such as phenolic acids, flavonoids, lignans, vitamin E, sterols, and folates [13]. These bioactive phytochemicals are known for their health benefits [14].

10.2.1 VITAMINS

The vitamins B1, B2, B3, E, and gamma-aminobutyric acid are among those found in barley, and their concentrations can differ based on the cultivar. Huang et al. [15] demonstrated that vitamins B1 and B2 were the most variable in a collection of barley from Tibet, China. In the collection, vitamin B1 was found in the highest proportion, followed by B2, E, and B3. Vitamin B3 was found in the lowest proportion. It was also shown that the content of gamma-aminobutyric acid was also highly variable. Furthermore, it was determined that the contents of vitamins B1, B2, and C, and proteins increased significantly after germination [16]. Barley vitamins promote sleep and have antidepressant, antioxidant, hypolipidemic, gastrointestinal, and anti-inflammatory effects, and prevent cardiovascular disease (CVD). Barley vitamins have very effective anti-stress capabilities by preventing the reduction of wheel spinning activity and hippocampal brain-derived neurotrophic factor mRNA in response to restraint stress [17]. Vitamin E has a unique therapy for chronic disease, antioxidant, and anti-inflammatory properties.

10.2.2 POLYPHENOLS AND FLAVONOIDS

Reducing the risk of age-related diseases like cancer and CVD is anticipated when the availability of plant-based phenolics is increased [18]. The majority of these effects result from their potent antioxidant and free radical scavenging properties [19]. Phenolic compounds, which can be found free or attached to fiber, are abundantly produced by barley [20].

Phenolic compounds in barley contain phenolic acids, flavonoids, monophenols, and other polyphenols [2, 17]. Flavanols, like prodelphinidins, procyanidins, and catechin, are the major compounds of the free phenolic fraction of barley grain, whereas phenolic acids, like coumaric and vanillic acids, and ferulic, are the main constituents of the bound phenolic portion [20–22].

The main phenolic group of barley phytochemicals is mostly found in the external part of the grain epidermis [23]. Two groups are distinguished from them: benzoic acid and cinnamic acid and their derivatives [24, 25]. Phenolic acids have been related to the prevention of chronic diseases, thanks to the existence of an unsaturated carboxylic group [26]. The high content of phenolic acids in barley, especially in the hulled variety, suggests that it could also be an excellent dietary supplement of natural antioxidants with antiradical and antiproliferative potential [24, 27]. In barley, the greatest levels of phenolic acids are present in the bound form, followed by the conjugated and free forms [28]. The free forms are frequently identified in the outer part of the pericarp [29].

Flavonoids are phytochemicals whose absorber is UV-B [30]. They are bioactive substances protecting against UV radiation when exposed to "excess light" stress [31, 32]. Clinical research suggests that flavonoids may be the bioactive substances in cereal grains that are responsible for the mitigation of many secondary effects, including cancer and coronary heart disease [33, 34]. The identification of flavonoids suggested that flavanols, anthocyanins, and proanthocyanidins are the major types of flavonoids present in barley grains. More specifically, flavanols and anthocyanins are found in the pericarp of barley grains, where they mainly exist as glycoside derivatives, namely delphinidin-3-glucoside, penidin-3-glucoside, and cyanidin-3-glucoside [22, 35].

10.2.3 CAROTENOIDS

Carotenoids are important phytonutrients responsible for plant color and are present in many fruits and vegetables. In barley, carotenoids are responsible for the yellow color of the grain endosperm. These carotenoids exhibit C40 isoprenoids composed of a long-conjugated polyene chain that is responsible for their color and biological activities. Carotenoids serve important functions in plants in both photosynthetic and non-photosynthetic tissues, where they assist in light harvesting and photoprotection, or serve as signaling molecules, precursors of volatile compounds, colorants, and attractants to pollinators [36]. The two classes of carotenoids are xanthophylls (oxygenated derivatives of carotenes) and carotenes (which are hydrocarbons). In terms of cereals, barley is rather low in total carotenoids, with contents similar to those of the classic bread wheat. Ndolo and Beta [37] found that yellow barley grain (the most typical form) contains 2.25 μg/g of total carotenoids, whereas its purple form contains double this content (4.54 μg/g). These authors have further noted that zeaxanthin is the principal carotenoid in barley (Table 10.1).

TABLE 10.1 Summary of the Natural Antioxidants of Barley

Origin	Phytochemical Class	Sous-Class	References
Barley	Flavonoids	Catechin, procyanidins, prodelphinidins, flavanols, anthocyanins, proanthocyanidins, prodelphinidin B3 anidin	[35]
	Phenolic acids	Ferulic acid, coumaric acid, vanillic acid, hydroxycinnamic acid	[20, 21, 23, 26, 38]
	Polyphenols	Carotenoids, catechin, 1,2-dihydroxybenzene, quercetin, isorhamnetin	[39]
	Carbohydrates and fiber	Starch, beta-glucan	[40, 41]
	Proteins	Prolamins, glutamine, proline, hordoindoline	[42, 43]
	Amino acids	Lysine, threonine, tryptophan	[44, 45]
	Lipids	phytosterols, tocols	[45]
	Vitamins	Vitamins B1, B2, B3, E, and gamma-aminobutyric acid; vitamin E	[15, 16, 46]

10.2.4 LIPIDS

Barley is a species that contains several lipids, such as fatty acids, tocols, and phytosterols. A recent study showed that linoleic acid is the most abundant unsaturated fatty acid in Irish barley varieties. Alpha-tocotrienol is the most abundant tocol counterpart, whereas beta-sitosterol is the most abundant sterol [45].

Studies have shown that tocopherols and tocotrienols, also referred to as tocols or Vitamin E, are found in barley and have a variety of chemical and physiological characteristics [47]. According to Bravi et al. [48], barley's total lipid content and fatty acid composition change depending on variety and decrease during the malting process. However, a study found that malt oil had 17.33% palmitic, 15.62% oleic, and 56.56% linoleic acids, whereas barley lipids had 18.53% palmitic, 19.94% oleic, and 51.74% linoleic acids. Moreover, the content of linoleic acid increased during malting, whereas that of oleic and palmitic acids decreased [39].

Tocopherols and tocotrienols, collectively referred to as tocols, are a group of lipid-soluble phytochemicals present in barley. These tocols are well-known for their antioxidant properties, particularly for their capacity to prevent lipid peroxidation in biological membranes [26, 49, 50]. Furthermore, the tocols found in cereals possess anticancer and cancer-suppressing properties [51, 52], boost the immune system [53], regulate CVD risk factors [54, 55], and encourage apoptosis induction [52, 56]. One of the most remarkable findings concerning tocols is their potential to clear atherosclerotic blockages (stenosis) in the carotid artery, which may reduce the risk of stroke [57].

Barley is considered one of the most abundant sources of tocols among cereals, with a high concentration and favorable distribution of all eight biologically active vitamers [58]. In their study, Moreau et al. [58] discovered that the total phytosterol and total tocotrienol levels in oils derived from both whole barley kernels and sacrificed fines ranged from 2911 mg/g to 6126 mg/g, which is significantly higher than the content in palm oil (530 mg/g) and rice bran oil (770 mg/g), both marketed as "high in tocotrienols." Temelli et al. [59] investigated the overall tocol content of whole-grain barley varieties, which varied between 40 mg/g and 151.1 mg/g. Among individual tocol isomers in whole grain barley, α-tocotrienol was the most prevalent, accounting for roughly 47.7% of the total tocol content [60], followed by α-tocopherol (17.7–33.9%), γ-tocotrienol (10.4–20.2%), γ-tocopherol (1.9–9.2%), β-tocotrienol (2.9–7.8%), and δ-tocotrienol (2.7–6.7%) [59]. The average tocotrienol content in barley is approximately 70.6–76.8% [60], making it one of the richest sources of tocotrienol among cereal grains.

Phytosterols, also known as plant sterols, are structurally similar to cholesterol and are important components of plant membranes. Recent studies have shown that dietary intake of phytosterols can decrease serum cholesterol levels, protect against CVDs, and prevent colon cancer [61–64]. Cereals, along with vegetable oils, are the main sources of phytosterols, contributing up to 40% of the daily intake of plant sterols [65].

Barley is regarded as a great source of phytosterols, even though its phytosterol level is moderate compared to other important grains [66]. Barley grains contain phytosterols in both free and bound forms, which can be esterified to fatty acids, phenolic acids, steryl glucosides, or acylatedsterol glycosides. The level of esterification varies among varieties and around different parts of the barley grain, with higher levels of phytosterols identified in the outer layers of the kernel [67]. Barley contains a variety of sterols, with sitosterol being the most predominant, accounting for 53–61% of total sterols, followed by campesterol (14–20%). Additionally, barley contains other forms of sterols such as brassicasterol, stigmasterol, d5-avenasterol, stigmastenol, stigmastadienol, and d7-avenasterol. The content of sterols in barley varies among varieties, with levels ranging from 820 μg/g to 1153 μg/g [68]. According to Piironen et al. [69], the highest plant sterol content was found in the rye (mean, 955 mg/g), followed by barley (761 mg/g), wheat (690 mg/g), and oat (447 mg/g).

10.2.5 NON-STARCH POLYSACCHARIDES

The composition and biological activity of polysaccharides found in flour vary depending on the type of grain. Plants and microorganisms are the primary natural sources of these bioactive compounds [70]. Non-starch polysaccharides, such as β-glucans, mannans, and fructans, have been shown to have positive effects on health and disease prevention [71–73]. The bran, husk, cereal endosperm cell walls, and aleurone layer contain the most important fiber components, which are non-starch polysaccharides [74, 75]. The endosperm cell walls of barley are particularly rich in mixed linkage β-glucans, whereas wheat cell walls contain low levels of β-glucans found in the bran [76]. β-glucans form viscous solutions in water, which limits the use of barley flour in baking. Mannans are also present in cell walls and are classified as dietary fibers; they have high viscosity and low palatability [77]. Recent studies showed that mannans make up 4–7% of all cell wall polysaccharides [75]. Fructans, which are linear or branched polymers, are also considered dietary fibers and are found in plant cell walls and microbial environments [78]. Fructans can have both positive and negative effects on human health. While they may cause bloating and stomach pain in individuals with functional gastrointestinal disorders like irritable bowel syndrome, their bacterial transformation into short-chain fatty acids (SCFAs) has prebiotic effects [79]. Polysaccharides also influence bread texture through their solubility and viscosity properties [80].

10.2.6 PROTEINS AND AMINO ACIDS

The protein content of barley grains is a crucial factor in determining their quality and value for various end-uses [42]. Prolamins make up as much as 70% of the total protein in barley, making them the primary storage protein, but they are deficient in essential amino acids like lysine, threonine, and tryptophan, and have high levels of glutamine and proline [44]. A recent study comparing six-row and two-row barley cultivars showed that they differ in the abundance of 20 proteins, with hordoindoline proteins being one of the key differences and potentially contributing to differences in seed hardness [43].

In a survey conducted in Western Canada in 2018, where barley is a significant source of feed for both ruminants and monogastric animals, it was found that barley has a higher crude protein content than corn, rye, and oats [81]. Another study found that barley has 4% more protein than corn grain [40]. Additionally, a study investigating the impact of future climate conditions on barley accessions reported that while there was an increase in grain protein concentration under elevated temperature and CO_2 levels, the decrease in grain yield resulted in a 23% decrease in harvestable protein. However, the response of different barley accessions varied, indicating the potential for exploitation [82].

10.3 PHYSICAL CHARACTERISTICS

Physical properties of barley grains were determined by the angle of internal friction, bulk density, porosity, true density, and static coefficient of friction. These physical characteristics are essential to design equipment involved in malting, such as cleaning, sorting, and separation [83], and eventually for the post-harvest processes such as transport ventilation, drying, and storage [84, 85]. Furthermore, the characteristic dimensions also allow calculating the surface area, the size of seeds, and the volume of grains, which are considered important aspects for the modeling of drying and ventilation [86, 87]. Therefore, physical measurements were considered advantageous commercial tools due to their rapidity and the early detection for the characterization of nutritional quality value compared to chemical tests used [88].

10.3.1 ROW TYPE

Barley can be categorized into two main types based on the number of rows of kernels on each spike: two-row barley and six-row barley.

Two-row barley has two rows of kernels, while six-row barley has six rows of kernels [89]. This is a simply inherited trait, readily discernible upon visual inspection, and is controlled by the Vrs1 and Int-c genes located on chromosomes 2(2H) and 4(4H), respectively [90, 91].

The two-rowed barleys have a central fertile floret and two sterile lateral florets, providing a single seed at each node and a flat head appearance. All kernels of two-rowed barleys are symmetrical, with an important variation in size for the smaller ones at the tip, middle, and base of the spike. Thus, the six-rowed barleys have three fertile spikelets, giving a tubular appearance and a shape to the kernels on the spike. One-third of the kernels of six-rowed varieties are symmetrical, as in two-rowed barley, and the other two-thirds are twisted. Therefore, the row type has an important effect on yield component and potential because six-rowed varieties usually provide more seed number per spike [92].

However, some studies have reported that two-rowed barleys contain larger kernels with slightly more total protein and amino acid composition compared to six-rowed varieties [93].

10.3.2 KERNEL SIZE, KERNEL WEIGHT, AND VOLUME WEIGHT

Barley kernels vary in size, weight, and volume. The size of barley kernels can range from 2–5 mm in length, while the weight can range from 20–50 mg per kernel. The volume weight of barley refers to the weight of a certain volume of barley, usually expressed in grams per liter [94]. Kernel size and shape contain relevant information to the end-use quality properties [88]. Moreover, the grain weight or kernel weight of barley, usually expressed as 1000 grain weight (Thousand kernel weight (TKW)), is one of the most important yield components, revealing the effect of environmental conditions during the grain-filling stages [95, 96]. Test weight or volume weight is considered an important trait due to the direct relationship of energy content and feeding value on foul feeling of the grain [88]. Grains of high-test weight reveal a high percentage of large and plump grains compared to lower test weight seed containing higher starch content [97]. Furthermore, according to several studies, 1000 grain weight and test weight seem to depend on row types, where two-row varieties have a higher value than six-row barley [88]. The removal of the hull of barley also has an important effect on TKW by reducing it from 15 to 20% for the two-row type of barley varieties. This effect was reported eventually on test weight, which may range from 52 to 72 Kg/hL, and could increase to 80 kg/hL for hulless barley [98].

10.3.3 KERNEL UNIFORMITY

Uniformity of kernel size, weight, and volume is an important quality trait for barley, as it can affect yield and overall grain quality [89]. This trait is determined by food processing, such as malting behavior and the pearling process. Kernels with different sizes could provide a malt with widely varying properties and specificity [99]. The six-rowed barley containing smaller, natural six-rowed seeds requires a longer pearling period compared to two-rowed varieties [100]. Thus, kernel uniformity is also needed to design equipment for cleaning, sorting, and separation, and for the creation of models of drying and ventilation processes [83, 87]. It could provide a further option to breeding programs for the choice of row type and barley cultivar [93] (Figure 10.1).

FIGURE 10.1 Barley varieties, including hulled (bottom left), hulless (bottom right), two-row (upper left), and six-row (upper right) types.

Source: The authors' personal photo collection.

10.4 HEALTH BENEFITS

Barley grains are actually receiving more attention compared to other cereal crops such as rice, maize, and wheat due to their chemical composition and health benefits. They are rich in important dietary fiber beneficial to human health, providing important protection against CVD, hypertension, and diabetes [101]. Therefore, in 2006, a final approval of a health claim for barley was given by the U.S. FDA to highlight the importance of regular consumption of barley grain in the protection from lowering blood cholesterol and CVD [101]. The European Food Safety Authority [102, 103], and more recently, Food Standards Australia and New Zealand [104] has also approved the barley health claims [102–104]. The presence of higher contents of dietary fiber, such as β-glucan and tocols in barley grains have contributed to the improvement of immunity, intestinal function, and protection from stroke, 90% of type 2 diabetes, 80% of coronary heart disease, and hypertension [105, 106]. For instance, the β-glucan content in barley grain is almost 10-fold higher than that of wheat on average [21]. The tocol content, such as alpha-tocopherol, alpha-tocotrienol, and polyunsaturated and fatty acids, was considered responsible for the control of degenerative diseases such as CVD and cancer by decreasing platelet aggregation and lipid peroxidation [20, 107]. Vanillic and coumaric acid have provided anticancer, anti-inflammatory, antioxidant, and antimicrobial effects by decreasing low-density lipoprotein (LDL) peroxidation [108, 109]. A therapeutic application of syringic acid present in the barley grain is used in reducing the risk of cerebral ischemia, CVDs, and preventing different types of diabetes [110]. Therefore, the regular consumption of barley wholegrain was considered by several studies to be responsible for the reduction of the risk of gallstones, chronic diseases, and cancer [111].

10.4.1 ANTI-INFLAMMATORY ACTIVITY

It has been discovered that barley contains anti-inflammatory qualities, which may be due to the presence of various bioactive compounds, including phenolic acids, flavonoids, and beta-glucans [112]. This crop exhibits anti-inflammatory effects by reducing inflammation markers. The consumption of barley grains or barley extracts can reduce the levels of inflammatory markers in the body, such as C-reactive protein (CRP), interleukin-6 (IL-6), and tumor necrosis factor-alpha (TNF-α). It is responsible for the regulation of immune response by the β-glucan content. They can activate immune cells such as macrophages and dendritic cells, which play a crucial role in regulating the inflammatory response. It protects against oxidative stress due to the antioxidant components, such as phenolic acids and flavonoids, that can protect against oxidative stress and inflammation by neutralizing free radicals [113, 114]. It helps in the control of inflammatory bowel disease (IBD), ulcerative colitis, and Crohn's disease by reducing inflammation. Moreover, it improves joint health by reducing inflammation associated with conditions like rheumatoid arthritis and osteoarthritis. Overall, the anti-inflammatory properties of barley make it a potentially useful dietary supplement for reducing chronic inflammation and associated health conditions [115].

10.4.2 ANTIMICROBIAL ACTIVITY

Barley has been shown to possess antimicrobial properties against a variety of microorganisms, including fungi, bacteria, and viruses. Some of the antimicrobial activities of barley have also been exploited for the preservation of food. Barley extracts have been shown to inhibit the growth of foodborne pathogens, including *Listeria monocytogenes* and *E. coli*, in various food products. Overall, the antimicrobial activities of barley make it a potentially useful natural antimicrobial agent for use in various applications, including food preservation and the development of novel antimicrobial agents. The β-glucan content is responsible for the formation of SCFAs such as propionate, acetate, and butyrate, fermented by intestinal bacteria in the colon [116]. Moreover, the resistant starch and β-glucan fermented in the large intestine also contribute to the production of SCFAs, especially butyrate and propionate, providing an energy source for the colonic mucosa's epithelial cells [117, 118].

10.4.3 ANTI-DIABETES

Numerous studies have demonstrated the beneficial impact of consistent barley consumption on human glucose metabolism [119]. The glycemic index (GI) of barley (34–70) is characterized by a lower level compared to other cereals such as wheat (52–75), maize (46–80), and rice (55–85) [120]. The high viscosity of β-glucan and the amylose/amylopectin ratio has contributed to lowering blood sugar and the reduction of the glycemic response [121]. This amylase component is only present in barley grains containing 99% amylose starch compared to other cereal crops [122].

This amylose/amylopectin ratio has also significantly affected the insulin responses by variation in blood sugar content [120, 121]. However, the resistant starch and β-glucan have contributed to slowing absorption and digestion, which is responsible for reducing blood glucose levels [123]. Minehira et al. [124] results have shown that β-glucan participates in the regulation of prandial-glucose metabolism and increases intestinal viscosity, which contributes to decreasing glucose absorption. In addition, these resistant starch components provide an important anti-digestive enzyme in the large intestine [125].

10.4.4 ANTICANCER ACTIVITY

A different physiological effect of β-glucan, as a soluble dietary fiber, is interacting with several mechanisms, such as fermentative formation of SCFA in the colon and gelation [117]. The main effect of SCFA is the induction of chemopreventive mechanisms in the colon responsible for the reduction of the initiation of cells in the colon into preneoplastic cells and the inhibition of further transformation to neoplastic cells. An important protection against DNA damage by inhibiting the further transformation of initiated cells and reduction of reactive oxygen was mainly induced by SCFA. Butyrate, one of the acetate and propionate components, provides an important chemopreventive effect by inducing apoptosis and inhibiting the growth of already degenerated cells. The main end product of dietary fiber fermentation, butyrate, contributes to the reduction of the risk of colon cancer development by acting as a histone deacetylase inhibitor [126, 127].

10.5 CONCLUSION

A versatile crop used for human nutrition and animal feed for energy and nutrients, barley is gaining attention again. This is because of its high content of biologically active constituents like phenolic compounds, dietary fiber (especially beta-glucan), and tocols (including tocopherols and toco-trienols). These widely dispersed components of barley are crucial for the biosynthesis of numerous metabolic compounds and have an impact on the color, flavor, and texture of food. With its anti-inflammatory, anticarcino-genic, antioxidant, and cardio- and neuroprotective properties, barley's nutritional qualities help prevent a variety of metabolic disorders. Overall, eating barley exhibited positive effects on preventing the development of chronic illnesses in both humans and animals. Future research should further focus on the phytochemical profile of different barley cultivars as well as on their antioxidant activity in relation to different types of barley.

KEYWORDS

- **antioxidant activity**
- **chronic disease prevention**
- **dietary fiber**
- **functional food**
- ***Hordeum vulgare***
- **phytochemicals**

REFERENCES

1. Ragaee, S., Abdel-Aal, E. M., Noaman, M., & Khalil, A. H. (2020). Barley: Production, processing, and health benefits. In *Grains for Food and Beverage Industries* (pp. 197–224). Academic Press.
2. Khan, I., Farooq, M., Ali, A., Khan, Z., & Hussain, M. (2020). Barley in food and feed: Health benefits, breeding strategies, and challenges. *Sustainability, 12*(8), 3356.
3. Zhang, G., Liu, X., Quan, C., Yan, J., & Li, J. (2020). Barley as a sustainable crop in a changing environment: Physiology, adaptation, and breeding. *Frontiers in Plant Science, 11*, 566.
4. Food and Agriculture Organization of the United Nations. (2020). *FAOSTAT Database: Crops*. http://www.fao.org/faostat/en/#data/QC (accessed on 1st January 2026).
5. United States Department of Agriculture. (2021). *World Agricultural Supply and Demand Estimates*. https://www.usda.gov/oce/commodity/wasde/ (accessed on 1st January 2026).
6. Bharadwaj, N., & Singh, V. (2021). Barley in India: Production, consumption, and challenges. *Indian Journal of Plant Genetic Resources, 34*(3), 389–395.
7. United Nations Framework Convention on Climate Change. (2022). *The Role of Agriculture in Climate Change Mitigation*. https://unfccc.int/topics/agriculture/the-big-picture/the-role-of-agriculture-in-climate-change-mitigation (accessed on 1st January 2026).
8. Newman, C. W., & Newman, R. K. (2006). A brief history of barley foods. *Cereal Foods World, 51*, 4–7.
9. Harwood, W. A. (2019). An introduction to barley: The crop and the model. *Methods in Molecular Biology, 1900*, 1–5.

10. Izydorczyk, M. S., McMillan, T., Bazin, S., Kletke, J., Dushnicky, L., Dexter, J., Chepurna, A., & Rossnagel, B. (2014). Milling of Canadian oats and barley for functional food ingredients: Oat bran and barley fibre-rich fractions. *Canadian Journal of Plant Science*, *94*, 573–586.
11. Nikkhah, A. (2012). Barley grain for ruminants: A global treasure or tragedy. *Journal of Animal Science and Biotechnology*, *3*(1), 1–9.
12. Verhulst, N., Govaerts, B., Verachtert, E., Castellanos-Navarrete, A., Mezzalama, M., Wall, P., & Sayre, K. D. (2010). Conservation agriculture, improving soil quality for sustainable production systems? In *Advances in Agronomy* (Vol. 109, pp. 1–78). Academic Press.
13. Guo, H., Lin, S., Lu, M., Gong, G. D. B., Wang, L., Zhang, Q., & Wu, D. T. (2018). Characterization, *in vitro* binding properties, and inhibitory activity on pancreatic lipase of β-glucans from different Qingke (Tibetan hulless barley) cultivars. *International Journal of Biological Macromolecules*, *120*, 2517–2522. https://doi.org/10.1016/j.ijbiomac.2018.09.023 (accessed on 1st January 2026).
14. Liu, J., Xu, D., Chen, S., Yuan, F., Mao, L., & Gao, Y. (2021). Superfruits en Chine: Composés phytochimiques bioactifs et leurs avantages potentiels pour la santé – Une revue. *Sciences Alimentaires et Nutrition*, *9*(12), 6892–6902.
15. Huang, H., Gao, X., Li, Y., Tian, P., Nima, Y., Laba, Z., Ci, Z., Wei, X., Qu, J., Guan, W., & Liao, W. (2020). Content analysis of vitamins, dietary fibers and amino acids in a wide collection of barley (*Hordeum vulgare* L.) from Tibet, China. *Bioinformation*, *16*, 314–322. https://doi.org/10.6026/97320630016314 (accessed on 1st January 2026).
16. Rico, D., Peñas, E., García, M. D. C., Martínez-Villaluenga, C., Rai, D. K., Birsan, R. I., Frias, J., & Martín-Diana, A. B. (2020). Sprouted barley flour as a nutritious and functional ingredient. *Foods*, *9*, 296. https://doi.org/10.3390/foods9030296 (accessed on 1st January 2026).
17. Food Standards Australia New Zealand (FSANZ). *Consultation About Beta-Glucan and Blood Cholesterol Health Claims*. http://www.foodstandards.gov.au/consumer/labelling/nutrition/Pages/Consultation-about-beta-glucan-and-blood-cholesterol-health-claims.aspx (accessed on 1st January 2026).
18. Fresco, P., Borges, F. I. G. M., Diniz, C., & Marques, M. P. M. (2006). New insights on the anticancer properties of dietary polyphenols. *Medicinal Research Reviews*, *26*(6), 747–766.
19. Rice-Evans, C., Miller, N., & Paganga, G. (1997). Antioxidant properties of phenolic compounds. *Trends in Plant Science*, *2*(4), 152–159.
20. Martínez, M., Motilva, M. J., de Las Hazas, M. C. L., Romero, M. P., Vaculova, K., & Ludwig, I. A. (2018). Phytochemical composition and β-glucan content of barley genotypes from two different geographic origins for human health food production. *Food Chemistry*, *245*, 61–70.
21. Holtekjølen, A. K., Kinitz, C., & Knutsen, S. H. (2006). Teneur en flavanols et en acides phénoliques liés dans différentes variétés d'orge. *Journal de Chimie Agricole et Alimentaire*, *54*(6), 2253–2260.
22. Abdel-Aal, E. S. M., Choo, T. M., Dhillon, S., & Rabalski, I. (2012). Acides phénoliques libres et liés et composés phénoliques totaux dans l'orge noire, bleue et jaune et leur contribution à la capacité de piégeage des radicaux libres. *Chimie céréalière*, *89*(4), 198–204.

23. Dykes, L., & Rooney, L. (2007). Phenolic compounds in cereal grains and their health benefits. *Cereal Foods World*, *52*(3), 105–111.
24. Quinde-Axtell, Z., & Baik, B. K. (2006). Phenolic compounds of barley grain and their implication in food product discoloration. *Journal of Agricultural and Food Chemistry*, *54*(26), 9978–9984.
25. Tang, Y., Li, X., Zhang, B., Chen, P. X., Liu, R., & Tsao, R. (2015). Characterization of phenolics, betanins and antioxidant activities in seeds of three *Chenopodium quinoa* Willd. genotypes. *Food Chemistry*, *166*, 380–388.
26. Pandey, K. B., & Rizvi, S. I. (2009). Plant polyphenols as dietary antioxidants in human health and disease. *Oxidative Medicine and Cellular Longevity*, *2*, 270–278.
27. Zhao, Z., & Moghadasian, M. H. (2008). Chemistry, natural sources, dietary intake and pharmacokinetic properties of ferulic acid: A review. *Food Chemistry*, *109*(4), 691–702.
28. Bonoli, M., Verardo, V., Marconi, E., & Caboni, M. F. (2004). Antioxidant phenols in barley (*Hordeum vulgare* L.) flour: Comparative spectrophotometric study among extraction methods of free and bound phenolic compounds. *Journal of Agricultural and Food Chemistry*, *52*(16), 5195–5200.
29. Gamel, T., & Abdel-Aal, E. S. M. (2012). Phenolic acids and antioxidant properties of barley wholegrain and pearling fractions. *Agricultural and Food Science*, *21*(2), 118–131.
30. Tossi, V., Lombardo, C., Cassia, R., & Lamattina, L. (2012). Nitric oxide and flavonoids are systemically induced by UV-B in maize leaves. *Plant Science*, 103–109.
31. Agati, G., Biricolti, S., Guidi, L., Ferrini, F., Fini, A., & Tattini, M. (2011). The biosynthesis of flavonoids is enhanced similarly by UV radiation and root zone salinity in *L. vulgare* leaves. *Journal of Plant Physiology*, *168*(3), 204–212.
32. Bashandy, T., Taconnat, L., Renou, J. P., Meyer, Y., & Reichheld, J. P. (2009). Accumulation of flavonoids in a *ntra ntrb* mutant leads to tolerance to UV-C. *Molecular Plant*, *2*(2), 249–258.
33. Tang, Y., Zhang, B., Li, X., Chen, P. X., Zhang, H., Liu, R., & Tsao, R. (2016). Bound phenolics of quinoa seeds released by acid, alkaline, and enzymatic treatments and their antioxidant and α-glucosidase and pancreatic lipase inhibitory effects. *Journal of Agricultural and Food Chemistry*, *64*(8), 1712–1719.
34. Gani, A., Wani, S., Masoodi, F., & Hameed, G. (2012). Whole-grain cereal bioactive compounds and their health benefits: A review. *Journal of Food Processing & Technology*, *3*(3), 146–156.
35. Abdel-Aal, E. S. M., Young, J. C., & Rabalski, I. (2006). Anthocyanin composition in black, blue, pink, purple, and red cereal grains. *Journal of Agricultural and Food Chemistry*, *54*(13), 4696–4704.
36. Rodriguez-Concepcion, M., Avalos, J., Bonet, M. L., Boronat, A., Gomez-Gomez, L., Hornero-Mendez, D., & Zhu, C. (2018). Une perspective globale sur les caroténoïdes: Métabolisme, biotechnologie et avantages pour la nutrition et la santé. *Progrès de la recherche sur les lipides*, *70*, 62–93.
37. Ndolo, V. U., & Beta, T. (2013). Distribution of carotenoids in endosperm, germ, and aleurone fractions of cereal grain kernels. *Food Chemistry*, *139*(1–4), 663–671.

38. Lempereur, I., Rouau, X., & Abecassis, J. (1997). Variation génétique et agronomique des teneurs en arabinoxylane et en acide férulique du grain de blé dur (*Triticum durum* L.) et de ses fractions meunières. *Journal of Cereal Science*, *25*(2), 103–110.
39. Özcan, M. M., Aljuhaimi, F., & Uslu, N. (2018). Effect of malt process steps on bioactive properties and fatty acid composition of barley, green malt and malt grains. *Journal of Food Science and Technology*, *55*(1), 226–232.
40. McAllister, M., & Meale, S. (2015). *Barley Grain—Feed Industry Guide*. Alberta Barley, *1*, 1–35. https://www.albertabarley.com/download-feed-guide/ (accessed on 1st January 2026).
41. Izydorczyk, M. S., & Dexter, J. E. (2008). Barley beta-glucan and arabinoxylans: Molecular structure, physicochemical properties, and uses in food products–A review. *Food Research International*, *41*, 850–868. https://doi.org/10.1016/j.foodres.2008.04.001 (accessed on 1st January 2026)
42. Young, V. R., & Pellett, P. L. (1994). Plant proteins in relation to human protein and amino acid nutrition. *American Journal of Clinical Nutrition*, *59*, 1203S–1212S. https://doi.org/10.1093/ajcn/59.5.1203S (accessed on 1st January 2026)
43. Mahalingam, R. (2017). Shotgun proteomics of the barley seed proteome. *BMC Genomics*, *18*, 44. https://doi.org/10.1186/s12864-016-3408-5 (accessed on 1st January 2026)
44. Lange, M., Vincze, E., Wieser, H., Schjoerring, J. K., & Holm, P. B. (2007). Suppression of C-hordein synthesis in barley by antisense constructs results in a more balanced amino acid composition. *Journal of Agricultural and Food Chemistry*, *55*, 6074–6081. https://doi.org/10.1021/jf0709505 (accessed on 1st January 2026)
45. Gangopadhyay, N., Rai, D. K., Brunton, N. P., Gallagher, E., & Hossain, M. B. (2016). Isolement guidé par les antioxydants et identification par spectrométrie de masse des principaux polyphénols du grain d'orge (*Hordeum vulgare*). *Chimie Alimentaire*, *210*, 212–220.
46. Ramakrishna, R., Sarkar, D., & Shetty, K. (2019). Functional bioactive from barley for human health benefits. In *Functional Foods and Biotechnology* (pp. 61–85). CRC Press. https://doi.org/10.1201/9781003003830-5 (accessed on 1st January 2026)
47 Azzi, A., & Zingg, J. M. (2006). Vitamin E textbooks require updating. *Biochemistry and Molecular Biology Education*, *33*, 184–187.
48. Bravi, E., Marconi, O., Perretti, G., & Fantozzi, P. (2012). Influence of barley variety and malting process on lipid content of malt. *Food Chemistry*, *135*, 1112–1117.
49. Tang, Y., Li, X., Chen, P. X., Zhang, B., Hernandez, M., Zhang, H., Marcone, M. F., Liu, R., & Tsao, R. (2014). Lipids, tocopherols, and carotenoids in leaves of amaranth and quinoa cultivars and a new approach to overall evaluation of nutritional quality traits. *Journal of Agricultural and Food Chemistry*, *62*(52), 12610–12619.
50. Hajieva, P., Abrosimov, R., Kunath, S., & Moosmann, B. (2023). Modulation antioxydante et prooxydante de la peroxydation lipidique par des protéines membranaires intégrales. *Recherche sur les Radicaux Libres*, 1–10.
51. Sen, C. K., Khanna, S., & Roy, S. (2007). Tocotrienols in health and disease: The other half of the natural vitamin E family. *Molecular Aspects of Medicine*, *28*(5–6), 692–728.
52. Nesaretnam, K., Yew, W. W., & Wahid, M. B. (2007). Tocotrienols and cancer: Beyond antioxidant activity. *European Journal of Lipid Science and Technology*, *109*, 445–452.

53. Becker, F. (2013). Tocopherols in wheat and rye: A review of research within the subject. *Institutionen för livsmedelsvetenskap*.
54. Tiwari, U., & Cummins, E. (2009). Nutritional importance and effect of processing on tools in cereals. *Trends in Food Science & Technology*, *20*, 511–520.
55. Nawawi, H. M. (2012). Tocotrienols and atherosclerosis: Potential in cardio protection. In *Tocotrienols: Vitamin E Beyond Tocopherols* (p. 163).
56. Suman, S., Datta, K., Chakraborty, K., Kulkarni, S. S., Doiron, K., Fornace, A. J., Kumar, K. S., Hauer-Jensen, M., & Ghosh, S. P. (2013). Gamma tocotrienol, a potent radioprotector, preferentially upregulates expression of anti-apoptotic genes to promote intestinal cell survival. *Food and Chemical Toxicology*, *60*, 488–496.
57. Upadhyay, J., & Misra, K. (2009). Towards the interaction mechanism of tocopherols and tocotrienols (vitamin E) with selected metabolizing enzymes. *Bioinformation*, *3*, 326–331.
58. Moreau, R. A., Flores, R. A., & Hicks, K. B. (2007). Composition of functional lipids in hulled and hulless barley in fractions obtained by scarification and in barley oil. *Cereal Chemistry*, *84*, 105.
59. Temelli, F., Stobbe, K., Rezaei, K., & Vasanthan, T. (2013). Tocol composition and supercritical carbon dioxide extraction of lipids from barley pearling flour. *Journal of Food Science*, *78*(11), C1643–C1650.
60. Andersson, A. A., Lampi, A. M., Nystrom, L., Piironen, V., Li, L., Ward, J. L., Gebruers, K., Courtin, C. M., Delcour, J. A., Boros, D., & Fras, A. (2008). Phytochemical and dietary fiber components in barley varieties in the HEALTHGRAIN diversity screen. *Journal of Agricultural and Food Chemistry*, *56*, 9767–9776.
61. Hendriks, H., Weststrate, J., Van Vliet, T., & Meijer, G. (1999). Spreads enriched with three different levels of vegetable oil sterols and the degree of cholesterol lowering in normocholesterolaemic and mildly hypercholesterolaemic subjects. *European Journal of Clinical Nutrition*, *53*, 319–327.
62. Hallikainen, M. A., Sarkkinen, E. S., & Uusitupa, M. I. (2000). Plant stanol esters affect serum cholesterol concentrations of hypercholesterolemic men and women in a dose-dependent manner. *Journal of Nutrition*, *130*, 767–776.
63. Awad, A. B., & Fink, C. S. (2000). Phytosterols as anticancer dietary components: Evidence and mechanism of action. *Journal of Nutrition*, *130*, 2127–2130.
64. Rao, A. V., & Janezic, S. A. (1992). The role of dietary phytosterols in colon carcinogenesis. *Nutrition and Cancer*, *18*, 43–52.
65. Valsta, L., Lemstrom, A., Ovaskainen, M. L., Lampi, A. M., Toivo, J., Korhonen, T., & Piironen, V. (2004). Estimation of plant sterol and cholesterol intake in Finland: Quality of new values and their effect on intake. *British Journal of Nutrition*, *92*, 671–678.
66. Frølich, W., Aman, P., & Tetens, I. (2013). Whole grain foods and health: A Scandinavian perspective. *Food & Nutrition Research*, *57*. https://doi.org/10.3402/fnr.v57i0.18503 (accessed on 1st January 2026).
67. Liu, K., & Moreau, R. (2008). Concentrations of functional lipids in abraded fractions of hulless barley and effect of storage. *Journal of Food Science*, *73*, C569–C576.
68. Lampi, A. M., Moreau, R. A., Piironen, V., & Hicks, K. B. (2004). Pearling barley and rye to produce phytosterol-rich fractions. *Lipids*, *39*, 783–787.
69. Piironen, V., Edelmann, M., Kariluoto, S., & Bedo, Z. (2008). Folate in wheat genotypes in the HEALTHGRAIN diversity screen. *Journal of Agricultural and Food Chemistry*, *56*, 9726–9731.

70. Yang, W., & Huang, G. (2021). Extraction methods and activities of natural glucans. *Trends in Food Science & Technology*, *112*, 50–57. https://doi.org/10.1016/j.tifs.2021.03.009 (accessed on 1st January 2026).
71. Belitz, H. D., Grosch, W., & Schieberle, P. (2008). *Food Chemistry*. Springer Science and Business Media.
72. Torbica, A., Radosavljević, M., Belović, M., Djukić, N., & Marković, S. (2022). Overview of nature, frequency and technological role of dietary fiber from cereals and pseudocereals from grain to bread. *Carbohydrate Polymers*, *290*, 119470. https://doi.org/10.1016/j.carbpol.2022.119470 (accessed on 1st January 2026).
73. Verma, D. K., Patel, A. R., Thakur, M., Singh, S., Tripathy, S., Srivastav, P. P., Chávez-González, M. L., Gupta, A. K., & Aguilar, C. N. (2021). A review of the composition and toxicology of fructans, and their applications in foods and health. *Journal of Food Composition and Analysis*, *99*, 103884. https://doi.org/10.1016/j.jfca.2021.103884 (accessed on 1st January 2026).
74. Ispiryan, L., Kuktaite, R., Zannini, E., & Arendt, E. K. (2021). Fundamental study on changes in the FODMAP profile of cereals, pseudocereals, and pulses during the malting process. *Food Chemistry*, *343*, 128549. https://doi.org/10.1016/j.foodchem.2020.128549 (accessed on 1st January 2026).
75. Verhertbruggen, Y., Falourd, X., Sterner, M., Guillon, F., Girousse, C., Foucat, L., Le Gall, S., Chateigner-Boutin, A. L., & Saulnier, L. (2019). Challenging the putative structure of mannan in wheat (*Triticum aestivum*) endosperm. *Carbohydrate Polymers*, *224*, 115063. https://doi.org/10.1016/j.carbpol.2019.115063 (accessed on 1st January 2026).
76. Comino, P., Shelat, K., Collins, H., Lahnstein, J., & Gidley, M. J. (2013). Separation and purification of soluble polymers and cell wall fractions from wheat, rye and hull-less barley endosperm flours for structure–nutrition studies. *Journal of Agricultural and Food Chemistry*, *61*, 12111–12122. https://doi.org/10.1021/jf402678t (accessed on 1st January 2026).
77. Singh, S., Singh, G., & Arya, S. K. (2018). Mannans: An overview of properties and application in food products. *International Journal of Biological Macromolecules*, *119*, 79–95. https://doi.org/10.1016/j.ijbiomac.2018.07.120 (accessed on 1st January 2026).
78. Verspreet, J., Dornez, E., Van den Ende, W., Delcour, J. A., & Courtin, C. M. (2015). Cereal grain fructans: Structure, variability and potential health effects. *Trends in Food Science & Technology*, *43*, 32–42. https://doi.org/10.1016/j.tifs.2015.01.002 (accessed on 1st January 2026).
79. Loponen, J., & Gänzle, M. G. (2018). Use of sourdough in low FODMAP baking. *Foods*, *7*, 96.
80. Tokusoglu, Ö., & Hall III, C. A. (2011). *Fruit and Cereal Bioactive: Sources, Chemistry, and Applications*. CRC Press.
81. Perrott, L. A., Strydhorst, S. M., Hall, L. M., Yang, R. C., Pauly, D., Gill, K. S., & Bowness, R. (2018). Advanced agronomic practices to maximize feed barley yield, quality, and standability in Alberta, Canada. I. Responses to plant density, a plant growth regulator, and foliar fungicides. *Agronomy Journal*, *110*(4), 1447–1457.
82. Ingvordsen, C. H., Gislum, R., Jørgensen, J. R., Mikkelsen, T. N., Stockmarr, A., & Jørgensen, R. B. (2016). Grain protein concentration and harvestable protein

under future climate conditions. A study of 108 spring barley accessions. *Journal of Experimental Botany*, *67*, 2151–2158. https://doi.org/10.1093/jxb/erw033 (accessed on 1st January 2026).

83. Kachru, R. P., Gupta, R. K., & Alam, A. (1994). *Physico-Chemical Constituents and Engineering Properties of Food Crops*. Scientific Publishers.
84. Kashaninejad, M., Mortazavi, A., Safekordi, A., & Tabil, L. G. (2006). Some physical properties of pistachio (*Pistacia vera* L.) nut and its kernel. *Journal of Food Engineering*, *72*(1), 30–38.
85. Ozturk, T., & Esen, B. (2008). Physical and mechanical properties of barley. *Agricultura Tropica et Subtropica*, *41*(3), 117–121.
86. Al-Mahasneh, M. A., & Rababah, T. M. (2007). Effect of moisture content on some physical properties of green wheat. *Journal of Food Engineering*, *79*, 1467–1473.
87. Sologubik, C. A., Campañone, L. A., Pagano, A. M., & Gely, M. C. (2013). Effect of moisture content on some physical properties of barley. *Industrial Crops and Products*, *43*, 762–767.
88. Bleidere, M. (2009). Characteristic of grain physical traits of spring barley. *Research for Rural Development*, 8–13.
89. Zeven, A. C., & De Wet, J. M. (1982). *Dictionary of Cultivated Plants and Their Regions of Diversity: Excluding Most Ornamentals, Forest Trees and Lower Plants*. Pudoc.
90. Kleinhofs, A., & Han, F. (2002). Molecular mapping of the barley genome. In G. A. Slafer, J. L. Molina-Cano, R. Savin, J. L. Araus, & I. Romagosa (Eds.), *Barley Science: Recent Advances from Molecular Biology to Agronomy of Yield and Quality* (pp. 31–63).
91. Fox, G. P., Kelly, A. M., Cakir, M., Bloustein, G., Poulsen, D. M. E., Inkerman, P. D., & Henry, R. J. (2006). Genetic impacts of the hull on barley grain quality. *Journal of the Institute of Brewing*, *112*(2), 101–107.
92. Briggs, D. E. G. (1978). The morphology, mode of life, and affinities of *Canadaspis perfecta* (Crustacea: Phyllocarida), Middle Cambrian, Burgess Shale, British Columbia. *Philosophical Transactions of the Royal Society of London. B, Biological Sciences*, *281*(984), 439–487.
93. Newman, R. K., & Newman, C. W. (2008). *Barley for Food and Health: Science, Technology, and Products*. John Wiley and Sons.
94. Jeroch, H., & Dänicke, S. (1995). Barley in poultry feeding: A review. *World's Poultry Science Journal*, *51*(3), 271–291.
95. Fox, G. P., Kelly, A., Poulsen, D., Inkerman, A., & Henry, R. (2006). Selecting for increased barley grain size. *Journal of Cereal Science*, *43*, 198–208.
96. Evers, T., & Millar, S. (2002). Cereal grain structure and development: Some implications in quality. *Journal of Cereal Science*, *36*, 261–284.
97. Christison, G. I., & Bell, J. M. (1975). An assessment of bulk weight and other simple criteria for predicting the digestible energy values of feed grains. *Canadian Journal of Animal Science*, *55*, 115–528.
98. Ullrich, S. E., Muir, C. E., Jitkov, V. A., Burns, J. W., Reisenauer, P. E., Wesenberg, D. M., & Hayes, P. M. (2002). Registration of 'Farmington' barley (Registrations of cultivars). *Crop Science*, *42*(6), 2209–2211.
99. Burger, W. C., & LaBerge, D. E. (1985). Malting and brewing quality. *Barley*, *26*, 367–401.

100. Edney, M. J., Rossnagel, B. G., Endo, Y., Ozawa, S., & Brophy, M. (2002). Pearling quality of Canadian barley varieties and their potential use as rice extenders. *Journal of Cereal Science*, *36*(3), 295–305.
101. Geng, L., Li, M., Zhang, G., & Ye, L. (2002). Barley: A potential cereal for producing healthy and functional foods. *Food Quality and Safety*, *6*.
102. European Food Safety Authority (EFSA). (2011). Scientific opinion on the substantiation of health claims related to beta-glucans from oats and barley and maintenance of normal blood LDL-cholesterol concentrations (ID 1236, 1299), increase in satiety leading to a reduction in energy intake (ID 851, 852), reduction of post-prandial glycaemic responses (ID 821, 824), and "digestive function" (ID 850) pursuant to Article 13(1) of Regulation (EC) No. 1924/2006. *EFSA Journal*, *9*, 2207.
103. N/A. (2012). Summary of Health Canada's assessment of a health claim about barley products and blood cholesterol lowering. https://www.canada.ca/en/healthcanada/services/food-nutrition/food-labelling/health-claims/assessments/assessment-health-claim-about-barley-products-blood-cholesterol-lowering.html (accessed on 1st January 2026).
104. N/A. (2017). FSANZ mid-year highlights. *Food Australia*, *69*(5), 21.
105. Chatterjee, S., & Bagchi, D. (2007). Eating habits in combating disease: Nutraceuticals and functional foods at the crossroads of immune health and inflammatory responses. In *Immunity and Inflammation in Health and Disease*. Academic Press. https://doi.org/10.1016/B978-0-12-805417-8.00032-9 (accessed on 1st January 2026).
106. Tosh, S. M., & Bordenave, N. (2020). Emerging science on benefits of whole grain oat and barley and their soluble dietary fibers for heart health, glycemic response, and gut microbiota. *Nutrition Reviews*, *78*, 13–20.
107. Jialal, I., & Devaraj, S. (2005). Scientific evidence to support a vitamin E and heart disease health claim: Research needs. *Journal of Nutrition*, *135*, 348–353. https://doi.org/10.1093/jn/135.2.348 (accessed on 1st January 2026).
108. Kumar, A., Mosa, K. A., Ji, L., Kage, U., Dhokane, D., Karre, S., Madalageri, D., & Pathania, N. (2018). Metabolomics-assisted biotechnological interventions for developing plant-based functional foods and nutraceuticals. *Critical Reviews in Food Science and Nutrition*, *58*, 1791–1807. https://doi.org/10.1080/10408398.2017.1285752 (accessed on 1st January 2026).
109. Brimson, J. M., Onlamoon, N., Tencomnao, T., & Thitilertdecha, P. (2019). *Clerodendrum petasites* S. Moore: The therapeutic potential of phytochemicals, hispidulin, vanillic acid, verbascoside, and apigenin. *Biomedicine and Pharmacotherapy*, *118*, 109319.
110. Srinivasulu, C., Ramgopal, M., Ramanjaneyulu, G., Anuradha, C. M., & Suresh Kumar, C. (2018). Syringic acid (SA)–A review of its occurrence, biosynthesis, pharmacological and industrial importance. *Biomedicine and Pharmacotherapy*, *108*, 547–557. https://doi.org/10.1016/j.biopha.2018.09.069 (accessed on 1st January 2026).
111. Idehen, E., Tang, Y., & Sang, S. (2017). Bioactive phytochemicals in barley. *Journal of Food and Drug Analysis*, *5*, 148–161. https://doi.org/10.1016/j.jfda.2016.08.002 (accessed on 1st January 2026).
112. Eid, O., Elkady, W. M., Ezzat, S., El Sayed, A., & Abdelsattar, E. (2023). Comprehensive overview: The effect of using different solvents for barley extraction

with its anti-inflammatory and antioxidant activity. *Chemistry and Biodiversity, 20*(3), e202200935.
113. Murtaugh, M. A., Jacobs, D. R., & Jacob, B. (2003). Epidemiological support for the protection of whole grains against diabetes. *Proceedings of the Nutrition Society, 62*, 143–149.
114. Etoh, H., Murakami, K., Yogoh, T., Ishikawa, H., Fukuyama, Y., & Tanaka, H. (2004). Antioxidative compounds in barley tea. *Bioscience, Biotechnology, and Biochemistry, 68*, 2616–2618.
115. Gul, S., Ahmed, S., Kifli, N., Uddin, Q. T., Batool Tahir, N., Hussain, A., & Zia-Ul-Haq, M. (2014). Multiple pathways are responsible for anti-inflammatory and cardiovascular activities of *Hordeum vulgare* L. *Journal of Translational Medicine, 12*(1), 1–8.
116. Mitsou, E. K., Panopoulou, N., & Turunen, K. (2010). Prebiotic potential of barley derived beta-glucan at low intake levels: A randomized, double-blinded, placebo-controlled clinical study. *Food Research International, 43*, 1086–1092.
117. Schlörmann, W., & Glei, M. (2017). Potential health benefits of beta-glucan from barley und oat. *Ernährungs Umschau, 64*, M555–M559.
118. Demartino, P., & Cockburn, D. W. (2020). Resistant starch: Impact on the gut microbiome and health. *Current Opinion in Biotechnology, 61*, 66–71.
119. Fuse, Y., Higa, M., Miyashita, N., et al. (2020). Effect of high β-glucan barley on postprandial blood glucose and insulin levels in type 2 diabetic patients. *Clinical Nutrition Research, 9*(1), 43–51.
120. Lal, M. K., Singh, B., Sharma, S., Singh, M. P., & Kumar, A. (2021). Glycemic index of starchy crops and factors affecting its digestibility: A review. *Trends in Food Science and Technology, 111*, 741–755.
121. Tosh, S. M. (2013). Review of human studies investigating the post-prandial blood-glucose lowering ability of oat and barley food products. *European Journal of Clinical Nutrition, 67*, 310–317.
122. Sagnelli, D., Chessa, S., Mandalari, G., et al. (2018). Low glycaemic index foods from wild barley and amylose-only barley lines. *Journal of Functional Foods, 40*, 408–416.
123. Hinata, M., Ono, M., Midorikawa, S., & Nakanishi, K. (2007). Metabolic improvement of male prisoners with type 2 diabetes in Fukushima Prison, Japan. *Diabetes Research and Clinical Practice, 77*, 327–332.
124. Minehira, K., Schwarz, J. M., Acheson, K., et al. (2001). Mechanisms of action of beta-glucan in postprandial glucose metabolism in healthy men. *European Journal of Clinical Nutrition, 55*(5), 327–333.
125. Lockyer, S., & Nugent, A. P. (2017). Health effects of resistant starch. *Nutrition Bulletin, 42*(1), 10–41.
126. Geier, M. S., Butler, R. N., & Howarth, G. S. (2006). Probiotics, prebiotics and symbiotic: A role in chemoprevention for colorectal cancer? *Cancer Biology & Therapy, 5*, 1265–1269.
127. Schlörmann, W., Hiller, B., Jahns, F., et al. (2012). Chemo preventive effects of *in vitro* digested and fermented bread in human colon cells. *European Journal of Nutrition, 51*, 827–839.

PART V
Industrial Crops

CHAPTER 11

Rapeseed/Canola (*Brassica napus* L.)

SANA MEDIMAGH,[1,*] SONIA MANSOURI,[1] HATEM ZGALLAI,[2] MOUNA MECHRI,[3] HELA MAHMOUDI,[4] and FAYSAL BEN JEDDI[5]

[1]*Laboratory of Field Crops, National Agricultural Research Institute of Tunisia (INRAT), University of Carthage, Tunis, Ariana Governorate, Tunisia*

[2]*National Agricultural Research Institute of Tunisia (INRAT), University of Carthage, Ariana Governorate, Tunisia*

[3]*National Institute of Field Crops (INGC), Bousalem, Jendouba Governorate, Tunisia*

[4]*Laboratory of Natural Substances, National Institute for Research and Physico-Chemical Analysis (INRAP), Biotechpole of Sidi Thabet, Ariana Governorate, Tunisia*

[5]*Laboratory of Horticultural and Environmental Sciences, National Institute of Agronomy of Tunisia (INAT), University of Carthage, Ariana Governorate, Tunisia*

*Corresponding author

ABSTRACT

Brassica napus L., commonly known as rapeseed or canola, is a common crop that is cultivated throughout much of the world. As a very important food crop in the form of oilseeds, forage, mulch, manure, and spices, canola has played a huge role in human history, contributing a large portion of food in one form or another. Rapeseed is used not only for human edible

Postharvest Quality Attributes of Fresh Vegetables and Grains.
Riadh Ilahy, Imen Tlili, Thouraya R'him, and Muhammad Azam (Eds.)
© 2027 Apple Academic Press, Inc. Co-published with CRC Press (Taylor & Francis)

DOI:10.1201/9781779642851-16

oil, but also for protein-rich feed for animals and renewable feedstock for biodiesel and industrial applications. It is also used as a cover crop to improve soil fertility and suppress weeds.

Rapeseed has been cultivated since ancient times and is now grown in many parts of the world, including Europe, North America, China, India, and Australia. Given that its output has increased more quickly than any other source of edible vegetable oil over the past 10 years, canola is one of the most significant oilseed crops in the world.

This chapter aims to summarize the botanical classification, history, production, chemical components, and health-beneficial properties of rapeseed or canola.

11.1 INTRODUCTION

11.1.1 BOTANICAL NAME, COMMON NAME

Rapeseed (*Brassica napus* L.), also known as oilseed rape, annual rape, colza, Hanover-salad, rape kale, rutabaga, Siberian kale, summer rape, Argentine rape swede, Swede rape, Swedish turnip, winter rape, and canola [1, 2], is an annual herbaceous species belonging to the Brassicaceae family, formerly named the Cruciferae family, and the subfamily Brassicoidae [3, 4]. The rapeseed with low erucic acid content in oil and low glucosinolate levels in seed meal, known as "Canola," was first bred in Canada and then in Europe in the 1970s and spread to other parts of the world as the gene source for "double-low" quality in breeding programs [5].

11.1.2 HISTORY

Rapeseed (*Brassica napus* L.) is one of the oldest oilseed crops in the world and has been cultivated for centuries. Due to its origin and evolution, rapeseed has a complex polyploid genome ($2n = 4x = 38$; genome AACC). It originated from a spontaneous interspecific hybridization between two diploid progenitors, *Brassica rapa* ($2n = 20$; genome AA) and *Brassica oleracea* ($2n = 18$; genome CC), less than 7500 years ago [4, 6].

Rapeseed has been used for centuries as an oilseed crop, providing a source of vegetable oil for cooking and other uses. The oil is high in monounsaturated fats, making it a healthier alternative to other cooking oils. In addition, rapeseed meal is a valuable source of protein for livestock feed [7].

The name rape comes from the Latin word "rapum," which means turnip. However, the exact time and place of rapeseed's origin remains uncertain, many of the earliest records of the crop's cultivation come from India, China, and Japan [8]. Early history shows that rapeseed has been cultivated for thousands of years and originated in Asia. The Chinese's oldest archaeological discoveries may date back as far as ca. 5000 BC [9, 10]. Sanskrit writings, the sacred language of Hinduism, of 2000–1500 BC, are directly related to the oleaginous form of *B. napus* (Sarson type) and mustard [8].

It is believed to have been in the Mediterranean region and was first used as a food crop by the Greeks and Romans. It was later introduced to Europe and North America in the 16th century, where it became a popular oilseed crop [4]. Although the crop has grown in many countries around the world, including Australia, Switzerland, Germany, China, India, Italy, France, Denmark, and the Netherlands. In 1976, much of the work to develop the modern premium rapeseed variety known as Canola was done in Canada. The emergence of canola, or double-low varieties, has elevated rapeseed to a major global vegetable oil source [11, 12].

Canada registered the appellation "canola" in 1979 to refer to this new seed, which was low in erucic acid and glucosinolates [13]. By definition, canola has a threshold for erucic acid (<2%) and glucosinolate (<30 umol/g), a specific edible material intended for consumption by both humans and animals [13, 14].

The original use of rapeseed was as lighting oil, but it was also sometimes used as cooking oil by the poor. In later centuries, it gradually developed into an important cooking oil and kindling oil. In recent years, rapeseed has become an important source of biofuel due to its high oil content and low carbon footprint. Biodiesel made from rapeseed is becoming increasingly popular as an alternative fuel source for cars and trucks.

Rapeseed of commercial interest is grown in the cooler areas of the world, covering mainly North America, the northern part of Europe, Canada, China, and India [8]. In addition to its use as an oilseed crop and biofuel, rapeseed is also used in the production of industrial lubricants, paints, soaps, detergents, plastics, and other products.

11.1.3 PRODUCTION (WORLD)

The production of rapeseed and oil quality has greatly increased since 1994. The most important growing regions for rapeseed/canola are Canada,

Europe, China, India, and Australia [15, 16]. The world production of rapeseed is around 27.64 M tons, which allows it to be an economically important agricultural crop [16]. Up until 2020, more than 66 countries grew rapeseed, with 35 million hectares of growth space globally [16]. Now, the EU, China, Australia, Ukraine, and Canada are the top five exporters of rapeseed worldwide. China, Japan, and EU nations rank among the market's top importers [17]. The total UK rapeseed area was 307,000 hectares (MHa) in marketing year (MY) 2021/22 [18]. Additionally, the production volume of rapeseed oil worldwide in 2021–2022 was just over 29 million metric tons [19].

11.1.4 BOTANICAL DESCRIPTION

Rapeseed (*Brassica napus* L.) is an annual plant in the Brassicaceae family. The Brassicaceae family is comprised of about 25 tribes, 338 genera, and 3709 species [20–22]. The species of genus Brassica are classified into two major groups commonly known as rapeseed and mustard. Rapeseed mainly includes *B. napus* L., *B. rapa* L., and *B. nappobrassica* L., while mustard includes *B. nigra* L., *B. carinata* L., *B. juncea* L., *B. hirta* L., and *B. arvensis* L. [23]. *Brassica napus* is included within the Brassiceae tribe, which consists of 9 subtribes, 48 genera, and 240 species [21].

Also, the Brassicaceae family includes the model plant *Arabidopsis thaliana* (mouse-ear cress), its weedy relative *Synapsis arvensis* (wild mustard), and *B. napus* (black cabbage, Siberian kale), *B. rapa* (Chinese cabbage, Pai-Tsai, mizuna, Takana, broccoli Raab, turnip), *B. oleracea* (cabbage, broccoli, cauliflower, Brussels sprouts, kohlrabi, cabbage, kale), and *Raphanus sativus* (radish). Such aromatic plants of the cruciferous family are *B. nigra* (black mustard), *B. carinata* (Ethiopian mustard), *B. juncea* (brown or Indian mustard), *Armoracea rusticana* (horseradish), and many other small potted herbs and salad vegetables [21, 24].

Rapeseed is a cool-season crop that grows best in temperate climates with well-drained soils. The plant grows from a taproot to a height of 30 to 150 cm and has a long, erect, thin, cylindrical stem that can reach up to 1.5 m in height. Stems, branched at the top, are glabrous or slightly hairy. Basal leaves are stalked, 5 to 25 cm long, 2 to 7 cm wide, pinnately lobed with large, rounded terminal lobes. The leaves are alternate, oblong-lanceolate, glaucous green in color, and have a waxy coating on the upper surface. Stem leaves are lanceolate to oblong, smaller than basal leaves, sessile, base surrounded or lobed, margin entire or slightly wavy. The petioles are usually winged and can be up to 15 cm long. Clusters of 20 to 40 flowers form at the ends of the stems. The flowers are yellow and borne in racemes at the ends of branches, and have four petals each. Petals are cream to pale yellow, broadly obovate, 10 to 16 mm long and 6 to 9 mm wide, with rounded tips. The buds are above or the same size as the open flowers. Fruits are narrow ascending pods or siliques (a dry fruit with two valves), 5–10 cm long, 3.5–5 mm wide, with a beak at the top, 7–16 mm long. The pod contains 15–30 seeds. Seeds are small, dark brown to gray-black in color, spherical in shape, 1.8 to 2.7 mm in diameter with a glossy surface. The weight of 1000 seeds is 3–4 g [15, 25–27]. It should be harvested when the pods are dry and the seeds are mature.

Varieties of canola that are divided into winter (requiring vernalization) and spring (non-vernalization) have different vernalization requirements for flowering, which affects yield and antioxidant content in seeds and oil [28, 29].

11.2 ANTIOXIDANT PROPERTIES

One of the most significant challenges that the world faces today is providing access to nutritious, sustainable, and sufficient food for the growing global population [30]. Every part of the rapeseed (flowers, seeds, leaves, stem, and root) can be used for food, medicinal, and cosmetic purposes [31]. The seeds are the most crucial component as they provide oil and protein [32].

11.2.1 FLOWERS

The blossoms of rapeseed can be used to make infused tea [33], and since the pollens help the body fight against illnesses like cancer, they can also be consumed. This might be explained by the steroids found in pollen, which may have an impact on the viability of cancer cells, particularly those found in the prostate [34–36].

11.2.2 SEEDS

The most significant component is the seeds, which can be used as spices, condiments, and sources of oil and protein [31]. Different cultivars of rapeseed have varying amounts of protein and oil, and the seeds also contain other substances like phenols, phytic acid, cellulose, sugars, and glucosinolates. Rapeseed is in competition with other crops despite being well-known for producing high-quality vegetable oil [1, 37].

11.2.3 LEAVES AND STEM

The leaves and stems of rapeseed can be utilized for either edible vegetables or as a seasoning potherb [31]. Essential minerals, protein, and vitamins can be found in green leaves in high amounts [38]. They are rich in antioxidant phenolic compounds and should be consumed regularly to prevent the risk of chronic diseases [39–44]. Rapeseed leaves are an excellent plant-based diet for weight loss, due to their low-fat composition, high calcium content (the highest among green leafy vegetables), and high vitamin C content [45, 46]. Vitamin C is a well-known antioxidant that enhances immunity and protects macromolecules, as oxidation is the cause of a number of acute and chronic diseases, such as colds, depression, allergies, and bad cholesterol [47]. Brassica is known to have a high level of folate (Vitamin B9), which is important in preventing birth defects, particularly in the brain and spine, during pregnancy [48]. Furthermore, rapeseed has the ability to reduce cholesterol and cholesteryl esters by containing iron, carotene, and dietary fiber [49]. Rapeseed has both phytates and tocopherols (vitamin E), which are antioxidants [50], and phytosterols could decrease serum and low-density lipoprotein (LDL) cholesterol levels [51, 52].

11.2.4 ROOTS

The growth of human Hep G2 cancer cells can be inhibited by rape root extract, which has an anticancer effect. Additionally, seed extract has been shown to contain tocopherol, which may have a potential impact on the risk of prostate cancer [53]. Rapeseed roots have various diuretic, anti-gout, anti-inflammatory, and anti-scurvy qualities that make them beneficial [54].

11.2.5 EDIBLE OIL

Rapeseed oil is one of the main vegetable oils used in the food industry in many countries. The proportion of saturated fatty acids in this oil is low, and the proportion of monounsaturated fatty acids is high compared to other edible oils. In addition, rapeseed oil contains linoleic acid, linolenic acid, and other essential fatty acids [55]. Rapeseed oil is used as refined, bleached, and deodorized oil because it is necessary to ensure no negative impact of the product on the oil with regard to color and taste. The original oil has some disadvantages because of the relatively high amount of polyunsaturated fatty acids (PUFAs), but by hydrogenation or interesterification, the stability of rapeseed oil for such processes is highly improved, with some higher amounts of trans-fatty acids in partially hydrogenated rapeseed oil. Recently, the use of high oleic, low linolenic rapeseed oil with more than 70% of oleic acid and less than 3% of linolenic acid is an interesting alternative to partially hydrogenated rapeseed oil as a frying medium. As of the absence of trans fatty acids and a comparable or better performance during frying, this oil can help to overcome the problems of rapeseed oil during frying [56]. A relatively new field for rapeseed oil is its use as cold-pressed oil, similar to native olive oil. Consumers appreciate the seedlike and sometimes nutty taste and smell, as well as the small processing of this product [57].

Canola oil exhibits a distinct fatty acid profile: It is a good source of monounsaturated fatty acids (MUFA, 68.6%) [58] and PUFA; cold-pressed canola seed oil (CNO) also significantly contains oleic acid (63.7%). It contains 11% omega-3 PUFAs, 53–59% MUFA, 22% omega-6 PUFAs, and 7.1% saturated fatty acids (SFA) [59, 60], and its ratio of omega-6 to omega-3 is appropriate [60–62]. There are trace amounts of stearic acid, eicosenoic acid, erucic acid, arachidic acid, behenic acid, and palmitoleic acid [62]. CNO has less saturated fat (SFA, 7.2%) than both sunflower oil (11%) and soybean oil (15%) [63].

11.2.6 RAPESEED-BASED PRODUCTS

Rapeseed has been grown since ancient times. It is being used for a variety of purposes that are advantageous to the economy, environment, and health. Rapeseed oil was inexpensive and used to lubricate engines, create soap, illuminate lights, and cook food. Moreover, rapeseed oil can be used in salad dressings, mayonnaise, and other emulsified products like biscuits and canning [64]. Canola oil is now recognized as one of the most nutritious edible vegetable oils due to its biological functions and its capacity to reduce risk factors for disease and enhance health [7]. Considerable interest in canola has been observed all over the world [65]. Rapeseed meal is the residual product from seed oil extraction [31]. It is characterized by a high fiber content (33–40%), protein content (31–34%), and the presence of carbohydrates (6–13%) and ash (5.5–6.8%) [66]. Rapeseed meal offers proteins with a balanced amino acid composition, which can be used as a substitute protein source and help meet the global demand for protein [32]. The rapeseed market in Europe is predominantly driven by the use of rapeseed oil in the food industry [67].

11.3 CHARACTERIZATION OF THE CHEMICAL COMPOUND(S) RESPONSIBLE FOR ANTIOXIDANT PROPERTIES

Numerous studies have revealed that the rapeseed (*Brassica napus* L.) is a valuable source of antioxidant compounds [68]. Moreover, rapeseed meal or its extracts can be used as a natural source of antioxidants [68]. Thus, the development of canola quality rapeseed has made rapeseed one of the major plant oil sources at the global level [11].

Canola oil (CNO) has phenolic compounds, tocopherols, phytosterols, and carotenoids that are bioactive and contribute to its properties [62]. Canola oil's antioxidant properties can be attributed to its hydroxyl group (–OH), which can scavenge free radicals through phenolic compounds, tocopherols, and phytosterols [62].

11.3.1 PHENOLIC COMPOUNDS

Phenolic compounds, such as flavonoids, phenolic acids, condensed tannins, and lignans, are the main compounds responsible for the antioxidant properties of canola. These compounds are found in the seed coat and oil of canola and have been shown to scavenge free radicals, reduce lipid oxidation, and inhibit the formation of advanced glycation end products (AGEs) [69].

The phenolic acid compounds in CNO are mainly sinapic acid, cinnamic acid, syringic acid, ferulic acid, 4-hydroxybenzoic acid, vanillic acid, and p-coumaric acid [70]. Studies indicate that the main phenolic compounds in rapeseed meal are sinapine, while sinapic acid was also found at one-tenth that of sinapine [71]. In order to obtain CNO from rapeseed cold-pressing, heat treatment is applied beforehand. This causes sinapic acid decarboxylation, which produces canolol [72], an oil-soluble bioactive phenolic compound that is simpler to extract into CNO. It's antimutagenic, anticarcinogenic, and antioxidant properties are noteworthy [73].

Canola flour had 11.74 and 0.40 mg/g sinapine and sinapic acid contents, representing 51.99 and 1.77% of the canola total phenolics, respectively, depending on the variety of oilseed plant and the oil-processing method [74–76]. The amount of canolol in crude canola oil was approximately 200 ppm [77, 78].

The total concentration of all detectable phenolic compounds is approximately 400 mg per kg of rapeseed meal. In rapeseed meal, sinapic acid is the most abundant phenolic compound, which accounts for over 85% of all quantified phenolic compounds (mean 357 ± 13 mg/kg, range 339–379 mg/kg). According to reports, rapeseed contains the most abundant bioactive compounds, including sinapine, which has antitumor, neuroprotective, antioxidant, and hepatoprotective properties that are beneficial for health [66].

Condensed tannins and flavonoids are mainly localized in seed skins, while simple phenolic compounds are more concentrated in seed almonds [74, 79]. Naczk and Shahidi [80] showed the presence of two major flavonoids in rapeseed cake: 3-(O-sinapoyl sophoroside)-7-O-glucoside kaempferol and 3-(O-sinapoyl glucoside) 7-O- sophoroside kaempferol. Auger et al. [81] identified 13 different flavonoid subunits and seven different flavonol subunits in rapeseed skins.

11.3.2 TOCOPHEROLS

Tocopherols of canola oil are γ-tocopherol and α-tocopherol, as the major tocopherols, while δ-tocopherol, plastochromanol-8 (PC-8), and β-tocopherol are the minor tocopherols [73]. Vitamin E is a fat-soluble vitamin found in canola oil. It is a powerful antioxidant that helps protect cells from oxidative damage. Studies have shown that canola oil is a rich source of vitamin E, providing up to 20% of the recommended daily intake.

11.3.3 PHYTOSTEROLS

Phytosterols, or plant sterols, are plant-derived compounds found in canola oil. They are known for their LDL cholesterol-lowering properties [82] and have been shown to reduce cholesterol levels and have antioxidant properties [83]. Studies have shown that canola oil is a rich source of phytosterols, providing up to 20% of the recommended daily intake. The main CNO phytosterols are cholesterol, brassicasterol (specific for rapeseed oil), campesterol, stigmasterol, sitosterol, and 5-avenasterol [73].

11.3.4 CAROTENOIDS

Carotenoids can be found in crude canola oil at concentrations that are enough to provide potential health and nutritional benefits to consumers [84, 85].

The chemical relationship between carotenoids and vitamin A is crucial for skin and eye health and the prevention of oxidative processes, including several types of cancer and cardiovascular diseases (CVDs) [86, 87]. β-carotene and canola oil carotenoids are β-carotene, zeaxanthin, and lutein [88]. The main carotenoid found in canola oil is lutein. Lutein helps prevent macular degeneration and filters harmful light to reduce the amount of light that reaches the retina, while carotene exhibits antioxidant properties and vitamin A activity [89].

11.4 HEALTH BENEFITS

Plant metabolites are the origin of a vast array of biological functions, including anti-inflammatory, antimicrobial, anticancer, and antianalgesic actions [7, 90].

11.4.1 ANTIOXIDANT ACTIVITY

Rapeseed is a rich source of antioxidants, which can help protect cells from oxidative stress, reduce inflammation, and damage caused by free radicals. Canola seeds contain a variety of bioactive compounds, including antioxidant vitamins such as phenolic acids (sinapic acid, canolol, sinapin), coenzyme Q (CoQ), flavonoids, and tocopherols. Sinapic acid is a phenylpropanoid and, with the exception of its antioxidant activities [91], these compounds have been shown to have antioxidant properties, anti-inflammatory, healthy metabolic, and physiological effects that can help reduce inflammation and protect against cell damage [92]. Additionally, rapeseed oil is high in omega-3 fatty acids, which are known for their anti-inflammatory effects.

Rapeseed is a good source of omega-3 PUFAs for humans, which are beneficial for heart health and cognitive function. The oil obtained from this plant contains 8 to 10% linolenic acid (ALA, 18:3 omega-3) and has a good ratio of omega-6/omega-3 acids [93]. Fatty acid consumption (especially PUFAs) leads to an increase in the rate of fatty acid oxidation, which produces hydrogen peroxide in peroxisomal and mitochondrial cells.

11.4.2 ANTICANCER ACTIVITY

Cancer is regarded as a disorder that affects basic cellular functions like growth signaling, anti-apoptotic signaling, gene stability, and immune response [94]. The agro-industrial sector produces rapeseed meal, which has a significant amount of anticancer proteins and peptides. Rapeseed plant metabolites are now becoming new sources of anticancer therapy. Bioactive compounds in these plants have anticancer properties for major types of tumors [95]. Efforts have been made to conduct numerous clinical studies to uncover effective cancer treatments. Radiation therapy and chemotherapy have significant adverse effects on healthy cells. Targeted therapies and immunotherapy are considered among the most effective cancer therapies, but they are limited to certain patients and are costly treatments. In this respect, combination therapy has been recently utilized to avoid their side effects [96]. In addition, efforts are being made to find new and efficient drugs that have less side effects and toxic effects. More than 60% of anticancer drugs are derived from plant materials, and Brassicaceae families are a major source of glucosinolates, which are widely used as biologically active compounds [97]. In various forms of cancer therapy, active metabolites of glucosinolates are employed. For example, sulforaphane helps inhibit the growth of prostate cancer tumors. It also helps treat breast cancer, ovarian cancer, and melanoma [98]. Breast cancer, hepatocellular carcinoma, and pancreatic tumors can all be successfully treated with erucin. Prostate, breast, colon, and hepatocellular carcinoma are all prevented by indole-3-carbinol [99]. It is now known that rapeseed may have chemoprotective and anticancer properties. According to certain research, different rapeseed extracts may effectively purify reactive oxygen species (ROS), which could be linked to the anticancer action [100]. Specifically, rapeseed extracts include brassinosteroids, which have anticancer properties. The steroid part of the rapeseed chloroform extract can kill some types of human cancer cells. Induction of caspase activity and suppression of Bcl-2 protein cause cell death, primarily in prostate cancer cells [101]. According to certain findings from different research, a diet high in omega-3 fatty acids can reduce testosterone and estradiol levels while also inhibiting the growth of prostate tumor cells in mice [102].

By preventing human Hep G2 cancer cells from proliferating, rape root extract has anticancer properties. According to additional information, rapeseed extract's tocopherol may help lower the risk of prostate cancer [53]. An enzymatic digest of rapeseed protein was shown in one study to stop a breast cancer cell line from proliferating. The alkaline rapeseed extract was used to extract the protein isolate. Consequently, every protein hydrolysate that was obtained showed an antiproliferative effect on MCF-7 [103].

11.4.3 ANTI-INFLAMMATORY ACTIVITY

Numerous research works have examined how a diet high in canola oil affects biomarkers of inflammation and antioxidant status in relation to the risk of coronary heart disease. Kwon et al. [104] looked at how diets high in MUFA (canola oil) affected the production of thromboxane and platelet aggregation. In comparison to the high-SFA baseline diet, the canola oil groups displayed reductions in platelet aggregation after three weeks. At week eight, canola oil affected platelet function by reducing ATP secretion. On platelet function, however, the canola oil-based diet had longer-lasting positive effects. Similarly, Gillingham et al. [105] note differences in plasma inflammatory markers amongst the high-oleic canola oil-containing experimental diets.

11.4.4 NEUROPROTECTIVE ACTIVITY

The neuroprotective effects of sinapic acid and sinapine have been reported [106–108]. According to Zou et al. [106], sinapic acid scavenges peroxynitrite, and this suggests that sinapic acid may be important for protecting neurons from disorders linked to peroxynitrite. The neuroprotective effects of sinapic acid on kainic acid (KA)-induced hippocampus brain injury in mice were examined by Kim et al. [108]. A pretreatment with sinapine significantly reduced the cell death of PC12 cells. Sinapine inhibited the loss of mitochondrial membrane potential and reduced the apoptosis rate [107].

11.4.5 ANTIMICROBIAL (ANTIVIRAL AND ANTIBACTERIAL) ACTIVITY

Rapeseed water extract under subcritical conditions has been investigated as an antiviral agent against the A/H1N1 virus. When compared to other extracts of n-hexane, ethanol, or hot water, the subcritical aqueous extract demonstrated antiviral efficacy against the influenza A/H1N1 virus at maximum non-toxic doses. When 0.5 mg/mL subcritical aqueous extract was added to the culture medium, the kidney cells of dogs infected with the influenza A/H1N1 virus showed 50.35% vitality. In order to prevent influenza virus infection, the extract exhibited antiviral activity. The information gathered suggests that rapeseed extract may be used as an antiviral agent [109].

One of the primary causes of chronic liver disease is the hepatitis C virus. With over 170 million cases worldwide, it is still a significant public health issue [110]. An extract of a transgenic hepatitis C virus core protein produced from rapeseed was studied for its antiviral properties. A robust humoral (IgG) and Th1-dependent cellular response was generated in mice immunized with the transgenic core protein oil. Intracellular cytokine labeling results demonstrated that transgenic core protein oil immunization in mice elicited the release of interferon from $CD4^+$ and $CD8^+$ T cells. The information gathered is crucial for the creation of a hepatitis C vaccination and shows the potential of the rapeseed antigen [111, 112].

Another study demonstrates the antibacterial properties of rapeseed alcohol extract on specific kinds of pathogenic bacteria. Disc diffusion agar and excellent diffusion agar were used in the investigation for *Pseudomonas aeruginosa*, *Bacillus cereus*, *Escherichia coli*, and *Staphylococcus aureus*. Using successive dilutions, an *in vitro* test was conducted to determine the minimal inhibitory concentration and minimum bactericidal concentration. Consequently, the rapeseed alcohol extract stopped the aforementioned harmful bacteria from growing. In Pakistan, rapeseed oil has been utilized to cure a variety of illnesses and skin conditions. N-hexane is used to extract the oil from rapeseed. A film with supercritical carbon dioxide-extracted rapeseed extract showed antibacterial efficacy against pathogenic microbes as well. The decimal reductive concentration and the minimum bactericidal concentration for microorganisms were found by an *in vitro* investigation. Using gas chromatography, terpenoids, diterpenes, and sesquiterpenes were identified in the composition of this extract. Four strains of pathogenic microorganisms (*S. aureus*, *E. coli*, and *Candida albicans*) were used in the investigation of the antibacterial and fungicidal properties of rapeseed extract. Consequently, the rapeseed extract showed biological activity against both *E. coli* and *C. albicans*. However, *P. aeruginosa* and *S. aureus* were not shown to have any inhibitory ability. Therefore, the information gathered suggests that rapeseed extract may be used as an antibacterial agent in the pharmaceutical industry [113].

11.4.6 ANTI-OBESITY

Diabetes and insulin resistance are primarily caused by being overweight or obese. Furthermore, the onset of diabetes is linked to an increase in adipose tissue, particularly visceral fat [114]. Studies have been conducted on the hypoglycemic properties of rapeseed because of its high flavonoid content. Wistar rats were given rapeseed extract orally for four weeks. Consequently, rapeseed extract helped lower serum triglyceride levels and dramatically decreased glucose levels in diabetic rats. Rapeseed extract, therefore, has a hypoglycemic effect and is advantageous for diabetic patients [115].

By modifying antioxidant activity and inflammation, concurrent antioxidant supplementation with omega-3 PUFAs is a suitable nutritional approach to lessen metabolic disturbances associated with obesity [116]. Preclinical trials were carried out by Capel et al. [117] to investigate the impact of rapeseed oil in rats' diets on obesity prevention. The diet consisted of an adequate ratio of omega-6/omega-3 acids and rapeseed oil supplemented with PUFAs. Thus, the rapeseed oil diet decreased triacylglycerol levels and prevented rats from developing glucose intolerance. Therefore, it has been demonstrated that enriched rapeseed oil with natural micronutrients enhances the metabolism of skeletal muscle and adipose tissue, enabling better control over excess fatty acid management and lowering glycemia, which may have long-term benefits [117].

Malgorzata Jamka et al. [118] carried out a clinical trial to investigate the impact of rapeseed oil on individuals who are overweight or obese. A chronic systemic or gastrointestinal illness causing amnesia, liver disease, exocrine pancreatic insufficiency, medications affecting the digestion or absorption of fat, pregnancy, and lactation were among the exclusion criteria. Every study participant was divided into groups at random. 20 milliliters of rapeseed oil were given each day. Tumor necrosis factor-alpha (TNF-α), adiponectin, oxidized LDL, apolipoproteins (Apo) A1, B, and E, as well as indicators of glucose and insulin homeostasis, were among the baseline variables that changed. The atherosclerosis markers were positively impacted by rapeseed oil [85].

11.4.7 ANTI-DIABETES

One of the most prevalent illnesses affecting people worldwide, diabetes is becoming more common each year [119]. The global epidemic of type 2 diabetes is increasing rapidly [120], and the number of people with diabetes has doubled in the past decade [121]. Currently, type 2 diabetes cannot be cured with conservative therapies [122]. It has been discovered in some studies that canola seeds contain inhibitory peptides for dipeptidase-IV (DPP-IV) and the angiotensin-converting enzyme (ACE) [123]. A serine protease called DPP-IV aids in the inactivation of glucagon-like peptide-1 upon its entry into the bloodstream. These peptides have a high rate of insulin secretion and glucose tolerance, according to preclinical research. As a result, *Brassica napus* L. seeds may have hypotensive and antidiabetic effects [124].

11.4.8 ANXIOLYTIC EFFECTS

Yoon et al. [91] characterized the putative anxiolytic-like properties of sinapic acid using an elevated plus-maze (EPM) and hole-board test. Interestingly, sinapic acid (4 mg/kg) significantly increased the percentages of time spent in the open arms of the EPM test.

11.4.9 CARDIOPROTECTIVE ACTIVITY

In many nations, CVD is one of the leading causes of premature mortality and disability. Atherosclerosis represents the most prevalent pathogenic process among cardiovascular illnesses. For the past 50 years, there has been widespread recognition of the connection between dietary fat and CVD [125]. Chronic inflammation, oxidative stress, and aberrant cholesterol levels are some of the variables that affect the course of atherosclerosis [126]. It is well recognized that a diet rich in cis-monounsaturated and PUFAs, as well as bioactive antioxidants like vitamins and phenolic compounds, can promote a healthy metabolism and have a cardioprotective effect [127].

The Food and Drug Administration (FDA) has approved a health claim for conventional canola oil and has confirmed the GRAS status of canola oil. A rule issued by the FDA in 2006 allows health claim labeling of food products made with canola oil. Moreover, canola is high in unsaturated fats, and consuming 1.5 tablespoons of canola oil per day may reduce the risk of coronary heart disease [128].

11.4.10 HYPOCHOLESTEROLEMIC ACTIVITY

Rapeseed oil has been shown in numerous studies to reduce LDL cholesterol and blood total cholesterol. By preventing the absorption of cholesterol, the phytosterols in rapeseed oil have a hypocholesterolemic effect [129]. The presence of trace components in rapeseed oil can stop atherosclerosis from happening. In one study, rapeseed oil's impact on atherosclerosis risk factors was investigated in laboratory rats. According to the study, triglyceride and plasma cholesterol levels in rats significantly dropped after they consumed oil. Therefore, by enhancing plasma oxidative stress, the lipid profile, and inflammation, rapeseed oil inhibits atherogenesis [130].

11.4.11 ANTIULCER ACTIVITY

A major factor in the etiopathology of gastric ulcers is stress [131]. Total antioxidant activity, reducing power, DPPH assay, metal chelating, and active oxygen species assays are most commonly used for the evaluation of antioxidant activities of extracts [132]. Rapeseed has antiulcer activity, which can help heal stomach ulcers and prevent them from recurring [131].

11.4.12 ANTI-HYPERTENSIVE ACTIVITY

High blood pressure, or hypertension, is a dangerous medical condition that greatly raises the risk of developing illnesses of the kidneys, brain, heart, and other organs. Adults in the modern world have an incidence of arterial hypertension of 30–45% [133]. The force that blood flow has on the arteries, the body's most vital blood vessels, is known as blood pressure. An excessive rise in blood pressure is the hallmark of hypertension. By 2025, 1.6 billion people worldwide are expected to have this illness [134]. Arterial hypertension may be treated with peptides produced from rapeseed. A study conducted *in vivo* on mice investigated the potential safety and antihypertensive effects of rapeseed peptides in combination with captopril. Additionally, in hypertensive rats, rapeseed peptides extended the period of antihypertensive activity by 20% and synergistically improved the amplitude of captopril's 9% blood pressure-lowering effect [135]. Rapeseed peptides can therefore be utilized to assist in lowering blood pressure through the development of an antihypertensive agent.

11.5 CONCLUSION

Rapeseed is a nutrient-dense food, containing high levels of bioactive components such as polypeptides, minerals, vitamins, alkaloids, flavonoids, condensed tannins, isoflavones, terpenes, anthroquinones, and glucosinolates. This plant contains essential fatty acids and antioxidants. It is an excellent source of vitamin E, which is important for healthy skin and eyesight. It also contains omega-3 fatty acids, which are beneficial for heart health. Additionally, rapeseed contains lutein and zeaxanthin, two carotenoids that help protect against age-related macular degeneration. The current investigation outlines the physicochemical properties, nutritional values, and health-promoting phytochemicals, as well as value-added products, of varieties that have dark-colored seed coats. In conclusion, the use of canola oil, which has been shown to have beneficial effects on circulating lipid levels, is now being supported by increasing scientific evidence as a health-promoting component of the diet.

KEYWORDS

- **bioactive compounds**
- ***Brassica napus***
- **canola oil**
- **functional food**
- **glucosinolates**
- **nutritional value**

REFERENCES

1. Gulden, R. H., Warwick, S. I., & Thomas, A. G. (2008). The biology of Canadian weeds. 137. *Brassica napus* L. and *B. rapa* L. *Canadian Journal of Plant Science*, *88*, 951–996.
2. N/A. (2017). Germplasm Resources Information Network (GRIN) Taxonomy for Plants. Taxon: *Brassica napus* L. United States Department of Agriculture, Agricultural Research Service, Beltsville Area. http://www.ars-grin.gov/cgi-bin/npgs/html/taxon.pl?7661 (accessed on 1st January 2026).
3. N/A. (2011). *The Biology of Brassica napus L. (Canola)*, Version 2.1. Department of Health and Aging, Office of the Gene Technology Regulator, p. 72.

4. Chalhoub, B., Denoeud, F., Liu, S., Parkin, I. A., Tang, H., Wang, X., et al. (2014). Early allopolyploid evolution in the post-Neolithic *Brassica napus* oilseed genome. *Science, 345*(6199), 950–953.
5. Dupont, J., White, P. J., Johnston, K. M., Heggtveit, A. H., & McDonald, B. E. (1989). Food safety and health effects of canola oil. *Journal of the American College of Nutrition, 5*, 360–375.
6. Nagaharu, U. (1935). Genome analysis in *Brassica* with special reference to the experimental formation of *B. napus* and peculiar mode of fertilization. *Japanese Journal of Botany, 7*, 389–452.
7. Lin, L., Allemekinders, H., Dansby, A., Campbell, L., Durance-Tod, S., Berger, A., & Jones, P. J. H. (2013). Evidence of health benefits of canola oil. *Nutrition Reviews, 71*(6), 370–385.
8. Gupta, S. K., & Aditya, P. (2007). History, origin, and evolution. In *Advances in Botanical Research Incorporating Advances in Plant Pathology: Volume 45–Rapeseed Breeding* (S. Gupta, Ed.), p. 464. https://doi.org/10.1016/S0065-2296(07)45001-7 (accessed on 1st January 2026).
9. Needham, J., & Bray, F. (1984). *Science and Civilization in China: Biology and Biological Technology, Part 2, Agriculture* (Vol. 6). Cambridge University Press, p. 768.
10. Yan, Z. (1990). Overview of rapeseed production and research in China. In *Proceedings of the International Canola Conference*. Potash and Phosphate Institute, Atlanta, Georgia, 29–35.
11. Downey, R. K., & Rakow, G. (1987). Rapeseed and mustard. In W. R. Fehr (Ed.), *Principles of cultivar development: Crop species* (Vol. 2, pp. 437–486). MacMillan, New York.
12. Busch, L., Gunter, V., Mentele, T., et al. (1994). Socializing nature–Technoscience and the transformation of rapeseed into canola. *Crop Science, 34*, 607–614.
13. Mag, T. (1983). Canola oil processing in Canada. *Journal of the American Oil Chemists' Society, 60*, 380–384.
14. Azadmard-Damirchi, S., Habibi-Nodeh, F., Hesari, J., Nemati, M., & Achachlouei, B. F. (2010). Effect of pretreatment with microwaves on oxidative stability and nutraceuticals content of oil from rapeseed. *Food Chemistry, 121*, 1211–1215.
15. Warwick, S. (2010). *Brassica napus* L. In Flora of North America Editorial Committee (Eds.), *Flora of North America North of Mexico* (Vol. 7, p. 422). New York and Oxford.
16. FAOSTAT. (2022). *FAO–Food and Agriculture Organization Statistical Database*. http://faostat.fao.org/ (accessed on 1st January 2026).
17. N/A. (2023). *Oil Crops Data Yearbook*. https://www.ers.usda.gov/data-products/oil-crops-yearbook.aspx (accessed on 1st January 2026).
18. N/A. (2022). *Oilseeds and Products Annual, United Kingdom*, Report Number: UK2022-0019, p. 17. https://apps.fas.usda.gov/newgainapi/api/Report/DownloadReportByFileName?fileName=Oilseeds%20and%20Products%20Annual_London_United%20Kingdom_UK2022-0019.pdf (accessed on 1st January 2026).
19. Shahbandeh, M. (2023). Production volume of rapeseed oil worldwide from 2012/13 to 2022/23. https://www.statista.com/aboutus/our-research-commitment/1239/m-shahbandeh (accessed on 1st January 2026).
20. Jessop, J. P., Toelken, H. R., & Black, J. M. (1986). *Flora of South Australia* (Ed. 4, 4 parts, p. 2248). South Australian Government, Printing Division Adelaide.

21. OECD. (2012). *Consensus Document on the Biology of the Brassica Crops (Brassica spp.)*. Series on Harmonization of Regulatory Oversight of Biotechnology, No. 54, p. 142. Organization for Economic Co-operation and Development, Paris.
22. Schlegel, R. H. J. (2020). *Dictionary of Plant Breeding* (3rd ed., p. 735). CRC Press. ISBN-13: 978-0-367-49216-8.
23. Zhou, W. J., Zhang, G. Q., Tuvesson, S., Dayteg, C., & Gertsson, B. (2006). Genetic survey of Chinese and Swedish oilseed rape (*Brassica napus* L.) by simple sequence repeats (SSRs). *Genetic Resources and Crop Evolution*, *53*(3), 443–447.
24. Rakow, G. (2004). Species origin and economic importance of *Brassica*. In E. C. Pua & C. J. Douglas (Eds.), *Brassica. Biotechnology in Agriculture and Forestry* (Vol. 54, pp. 3–11). Springer, Berlin, Heidelberg. https://doi.org/10.1007/978-3-662-06164-0_1 (accessed on 1st January 2026).
25. Hultén, E. (1968). *Flora of Alaska and Neighboring Territories* (p. 1008). Stanford University Press, Stanford, CA.
26. Artemov, I. V. (1989). *Rape* (p. 43). Agropromizdat, Moscow.
27. Terekhina, N. (2009). Crops, *Brassica napus* L. ssp. *oleifera* Moench–Annual rape. *Agro Atlas: Interactive Agricultural Ecological Atlas of Russia and Neighboring Countries*. http://www.agroatlas.ru/en/content/cultural/Brassica_napus_K/ (accessed on 1st January 2026).
28. Friedt, W., Snowdon, R., Ordon, F., & Ahlemeyer, J. (2007). Plant breeding: Assessment of genetic diversity in crop plants and its exploitation in breeding. *Progress in Botany*, *68*, 151–178.
29. Maltas, E., & Yildiz, S. (2011). Distribution of secondary metabolites in *Brassica napus* genotypes. *Journal of Food Biochemistry*, *35*, 1071–1082. https://doi.org/10.1111/j.1745-4514.2010.00435.x (accessed on 1st January 2026).
30. Froggatt, A., & Wellesley, L. (2019). *Meat Analogs: Considerations for the EU* (p. 42). Energy, Environment and Resources Department.
31. Raboanatahiry, N., Li, H., Yu, L., & Li, M. (2021). Rapeseed (*Brassica napus*): Processing, utilization, and genetic improvement. *Agronomy*, *11*(9), 1776.
32. Arrutia, F., Binner, E., Williams, P., & Waldron, K. W. (2020). Oilseeds beyond oil: Press cakes and meals supplying global protein requirements. *Trends in Food Science & Technology*, *100*, 88–102.
33. Lim, Y. H., Chun, J. H., Lee, K. T., Hong, S. T., Lee, Y. H., & Kim, S. J. (2017). Changes in composition and content of flavonoids by processing type in rapeseed (*Brassica napus*) flowers. *Korean Society of Environmental Agriculture*, *36*, 7–16.
34. Wu, L. H., Luo, Y. M., Xing, X. R., & Christie, P. (2004). EDTA-enhanced phytoremediation of heavy metal contaminated soil with Indian mustard and associated potential leaching risk. *Agriculture, Ecosystems & Environment*, *102*, 307–318.
35. Feng, H., Jiao, S., Tang, Y., Tang, P., & Liu, J. (2011). Advance studying of rape pollen. *Food Industry*, *9*, 108–111.
36. Zhu, C., Du, L., Gao, Y., Qian, L., & Gu, L. (2011). Suppression effect of *Brassica campestris* pollen extract on the growth of prostate cancer cell *in vitro*. *Zhejiang Journal of Integrative Traditional Chinese and Western Medicine*, *21*, 84–86.
37. Beyzi, E., Gunes, A., Buyukkilic, B. S., & Konca, Y. (2019). Changes in fatty acid and mineral composition of rapeseed (*Brassica napus* ssp. *oleifera* L.) oil with seed sizes. *Industrial Crops and Products*, *129*, 10–14. https://doi.org/10.1016/j.indcrop.2018.11.064 (accessed on 1st January 2026).

38. Martinez, S., Losada, P., Franco, I., & Carballo, J. (2010). Protein, amino acid, ash and mineral contents in *Brassica* spp. grown in northwest Spain. *International Journal of Food Science & Technology, 46*, 146–153.
39. Van Duyn, M. A., & Pivonka, E. (2000). Overview of health benefits of fruit and vegetable consumption for the dietetics professional: Selective literature. *Journal of the American Dietetic Association, 100*, 1511–1521.
40. Kris-Etherton, P. M., Hecker, K. D., Bonanome, A., Coval, S. M., Binkoski, A. E., Hilpert, K. F., Griel, A. E., & Etherton, T. D. (2002). Bioactive compounds in foods: Their role in the prevention of cardiovascular disease and cancer. *American Journal of Medicine, 113*, 71–88.
41. Akubugwo, I. E., Obasi, N. A., Chinyere, G. C., & Ugbogu, A. E. (2007). Nutritional and chemical value of *Amaranthus hybridus* L. leaves from Nigeria. *African Journal of Biotechnology, 6*, 2833–2839.
42. Boffetta, P., Couto, E., Wichmann, J., Ferrari, P., Trichopoulos, D., Bueno-de-Mesquita, H. B., van Duijnhoven, F. J., Büchner, F. L., Key, T., Boeing, H., et al. (2010). Fruit and vegetable intake and overall cancer risk in the European prospective investigation into cancer and nutrition (EPIC). *JNCI: Journal of the National Cancer Institute, 102*, 529–537.
43. Brandt, K. (2016). Vegetables and fruit in the prevention of chronic age-related diseases. In *Molecular Basis of Nutrition and Aging. Molecular Nutrition* (pp. 707–722). Academic Press: Cambridge, MA, USA.
44. Neugart, S., Baldermann, S., Ngwene, B., Wesonga, J., & Schreiner, M. (2017). Indigenous leafy vegetables of Eastern Africa—A source of extraordinary secondary plant metabolites. *Food Research International, 100*, 411–422.
45. Shi, Y., Zhang, P., Shi, H., Zhou, Y., Huang, H., & Tang, S. (2009). Preliminary results of yield, income and nutrition of double low rapeseed (*Brassica napus* L.) cultivated for oilseed vegetable-dual-purpose. *Chinese Agricultural Science Bulletin, 25*, 224–227.
46. Domínguez-Perles, R., Moreno, D. A., & García-Viguera, C. (2012). Analysis of the tumoral cytotoxicity of green tea-infusions enriched with broccoli. *Food Chemistry, 132*, 1197–1206.
47. Michels, A. J., & Frei, B. (2012). Vitamin C. In M. H. Stipanuk & M. A. Caudill (Eds.), *Biochemical, Physiological, and Molecular Aspects of Human Nutrition* (pp. 626–654). Elsevier/Saunders: St. Louis, MO, USA.
48. Naderi, N., & House, J. D. (2018). Chapter 5–Recent developments in folate nutrition. *Advances in Food and Nutrition Research, 83*, 195–213.
49. Zhang, Y., Zhang, N., Wang, C., & Yu, G. (2010). Research on functional properties and modification of dietary fiber. *Grain Processing, 35*, 57–59.
50. Khattab, R., Goldberg, E., Lin, L., & Thiyam, U. (2010). Quantitative analysis and free-radical-scavenging activity of chlorophyll, phytic acid, and condensed tannins in canola. *Food Chemistry, 122*, 1266–1272.
51. Ling, W. H., & Jones, P. J. H. (1995). Dietary phytosterols: A review of metabolism, benefits, and side effects. *Life Sciences, 57*, 195–206.
52. Venkatramesh, M., Karunanandaa, B., Sun, B., Grunter, C. A., Boddupalli, S., & Kishore, G. M. (2003). Expression of a *Streptomyces* 3-hydroxysteroid oxidase gene in oilseeds for converting phytosterols to phytostanols. *Phytochemistry, 62*, 39–46.

53. Takahashi, S. I., Takeshita, K., Seeni, A., Sugiura, S., Tang, M., Sato, S. Y., Kuriyama, H., Nakadate, M., Abe, K., Maeno, Y., et al. (2009). Suppression of prostate cancer in a transgenic rat model via gamma-tocopherol activation of caspase signaling. *Prostate*, *69*, 644–651.
54. Saeidnia, S., & Gohari, A. R. (2012). Importance of *Brassica napus* as a medicinal food plant. *Journal of Medicinal Plants Research*, *6*, 2700–2703.
55. Seppanen-Laakso, T., Laakso, I., Lehtimaki, T., Rontu, R., Moilanen, E., Solakivi, T., Seppo, L., Vanhanen, H., Kiviranta, K., & Hiltunen, R. (2010). Elevated plasma fibrinogen caused by inadequate alpha-linolenic acid intake can be reduced by replacing fat with canola-type rapeseed oil. *Prostaglandins, Leukotrienes and Essential Fatty Acids*, *83*, 45–54.
56. Matthäus, B. (2006). Utilization of high-oleic rapeseed oil for deep-fat frying of French fries compared to other commonly used edible oils. *European Journal of Lipid Science and Technology*, *108*, 200–211.
57. Singh, S. K., Chakrabarty, S. K., Kumar, M., Prakash, K., & Kumari, S. (2023). Response of early, timely & late genotypes of *Brassica* sps. grown under different temperature conditions for oil content & fatty acid composition. *International Journal of Plant & Soil Science*, *35*(3), 170–179. https://doi.org/10.9734/ijpss/2023/v35i32789 (accessed on 1st January 2026).
58. Tarrago-Trani, M. T., Phillips, K. M., Lemar, L. E., & Holden, J. M. (2006). New and existing oils and fats used in products with reduced trans-fatty acid content. *Journal of the American Dietetic Association*, *106*, 867–880. https://doi.org/10.1016/j.jada.2006.03.010 (accessed on 1st January 2026).
59. Patade, A., Devareddy, L., Lucas, E., Korlagunta, K., Daggy, B. P., & Arjmandi, B. H. (2008). Flaxseed reduces total and LDL cholesterol concentrations in Native American postmenopausal women. *Journal of Women's Health (Larchmt)*, *17*, 355–366. https://doi.org/10.1089/jwh.2007.0359 (accessed on 1st January 2026).
60. Gaziano, J. M., Hennekens, C. H., O'Donnell, C. J., Breslow, J. L., & Buring, J. E. (1997). Fasting triglycerides, high-density lipoprotein, and risk of myocardial infarction. *Circulation*, *96*, 2520–2525. https://doi.org/10.1161/01.CIR.96.8.2520 (accessed on 1st January 2026).
61. Quiles, J. L., Ramírez-Tortosa, M. C., & Yaqoob, P. (2006). *Olive Oil and Health* (p. 351). CABI. ISBN: 1845930681.
62. Chew, S. C. (2020). Cold-pressed rapeseed (*Brassica napus*) oil: Chemistry and functionality. *Food Research International*, *131*, 108997.
63. McDowell, D., Elliott, C. T., & Koidis, A. (2017). Characterization and comparison of UK, Irish, and French cold-pressed rapeseed oils with refined rapeseed oils and extra virgin olive oils. *European Journal of Lipid Science and Technology*, *119*, 1600327. https://doi.org/10.1002/ejlt.201600327 (accessed on 1st January 2026).
64. Perrier, A., Delsart, C., Boussetta, N., Grimi, N., Citeau, M., & Vorobiev, E. (2017). Effect of ultrasound and green solvents addition on the oil extraction efficiency from rapeseed flakes. *Ultrasonics Sonochemistry*, *39*, 58–65.
65. Tileuberdi, N., Turgumbayeva, A., Yeskaliyeva, B., Sarsenova, L., & Issayeva, R. (2022). Extraction, isolation of bioactive compounds and therapeutic potential of rapeseed (*Brassica napus* L.). *Molecules*, *27*, 8824. https://doi.org/10.3390/molecules27248824 (accessed on 1st January 2026).

66. Yates, K., Pohl, F., Busch, M., Mozer, A., Watters, L., Shiryaev, A., & Kong Thoo Lin, P. (2019). Determination of sinapine in rapeseed pomace extract: Its antioxidant and acetylcholinesterase inhibition properties. *Food Chemistry*, *276*, 768–775.
67. Future Market Insights. (2018). *Rapeseed Oil Market: Industrial Applications of Rapeseed Oil in Biodiesel Production to Compete with its Use in Food Processing: Global Industry Analysis (2013–2017) & Opportunity Assessment (2018–2027).* https://www.futuremarketinsights.com/reports/rapeseed-oil-market (accessed on 1st January 2026).
68. Cvjetko, M., Lepojević, Ž., Zeković, Z., Vidović, S., & Milošević, S. (2009). Antioxidant properties of rapeseed. *Zbornik radova Tehnološkog fakulteta u Leskovcu*, *19*, 27–33.
69. Sosulski, F. (1979). Organoleptic and nutritional effects of phenolic compounds on oilseed protein products: A review. *Journal of the American Oil Chemists' Society*, *56*, 711–715.
70. Sobota, A., Wirkijowska, A., & Zarzycki, P. (2020). Application of vegetable concentrates and powders in colored pasta production. *International Journal of Food Science & Technology*, *55*, 2677–2687. https://doi.org/10.1111/ijfs.14521 (accessed on 1st January 2026).
71. Naczk, M., Amarowicz, R., Sullivan, A., & Shahidi, F. (1998). Current research developments on polyphenolics of rapeseed/canola: A review. *Food Chemistry*, *62*, 489–502.
72. Kraljić, K., Škevin, D., Barišić, L., Kovačević, M., Obranović, M., & Jurčević, I. (2015). Changes in 4-vinylsyringol and other phenolics during rapeseed oil refining. *Food Chemistry*, *187*, 236–242. https://doi.org/10.1016/j.foodchem.2015.04.039 (accessed on 1st January 2026).
73. Siger, A., Kaczmarek, A., & Rudzińska, M. (2015). Antioxidant activity and phytochemical content of cold-pressed rapeseed oil obtained from roasted seeds. *European Journal of Lipid Science and Technology*, *117*, 1225–1237. https://doi.org/10.1002/ejlt.201400378 (accessed on 1st January 2026).
74. Kozlowska, H., Rotkiewicz, D. A., & Zademowski, R. (1983). Phenolic acids in rapeseed and mustard. *Journal of the American Oil Chemists' Society*, *60*, 1191–123.
75. Shahidi, F., & Naczk, M. (1995). *Food Phenolics: Sources, Chemistry, Effects, Applications* (p. 331). Technomic Publishing.
76. Cai, R., & Arntfield, S. (2001). A rapid high-performance liquid chromatographic method for the determination of sinapine and sinapic acid in canola seed and meal. *Journal of the American Oil Chemists' Society*, *78*, 903–910. https://doi.org/10.1007/s11746-001-0362-4 (accessed on 1st January 2026).
77. Tsunehiro, J., Yasuda, F., Wakamatsu, D., & Nakai, C. (2002). Isolation and characterization of lipid-radical scavenging component in crude canola oil. In *Third International Conference and Exhibition on Nutraceuticals and Functional Foods: From Laboratory to the Real World and the Marketplace*, Nov. 17–20, San Diego, CA. American Oil Chemists' Society: Champaign, IL.
78. Wakamatsu, D., Morimura, S., Sawa, T., Kida, K., Nakai, C., & Maeda, H. (2005). Isolation, identification, and structure of a potent alkyl-peroxyl radical scavenger in crude canola oil, canolol. *Bioscience, Biotechnology, and Biochemistry*, *69*(8), 1568–1574.

79. Durkee, A. B. (1971). The nature of tannin in rapeseed (*Brassica campestris*). *Phytochemistry, 10*, 1583–1585.
80. Naczk, M., & Shahidi, F. (2004). Extraction and analysis of phenolics in food. *Journal of Chromatography A, 1054*, 95–111.
81. Auger, B., Marnet, N., Gautier, V., Maia-Grondard, A., Leprince, F., Renard, M., Guyot, S., Nesi, N., & Routaboul, J. M. (2010). A detailed survey of seed coat flavonoids in developing seeds of *Brassica napus* L. *Journal of Agricultural and Food Chemistry, 58*, 6246–6256.
82. Malik, R. (1990). Prospects for *Brassica carinata* as an oilseed crop in India. *Experimental Agriculture, 26*, 125–129.
83. Alander, J., Andersson, A., Bagge, C., Bringsarve, K., Hjorth, M., Johansson, M., Granroth, B., Norberg, S., Pedersen, M., Persson, M., et al. (2007). Raw materials. In *Handbook of Vegetable Oils and Fats* (2nd ed., Lidefelt, J., Ed.). Alfaprint: Sundbyberg, Sweden.
84. Flakelar, C. L., Luckett, D. J., Howitt, J. A., Doran, G., & Prenzler, P. D. (2015). Canola (*Brassica napus*) oil from Australian cultivars shows promising levels of tocopherols and carotenoids, along with good oxidative stability. *Journal of Food Composition and Analysis, 42*, 179–186.
85. Bowen, K. J., Kris-Etherton, P. M., West, S. G., Fleming, J. A., Connelly, P. W., Lamarche, B., et al. (2019). Diets enriched with conventional or high-oleic acid canola oils lower atherogenic lipids and lipoproteins compared to a diet with a western fatty acid profile in adults with central adiposity. *Journal of Nutrition, 149*, 471–478.
86. Fernández-García, E., Carvajal-Lérida, I., Jarén-Galán, M., Garrido-Fernández, J., Pérez-Gálvez, A., & Hornero-Méndez, D. (2012). Carotenoids bioavailability from foods: From plant pigments to efficient biological activities. *Food Research International, 46*, 438–450.
87. Delgado-Vargas, F., Jiménez, A. R., & Paredes-López, O. (2000). Natural pigments: Carotenoids, anthocyanins, and betalains—Characteristics, biosynthesis, processing, and stability. *Critical Reviews in Food Science and Nutrition, 40*, 173–289.
88. Tańska, M., Mikołajczak, N., & Konopka, I. (2018). Comparison of the effect of sinapic and ferulic acid derivatives (4-vinylsyringol vs. 4-vinylguaiacol) as antioxidants of rapeseed, flaxseed, and extra virgin olive oils. *Food Chemistry, 240*, 679–685. https://doi.org/10.1016/j.foodchem.2017.08.007 (accessed on 1st January 2026).
89. Granado, F., Olmedilla, B., & Blanco, I. (2003). Nutritional and clinical relevance of lutein in human health. *British Journal of Nutrition, 90*, 487–502.
90. Gordaliza, M. (2007). Natural products as leads to anticancer drugs. *Clinical and Translational Oncology, 9*, 767–776.
91. Yoon, B. H., Jung, J. W., Lee, J. J., Cho, Y. W., Jang, C. G., Jin, C., Oh, T. H., & Ryu, J. H. (2007). Anxiolytic-like effects of sinapic acid in mice. *Life Sciences, 81*(3), 234–240. https://doi.org/10.1016/j.lfs.2007.05.007 (accessed on 1st January 2026).
92. Xu, J., Ma, C., Han, L., Gao, H., Zhou, Q., Yang, M., Chen, C., Deng, Q., Huang, Q., & Huang, F. (2014). Optimized rapeseed oils rich in endogenous micronutrients ameliorate risk factors of atherosclerosis in high fat diet fed rats. *Lipids in Health and Disease, 13*, 166.

93. Fedor, D., & Kelley, D. S. (2009). Prevention of insulin resistance by n-3 polyunsaturated fatty acids. *Current Opinion in Clinical Nutrition and Metabolic Care*, *12*, 138–146.
94. Lee, C. S., Baek, J., & Han, S. Y. (2017). The role of kinase modulators in cellular senescence for use in cancer treatment. *Molecules*, *22*, 1411.
95. Mandrich, L., & Caputo, E. (2020). Brassicaceae-derived anticancer agents: Towards a green approach to beat cancer. *Nutrients*, *12*, 868.
96. Morrissey, K. M., Yuraszeck, T. M., Li, C. C., Zhang, Y., & Kasichayanula, S. (2016). Immunotherapy and novel combinations in oncology: Current landscape, challenges, and opportunities. *Clinical and Translational Science*, *9*, 89–104.
97. Johson, I. T. (2018). Cruciferous vegetables and risk of cancer of the gastrointestinal tract. *Molecular Nutrition & Food Research*, *62*, e1701000.
98. Arcidiacono, P., Kuligina, E., Crisanti, A., Ragonese, F., Rende, M., Spaccapelo, R., Stabile, A., Bottoni, U., Pistilli, A., & Calvieri, S. (2018). Antitumor activity and expression profiles of genes induced by sulforaphane in human melanoma cells. *European Journal of Nutrition*, *57*, 2547–2569.
99. Bao, Y., Wang, W., Zhou, Z., & Sun, C. (2014). Benefits and risks of the hormetic effects of dietary isothiocyanates on cancer prevention. *PLoS ONE*, *9*, e114764.
100. Nagai, T., Inoue, R., Inoue, H., & Suzuki, N. (2002). Scavenging capacities of pollen extracts from *Cistus ladaniferus* on autoxidation, superoxide radicals, hydroxyl radicals and DPPH radicals. *Nutrition Research*, *22*, 519–526.
101. Steigerová, J., Oklešťková, J., Levková, M., Rárová, L., Kolář, Z., & Strnad, M. (2010). Brassinosteroids cause cell cycle arrest and apoptosis of human breast cancer cells. *Chemico-Biological Interactions*, *188*, 487–496.
102. Akinsete, J. A., Ion, G., Witteand, T. R., & Hardman, W. E. (2012). Consumption of high ω-3 fatty acid diet suppressed prostate tumorigenesis in C3(1) Tag mice. *Carcinogenesis*, *33*, 140–148.
103. Ferrero, R. L., Soto-Maldonado, C., Weinstein-Oppenheimer, C., Cabrera-Muñoz, Z., & Zúñiga-Hansen, M. E. (2021). Antiproliferative rapeseed defatted meal protein and their hydrolysates on MCF-7 breast cancer cells and human fibroblasts. *Foods*, *10*, 309.
104. Kwon, J., Snook, J., Wardlaw, G., et al. (1991). Effects of diets high in saturated fatty acids, canola oil, or safflower oil on platelet function, thromboxane B2 formation, and fatty acid composition of platelet phospholipids. *American Journal of Clinical Nutrition*, *54*, 351–358.
105. Gillingham, L. G., Gustafson, J. A., Han, S. Y., et al. (2011). High-oleic rapeseed (canola) and flaxseed oils modulate serum lipids and inflammatory biomarkers in hypercholesterolemia subjects. *British Journal of Nutrition*, *105*, 417–427.
106. Zou, Y., Kim, A. R., Kim, J. E., Choi, J. S., & Chung, H. Y. (2002). Peroxynitrite scavenging activity of sinapic acid (3,5-dimethoxy-4-hydroxycinnamic acid) isolated from *Brassica juncea*. *Journal of Agricultural and Food Chemistry*, *50*, 5884–5890.
107. Yang, C. Y., & He, L. (2008). Neuroprotective effects of sinapine on PC12 cells apoptosis induced by sodium dithionite. *Chinese Journal of Natural Medicines*, *6*(3), 205–209.
108. Kim, D. H., Yoon, B. H., Jung, W. Y., Kim, J. M., Park, S. J., Park, D. H., Huh, Y., et al. (2010). Sinapic acid attenuates kainic acid-induced hippocampal neuronal damage in mice. *Neuropharmacology*, *59*, 20–23.

109. Lee, N. K., Lee, J. H., Lim, S. M., Lee, K. A., Kim, Y. B., Chang, P. S., & Paik, H. D. (2014). Short communication: Antiviral activity of subcritical water extract of *Brassica juncea* against influenza virus A/H1N1 in nonfat milk. *Journal of Dairy Science*, *97*, 5383–5386.
110. Hajarizadeh, B., Grebely, J., & Dore, G. J. (2013). Epidemiology and natural history of HCV infection. *Nature Reviews Gastroenterology & Hepatology*, *10*, 553–562.
111. Mohammadzadeh, S., Roohvand, F., Ehsani, P., Salmanian, A. H., & Ajdary, S. (2020). Canola oilseed- and *Escherichia coli*-derived hepatitis C virus (HCV) core proteins adjuvanted with oil bodies induced robust Th1-oriented immune responses in immunized mice. *APMIS: Acta Pathologica, Microbiologica et Immunologica Scandinavica*, *128*, 593–602.
112. Heterologous expression of hepatitis C virus core protein in oil seeds of *Brassica napus* L. (2015). *Jundishapur Journal of Microbiology*, *8*, e25462.
113. Turgumbayeva, A., Tileuberdi, N., Zhakipbekov, K., Tulemissov, S., Umurzakhova, G., & Utegenova, G. (2021). Antimicrobial efficacies of *Brassica napus* L. essential oils/nanoparticles composites. *Journal of Nanostructures*, *11*, 524–533.
114. Vosough-Ghanbari, S., Rahimi, R., Kharabaf, S., Zeinali, S., Mohammadirad, A., & Amini, S. (2010). Effects of *Satureja khuzestanica* on serum glucose, lipids and markers of oxidative stress in patients with type 2 diabetes mellitus: A double-blind randomized controlled trial. *Evidence-Based Complementary and Alternative Medicine*, *7*, 465–470.
115. Akbari, F., Khodadadi, S., Asgari, S., Shirzad, H., Mirhoseini, M., Shahinfard, N., & Rafieian-Kopaei, M. (2015). A comparative study on hypoglycemic properties, lipid profile and bioactive components of hydro-alcoholic extracts of cooked and raw *Brassica napus*. *Journal of Nephropharmacology*, *5*, 86–90.
116. Valdecantos, M. P., Perez-Matute, P., & Obesity, M. J. A. (2009). Oxidative stress: Role of antioxidant supplementation. *Revista de Investigación Clínica*, *61*, 127–139.
117. Capel, F., Geloen, A., Vaysse, C., Pineau, G., Demaison, L., Chardigny, J. M., Michalski, M. C., & Malpuech-Brugère, C. (2018). Rapeseed oil fortified with micronutrients can reduce glucose intolerance during a high fat challenge in rats. *Nutrition & Metabolism*, *15*, 22.
118. Jamka, M., Morawska, A., Krzyżanowska-Jankowska, P., Bajerska, J., Przysławski, J., Walkowiak, J., & Lisowska, A. (2021). Comparison of the effect of amaranth oil vs. rapeseed oil on selected atherosclerosis markers in overweight and obese subjects: A randomized double-blind cross-over trial. *International Journal of Environmental Research and Public Health*, *18*, 8540.
119. Forouhi, N. G., & Wareham, N. J. (2019). Epidemiology of diabetes. *Medicine*, *47*, 22–27.
120. Rahimi, M., & Naghizadeh, M. (2014). The effect of L-Arginin supplementation on lipid profiles in patients with type 2 diabetes. *Journal of Fasa University of Medical Sciences*, *4*, 99–110.
121. Aram-Ahmaddi, M., Bahrampour, A., & Najafipour, H. (2014). Predicting type two diabetes and determination of effectiveness of risk factors applying logistic regression model. *Journal of Kerman University of Medical Sciences*, *21*, 103–113.
122. Ballantyne, G. H., Wasielewski, A., & Saunders, J. K. (2009). The surgical treatment of type II diabetes mellitus: Changes in HOMA insulin resistance in the first year

following laparoscopic Roux-en-Y gastric bypass (LRYGB) and laparoscopic adjustable gastric banding (LAGB). *Obesity Surgery, 19*, 1297–1303.
123. Xu, F., de Mejia, E. G., Chen, H., Rebecca, K., Pan, M., He, R., Yao, Y., Wang, L., & Ju, X. (2020). Assessment of the DPP-IV inhibitory activity of a novel octapeptide derived from rapeseed using Caco-2 cell monolayers and molecular docking analysis. *Journal of Food Biochemistry, 44*, e13406.
124. Hussain, S., Rehman, A. U., Luckett, D. J., Naqvi, S., & Blanchard, C. L. (2021). Protease inhibitors purified from the canola meal extracts of two genetically diverse genotypes exhibit antidiabetic and antihypertension properties. *Molecules, 26*, 2078.
125. McDonald, B. E. (2013). In *Canola and Rapeseed: Production, Processing, Food Quality, and Nutrition* (U. Thiyam-Holländer, N. A. Michael Eskin, & B. Matthäus, Eds.). CRC Press, Taylor & Francis Group. ISBN-13: 978-1-4665-1388-4 (eBook–PDF).
126. Hansson, G. K. (2005). Inflammation, atherosclerosis, and coronary artery disease. *New England Journal of Medicine, 352*, 1685–1695.
127. Lorente-Cebrian, S., Costa, A. G., Navas-Carretero, S., Zabala, M., Martinez, J. A., & Moreno-Aliaga, M. J. (2013). Role of omega-3 fatty acids in obesity, metabolic syndrome, and cardiovascular diseases: A review of the evidence. *Journal of Physiology and Biochemistry, 69*, 633–651.
128. Dunford, N. (2018). Canola oil properties. *Food Technology Fact Sheet, 405*, 744–6071.
129. Ostlund, R. E., Racette, S. B., Okeke, A., & Stenson, W. F. (2002). Phytosterols that are naturally present in commercial corn oil significantly reduce cholesterol absorption in humans. *American Journal of Clinical Nutrition, 75*, 1000–1004.
130. Xu, J., Zhou, X., Deng, Q., Huang, Q., Yang, J., & Huang, F. (2011). Rapeseed oil fortified with micronutrients reduces atherosclerosis risk factors in rats fed a high-fat diet. *Lipids in Health and Disease, 10*, 96.
131. Favier, L. S., María, A. O. M., Wendel, G. H., Borkowski, E. J., Giordano, O. S., Pelzer, L., & Tonn, C. E. (2005). Anti-ulcerogenic activity of xanthanolide sesquiterpenes from *Xanthium cavanillesii* in rats. *Journal of Ethnopharmacology, 100*, 260–267.
132. Amarowicz, R., Naczk, M., & Shahidi, F. (2000). Antioxidant activity of crude tannins of canola and rapeseed hulls. *Journal of the American Oil Chemists' Society, 77*, 957–961.
133. Frolova, E. V. (2016). Arterial hypertension. *Russian Family Doctor, 20*, 1–18.
134. Kearney, P. M., Whelton, M., Reynolds, K., Muntner, P., Whelton, P. K., & He, J. (2005). Global burden of hypertension: Analysis of worldwide data. *Lancet, 365*, 217–223.
135. Wang, Y., Li, Y., Ruan, S., Lu, F., Tian, W., & Ma, H. (2021). Antihypertensive effect of rapeseed peptides and their potential in improving the effectiveness of captopril. *Journal of the Science of Food and Agriculture, 101*, 3049–3055.

Index

F

G

H

I

Q

R

S

T

Z

For Product Safety Concerns and Information please contact our EU representative GPSR@taylorandfrancis.com Taylor & Francis Verlag GmbH, Kaufingerstraße 24, 80331 München, Germany

www.ingramcontent.com/pod-product-compliance
Ingram Content Group UK Ltd.
Pitfield, Milton Keynes, MK11 3LW, UK
UKHW020904190926
475560UK00011B/334

9 781793 642554